FABRICATION AND FUNCTIONALIZATION OF ADVANCED TUBULAR NANOFIBERS AND THEIR APPLICATIONS

The Textile Institute Book Series

FABRICATION AND FUNCTIONALIZATION OF ADVANCED TUBULAR NANOFIBERS AND THEIR APPLICATIONS

Edited by

BAOLIANG ZHANG

School of Chemistry and Chemical Engineering, Northwestern Polytechnical University, Xi'an, China; Xi'an Key Laboratory of Functional Organic Porous Materials, Northwestern Polytechnical University, Xi'an, China

MUDASIR AHMAD

School of Chemistry and Chemical Engineering, Northwestern Polytechnical University, Xi'an, China; Xi'an Key Laboratory of Functional Organic Porous Materials, Northwestern Polytechnical University, Xi'an, China

WP
WOODHEAD PUBLISHING

An imprint of Elsevier
elsevier.com/books-and-journals

ELSEVIER

Woodhead Publishing is an imprint of Elsevier
50 Hampshire Street, 5th Floor, Cambridge, MA 02139, United States
The Boulevard, Langford Lane, Kidlington, OX5 1GB, United Kingdom

Notices
Knowledge and best practice in this field are constantly changing. As new research and
experience broaden our understanding, changes in research methods, professional
practices, or medical treatment may become necessary.

Practitioners and researchers must always rely on their own experience and knowledge in
evaluating and using any information, methods, compounds, or experiments described
herein. In using such information or methods they should be mindful of their own safety
and the safety of others, including parties for whom they have a professional responsibility.

To the fullest extent of the law, neither the Publisher nor the authors, contributors, or
editors, assume any liability for any injury and/or damage to persons or property as a
matter of products liability, negligence or otherwise, or from any use or operation of any
methods, products, instructions, or ideas contained in the material herein.

ISBN: 978-0-323-99039-4

For Information on all Woodhead Publishing publications visit our
website at https://www.elsevier.com/books-and-journals

Publisher: Matthew Deans
Acquisitions Editor: Brian Guerin
Editorial Project Manager: Franchezca A. Cabural
Production Project Manager: Kamesh R
Cover Designer: Vicky Pearson

Working together
to grow libraries in
developing countries

www.elsevier.com • www.bookaid.org

Typeset by Aptara, New Delhi, India
Transferred to Digital Printing 2023

Contents

Contributors   *ix*

Preface   *xi*

1. The review and introduction of hypercrosslinked polymer   1
Tariq Shah, Yanting Lyu and Baoliang Zhang

    1.1 Introduction   1
    1.2 Synthesis of different types of hypercrosslinked polymers   3
    1.3 Hypercrosslinked polymer by polycondensation   6
    1.4 Properties of the hypercrosslinked polymers   9
    1.5 Morphology of hypercrosslinked polymers   16
    1.6 Applications of hypercrosslinking polymers   20
    1.7 Future consideration and challenges   23
    References   24

2. Novel synthetic method for tubular hypercrosslinked polymer nanofibers and its mechanism   29
Jiqi Wang, Yihao Fan and Baoliang Zhang

    2.1 Introduction   29
    2.2 The study on hypercrosslinked microporous materials with regular morphology   30
    2.3 Preparation and formation mechanism of hypercrosslinked tubular nanofibers   35
    2.4 Application of hypercrosslinked tubular nanofibers   39
    2.5 Summary and prospect   44
    References   44

3. Design and preparation of self-driven BSA surface imprinted tubular nanofibers and their specific adsorption performance   47
Zuoting Yang and Baoliang Zhang

    3.1 Introduction   47
    3.2 Surface protein imprinting technique   49
    3.3 Concept of self-driven protein imprinting   71
    3.4 Self-driven surface imprinted tubular carbon nanofibers   71
    3.5 Self-driven surface imprinted magnetic tubular carbon nanofibers   75

3.6 Surface imprinted manganese dioxide-loaded tubular carbon fibers   80
3.7 Summary and prospect   85
References   85

4. **Functional carbon nano-material as efficient metal ion adsorbent from wastewater   91**

Mudasir Ahmad and Baoliang Zhang

4.1 Introduction   91
4.2 Functional group modifications and synthetic methods   92
4.3 Preparation and functionalization of adsorbents from biowastes   100
4.4 Preparation and functionalization of inorganic/organic adsorbents   104
4.5 Our contribution   105
References   106

5. **Preparation of carbon tubular nanofibers and their application for efficient enrichment of uranium from aqueous solution   115**

Mudasir Ahmad and Baoliang Zhang

5.1 Introduction   115
5.2 Synthesis of CNF   117
5.3 Applications of carbon tubular nanofibers   124
5.4 Our contribution   126
References   128

6. **Uranium adsorption property of carboxylated tubular carbon nanofibers enhanced chitosan microspheres   133**

Mudasir Ahmad and Baoliang Zhang

6.1 Introduction   133
6.2 Preparation and functionalization of tubular carbon nanofibers   135
6.3 Chitosan enhanced COOH-TCN for uranium adsorption   144
References   147

7. **Oil adsorption performance of tubular hypercrosslinked polymer and carbon nanofibers   153**

Ke Yang, Yuhong Cui and Baoliang Zhang

7.1 Introduction   153
7.2 Introduction of oil-absorbing materials   154
7.3 Research progress of one-dimensional oil-absorbing materials   161
7.4 Tubular hypercrosslinked polymer   169

7.5 Length controllable hydrophobically modified tubular carbon nanofibers   173
7.6 Porous biomass tubular carbon fibers from platanus orientalis   175
7.7 Summary and outlook   177
References   178

8. Fabrication of carboxylated tubular carbon nanofibers as anode electrodes for high-performance lithium-ion batteries   183

Yu Huyan, Junjie Chen and Baoliang Zhang

8.1 Introduction   183
8.2 Introduction of anode materials for lithium-ion batteries   184
8.3 Research status of carbon nanotube-based anode materials   190
8.4 Preparation of carboxyl modified carbon nanotube anode materials   191
8.5 Fabrication of acidified tubular carbon nanofibers   192
8.6 Summary and prospect   204
References   206

9. Tubular carbon nanofibers loaded with different MnO_2: Preparation and electrochemical performance   211

Yu Huyan, Mengmeng Wei and Baoliang Zhang

9.1 Introduction   211
9.2 Research progress of MnO_2-based anode materials   212
9.3 One-dimensional carbon nanomaterials and MnO_2 composite electrode materials   221
9.4 Tubular carbon nanofibers loaded with worm-like MnO_2 222
9.5 Tailoring carboxyl tubular carbon nanofibers/MnO_2 composites   240
9.6 Summary and prospect   242
References   244

10. Preparation and microwave absorption properties of tubular carbon nanofibers and magnetic nanofibers   249

Jiqi Wang, Fei Wu and Baoliang Zhang

10.1 Introduction   249
10.2 One-dimensional carbon nanomaterials   250
10.3 Microwave absorbers and electromagnetic stealth   257
10.4 1D carbon-based microwave absorbers   262
10.5 Tubular carbon nanofibers   276
10.6 Tubular magnetic carbon nanofibers with hierarchical pore structure   279
10.7 Helical/chiral biomass-derived 3D magnetic porous carbon fibers   282

10.8 Ultralight helical porous carbon fibers with CNTs-confined
Ni nanoparticles 286
10.9 Summary and outlook 289
References 290

11. Multiple composite tubular carbon nanofibers: Synthesis, characterization, and applications in microwave absorption 299

Fei Wu, Jiqi Wang and Baoliang Zhang

11.1 Introduction 299
11.2 1D multishell composite materials 300
11.3 Magnetic tubular fiber with multilayer heterostructure 304
11.4 Intertwined one-dimensional heterostructure obtained by
Mxene and MOF 311
11.5 Core-shell MnO_2@NC@MoS_2 nanowires 314
11.6 Summary and outlook 319
Reference 319

12. Biomedical applications of multifunctional tubular nanofibers 323

Idrees Khan and Baoliang Zhang

12.1 Introduction 323
12.2 Tissue regeneration 327
12.3 Bio-sensing applications 332
12.4 Drug delivery applications 333
12.5 Antimicrobial applications 334
12.6 Applications in blood dialysis 336
12.7 Applications in medical devices 337
12.8 Other medical applications 338
12.9 Conclusion 339
References 339

Index 345

Contributors

Mudasir Ahmad
School of Chemistry and Chemical Engineering, Northwestern Polytechnical University, Xi'an, China; Xi'an Key Laboratory of Functional Organic Porous Materials, Northwestern Polytechnical University, Xi'an, China

Junjie Chen
School of Chemistry and Chemical Engineering, Northwestern Polytechnical University, Xi'an, China

Yuhong Cui
School of Chemistry and Chemical Engineering, Northwestern Polytechnical University, Xi'an, China

Yihao Fan
School of Chemistry and Chemical Engineering, Northwestern Polytechnical University, Xi'an, China

Yu Huyan
School of Chemistry and Chemical Engineering, Northwestern Polytechnical University, Xi'an, China

Idrees Khan
School of Chemistry and Chemical Engineering, Northwestern Polytechnical University, Xi'an, China

Yanting Lyu
School of Chemistry and Chemical Engineering, Northwestern Polytechnical University, Xi'an, China

Tariq Shah
School of Chemistry and Chemical Engineering, Northwestern Polytechnical University, Xi'an, China

Jiqi Wang
School of Chemistry and Chemical Engineering, Northwestern Polytechnical University, Xi'an, China

Mengmeng Wei
School of Chemistry and Chemical Engineering, Northwestern Polytechnical University, Xi'an, China

Fei Wu
School of Chemistry and Chemical Engineering, Northwestern Polytechnical University, Xi'an, China

Zuoting Yang
School of Chemistry and Chemical Engineering, Northwestern Polytechnical University, Xi'an, China

Ke Yang
School of Chemistry and Chemical Engineering, Northwestern Polytechnical University, Xi'an, China

Baoliang Zhang
School of Chemistry and Chemical Engineering, Northwestern Polytechnical University, Xi'an, China; Xi'an Key Laboratory of Functional Organic Porous Materials, Northwestern Polytechnical University, Xi'an, China

Preface

Carbon nanomaterial fabrication leads sustainable green industrial materials toward a new horizon. Carbon tubular nanofibers are playing a vital role in the field of energy and environmental safety. This book introduces the knowledge of carbon-based tubular nanofibers. Therefore, researchers, scientists, and academia can learn and synthesize the cutting-edge materials and techniques for the fabrication of tubular carbon nanofibers for advanced applications.

This book starts with a depth introductory section on the crosslinking polymers along with different synthetic routes, morphological features, and applications in various fields (Chapter 1). Chapter 2 is a novel synthetic method for tubular hypercrosslinked polymer nanofibers and their mechanism. This chapter describes the potential of carbon tubular nanofibers in polymers, and the common synthetic procedures and formation of mechanism procedure. Chapter 3 is the design and preparation of self-driven BSA surface imprinted tubular nanofibers and their specific adsorption performance, which summarizes the surface protein imprinting technique and its applications as a functional carrier, carrier material surface design and construction, and the composition of the imprinted layer polymer. Chapter 4 introduces functional carbon nanomaterial as an efficient metal ion adsorbent from wastewater. In this chapter, carbon-based nanomaterial synthesis such as magnetic nanomaterials, selectively functionalized nanomaterials and their application for wastewater treatment. Chapter 5 demonstrates the synthesis of carbon tubular nanofibers by different methods from various sources and functionalized for uranium extraction. Chapter 6 is the uranium adsorption property of carboxylated tubular carbon nanofibers enhanced chitosan microspheres. In this chapter, experimental and theoretical approaches include carboxyl group (COOH), and the mechanism of composite formation with biopolymers. Chapter 7 utilizes hypercrosslinked tubular nanofibers and biomass-derived nanofibers that possess different diameters, lengths, and surface pore structures for oil separation application. Chapter 8 is a fabrication of carboxylated tubular carbon nanofibers as anode electrodes for high-performance lithium-ion batteries. This chapter introduces carboxyl-modified tubular carbon nanofibers that are prepared using carbonization and liquid phase oxidation technology for energy application. Chapter 9 discusses the synthesis of tubular carbon nanofibers loaded with different MnO_2 as composite electrode materials. Chapter 10 is the preparation and microwave absorption properties of tubular carbon nanofibers and magnetic nanofibers. This chapter summarizes the synthesis of one-dimensional carbon nanomaterials from synthetic polymers or natural polymers using various preparation methods such as the arc discharge method, laser ablation method, chemical vapor deposition method, and flame method for microwave application. Chapter 11 discusses synthesis and microwave absorption applications of multiple composite tubular carbon nanofibers. The 1D multishell composite absorbers include fiber-based multishell absorbers, magnetic nanochain-based multishell absorbers, inorganic nanowire-based multishell absorbers, magnetic tubular fiber with multi-layer heterostructure and core-shell $MnO_2@NC@MoS_2$ nanowires. Chapter 12 is biomedical applications of multifunctional tubular nanofibers. This chapter introduces the use of tubular carbon nanofibers in medical field applications such as vascular, urethral, nerve tissues, cardiac tissues, and soft bone engineering.

CHAPTER 1

The review and introduction of hypercrosslinked polymer

Tariq Shah[a], Yanting Lyu[a] and Baoliang Zhang[a,b]
[a] School of Chemistry and Chemical Engineering, Northwestern Polytechnical University, Xi'an, China
[b] Xi'an Key Laboratory of Functional Organic Porous Materials, Northwestern Polytechnical University, Xi'an, China

1.1 Introduction

Organic polymers with microporous and mesoporous structures have been highly investigated in few decades due to their various potential applications such as chromatographic separation, drug delivery, gas storage, catalyst, and sensor substrates [1–4]. The term "nanoporous" is also used for mesoporous and microporous polymers. Nanoporous organic polymers were divided into different subclasses such as hypercrosslinked polymers (HCPs), covalent triazin frameworks (CTFs), porous aromatic frameworks (PAFs), covalent organic frameworks (COFs), and conjugated microporous frameworks on the bases of their structure and synthetic methods [5].

Hypercrosslinked polymers were first invented by Davankov and Tsyurupa in the early 1970s, and had grown in popularity over the last few decades. HCPs are the series of microporous polymer materials which possess more extensive crosslinked network than conventional crosslinked polystyrene, and thus have attracted more researchers into this area. In recent years, HCPs have undergone rapid development due to its remarkable advantages such as high surface area, low cost reagents, easy functionalization, synthetic methods, and mild operation conditions. Synthesis of HCPs is mainly based on Friedel-Crafts chemistry, it provides fast kinetics which can form strong linkage producing highly crosslinked network with leading porosity [6]. The rigid networks and high level crosslinked nanoporous structure prevent it from collapse. HCPs require low–cost reagents such as catalysts, monomers, and reaction conditions in conventional synthetic methodologies; it also can produce high yield products by controlling the reaction conditions. Moreover, the products can compete with inexpensive materials such as activated carbon. HCPs are prepared by three synthetic approaches which are (1) postcrosslinking polymers precursor, (2) one step of

Fabrication and Functionalization of Advanced Tubular Nanofibers and their Applications.
DOI: https://doi.org/10.1016/B978-0-323-99039-4.00003-6

1

poly-condensation of functional monomer, and (3) knitting aromatic building blocks. The varieties of blocks mixed with the synthetic approaches to make HCPs valuable for exploring new organic porous materials which help in many fields like energy and environment. The first and only representative of hypercrosslinked polymeric material is polystyrene. Other materials could have same properties, if the formation process could correspond to hypercorsslinked polymer synthesis. Davankov and Tsyurupa proposed new approach for the formation of homogeneous and highly rigid network with high expanding capacity [7,8]. They used large number of rigid bridges to perform the intensive postcrosslinking of polystyrene chain in solution. The initial polystyrene chain connects all phenly rings to another by using active biofunctional crosslinking agents, thus produce homogeneous, rigid and expanded open network structure. Polystyrene-type material is new type of three dimensional (3D) networks which is the "third generation" with unusual and peculiar properties. The hypercrosslinked rigid network exhibits high mobility. For illustration, hypercrosslinked polystyrene swells up in excellent solvents with significant increase 3–5 times in volume. The conformation rearrangements of all interconnected networks are accountable for the significant change of total volume of the material. The interconnected material has similar aromatic chemical property of polystyrene precursor. Linear polystyrene is failed to dissolve but the hypercrosslinked polymer swells to high extent in some solvent. After the polymer-polymer interaction reduced energy, the dry polymer with rigid network and strong stress will be responsible for hypercrosslinked polymer network to swell in nonsolvents. The hypercrosslinked polymer swelling behavior can be analyzed by revising and re-examining the factors which govern the swelling behavior in 3D networks. The meshes mutual arrangement in space, their interpenetration, and interconnection are the topology of network, as well as the meshes conformational flexibility determines the network mobility. Under certain conditions, the parameters like ratio between inter- and intra- crosslinking, even the density of crosslinking will become unimportant. There is certain condition in which the strong network topology is depending on the network formation, unexpected things can be observed like the rise in hypercrosslinking swelling ability with an increase in its degree of crosslinking. Before the evaporation of residual water from the materials, during the drying process, the hydrophobic porous polymer swollen bed may exhibit 12% of decrease in volume below the dry polymer final equilibrium. The polymer contracts to the significant extent when immersed in phosphoric acid and LiCl concentrated solution. HCPs have the high deforming ability under loading, which is the characteristic of conventional linear polymers or slightly crosslink networks in the state of rubber elasticity, HCPs also retain their mobility under low temperatures. HCPs do not exhibit the characteristics of Elastomers nor are they polymeric glasses. However, the HCPs networks reveal unique deformation, relax properties and distinct difference.

Hypercrosslinked polystyrene is considered to be excellent sorbing material from practical point of view, and referred as *"Styrosorbs"* which have the ability to absorb large number of organic substances, polar and nonpolar, liquid and gaseous media. Their absorption capacity is much higher than the macroporous polystyrene and gel sorbents. Styrosorbs can be used in catalysis system, fabrication of nanocomposites and functional polymer, and extraction of solid phase contaminants from water and air. Styrosorbs are the first representative of the networks and sorbent for the third generation. The use of styrosorbs can improve separation process, concentration, traditional absorption efficiency [9].

1.2 Synthesis of different types of hypercrosslinked polymers

1.2.1 Macroporous and gel type polystyrene

One of the most common used polymeric adsorbents is crosslinked polystyrene due to its diversified modification and simplicity. In 1930, Heuer and Staudinger first reported divinylbenzene (DVB) styrene network which was the earliest crosslinked polystyrene [10]. In 1945, D Alelio registered a patent on synthesizes ion exchange resin as application of DVB–styrene matrices [11]. These networks were referred as gel-type polystyrene and considered to be the first generation of crosslinked copolymer. Bulk styrene is used for the copolymerization of homogenous gels with DVB of 5%–8% [12,13]. Gel-type styrene can swell in organic nonpolar solvents such as dichlroethane and toluene. In order to swell in water and polar alcohols, the ammonium and sulfonic were introduced as functional groups to gels–type styrene. In 1952, McBurney registered a patent on the second generation of gel-type polystyrene networks which is also known as macroporous polymers. The second-generation networks have more porosity compared to the first generation. The copolymerization of DVB (6%–12%) and styrene can generate high porosity during the induced phase polymerization in the presence of monomer porogen, and high percentage of DVB can maintain the structure in the synthesis process. Different pore sizes of second-generation networks were commercialized by the Haas and Roham. Pore volume and surface area of the heterogeneous are about 1 mL g^{-1} and

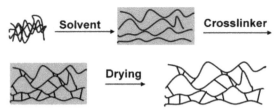

Figure 1.1 Schematic representation of HCPs polystyrene network by postcrosslinkg method. Reproduced with permission, copyright 2007, Royal Society of Chemistry [10].

20–300 m^2 g^{-1}, respectively. In chromatography, hypercrosslinked networks are used as stationary phase [10]. In Manhattan nuclear project, the first-generation DVB and styrene as ion exchange resin can be used to separate the rare earth element, gel-type copolymer cannot meet all the requirements. High osmotic pressure, sorption capacity and permeability are the requirements for excellent column packing materials, and they will be able to preserve the change in conditions such as mobile phase pressure, high mechanical strength, concentration, polarity and pH value. The change in external environment results copolymer rapid swelling and deswelling under lower osmotic pressure, and also shorten the material life. For industrial applications, the macroporous sorption capacity is insufficient due to the functional sites are on the pores surface [14].

1.2.2 Third generation hypercrosslinked polystyrene

Hypercrosslinked polystyrene can be prepared through suitable solvent dichloroethane with highly crosslinked precursor. External crosslinkers and catalysts SnCl$_3$ or FeCl$_3$ are used for generation of crosslinks, in the swollen state chain of polymer will be locked. Micropores and microporous of HCPs polystyrene networks are generated by the results of solvent removing (Fig. 1.1).

Hypercrosslinked polystyrene with high density can be synthesized through Friedel–Crafts alkylation. Davankov used monochlorodimethyl ether (MCDE), xylene dichloride (XDC), 1,4-bis chloromethyl diphenyl (CMPD) and tris-chloromethyl compound to synthesize the polystyrene chain. Methyl bridges are formed by introducing choloromethyl groups to phenyl ring. Dicholro ethylene (DCE), carbon tetra chloride (CCl$_4$), formaldehyde dimethyl acetal (FDA) are replaced for croslinking polystyrene through alkylation of two phenyl rings [15,16]. MCDE is carcinogenic but FDA is widely used for the crosslinking polymer chain. HCPs precursors such as highly crosslinked polystyrene and linear polystyrene with precursor (DVB 30%) were also used for hypercrosslinking. The density of highly crosslinked precursors was higher than the obtained crosslinking polymer chain. Precholromethylation introduced internal electrophiles before crosslinking while Davankon used external electrophiles as crosslinkers. The rigidity of network was increased, and the conversion rate of choloromethyl to methylene bridges was decreased. The surface area of 1000 m^2 g^{-1} is the result of the hypercrosslinking via Friedel-Crafts reaction with 5% of DVB as precursor. Sherrington synthesized the HCPs polystyrene with high surface area by using new modified route [16]. Suspension polymerization technique was used for the copolymerized styrene with small amount of DVB and VBC. Poly (S-co-VBC-co-DVB) highly crosslinked precursor was swollen in 1, 2-dichloroethane before the postlinking by Lewis acid catalysis [17,18]. The prepared HCPs had high surface area of 2029 m^2 g^{-1}. Lower surface area of HCPs were produced by higher content of DVB (20% content of DVB) because the catalyst and solvent limited accessibilities. With the increase in the styrene contents, the surface area was decreased because of the decrease in the choloromethyl group. The amount of micropores was increased with the increase of DVB content. HCPs showed the solely micropores when the DVB content higher than 7% [19–22].

1.2.3 Other DVB highly crosslinked polymer networks

The copolymerized styrene polymers will become insoluble after treated with 0.5% of DVB because DVB provides rigidity and short links. The void spaces become permanent porous when the diluents are removed in the synthesis of macroporous DVB crosslinked styrene. Copolymer synthesis on the industrial scale is 80% technical grade. Isomers para and meta mixture of DVB is about 80% on commercially scale and are available, while the ortho-DVB isomers are negligible because its cyclization to the naphthalene, diethyl benzene, and ethylstyrene para and meta isomers are impure. Styrene and meta-DVB have similar and low reactivity while para-DVB has high reactivity. Therefore, the crosslinks distribution is extremely inhomogeous during the copolymerization because the para-DVB is consumed first. Such copolymer inhomgeiousity may affect ion exchange/adsorption and result poor swelling properties [22–24]. The copolymerization of maleic anyhydride (electron deficient) with styrene polymer (electron rich) is an alternating way. For reducing the anhydride hydrolysis, glycerol is used instead of water. Dispersion, precipitation, suspension polymerization techniques

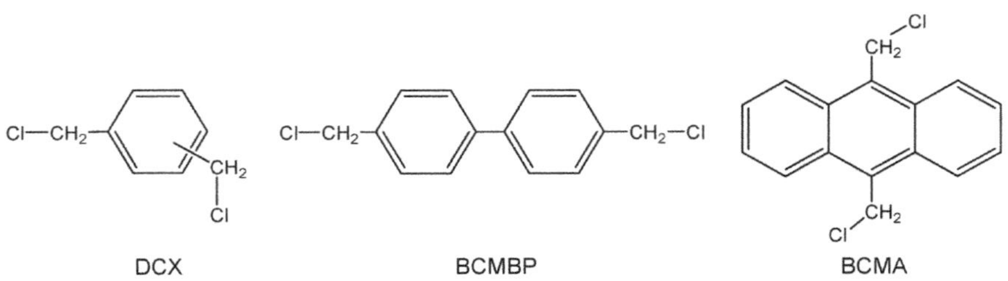

Cl—CH₂   Cl—CH₂   CH₂—Cl

DCX   BCMBP   BCMA

Figure 1.2 Monomers used for the synthesis of the hypercrosslinked polymer networks [31]. Reproduced with permission, copyright 2007, American Chemical Society.

are used for the preparation of DVB maleic anhydride particles. Amines react with the anyhydride groups in hydrolysis to form amide ester. For the potential applications such as gas sorption, heavy metal adsorption, and magnetite supporting matrix, highly crosslinked DVB maleic anhydride networks were investigated. Porosity of polymer particles can be achieved by etching block copolymer components with the degradable blocks such as poly tert-butyl acrylate (PtBA) and polylactide (PLA). Porous structure can be maintained from collapsing because the remaining matrix rigidity is enough though for etchable block accessible to the etching agent [25–27].

Hillmyer and Seo had employed crosslinking styrene and DVB for the polymerization of inducing phase separation, in the presence of RAFT agent containing the PLA segment. Styrene rich domain and PLA rich domain are the separate microphase of the resulting copolymer. The resultant mesopores showed 3D structure with poly(styrene-co-DVB) after removing the PLA segment by hydrolysis. Hierarchically network can be achieved by hypercrosslinking using the same precursor PLA-b-poly(VBzCl-co-DVB) and then etching the copolymer component. Polyethylene oxide (PEO) fraction and molecular weight-controlled micro and meso phase separation [28–30].

1.3 Hypercrosslinked polymer by polycondensation

Polycondensation technique can be employed for the preparation of HCPs without monomer precursors. Bischloro-methyl-unthtracene, dicholrox-ylene (DCX) and bis choloromethyl biphenyl (BCMBP) are the aromatic monomers used for preparation of macroporuous organic networks (Fig. 1.2). Cooper and coworkers used different amount of copolymer and Lewis acid, different isomers of DCX and BCMBP to synthesize organic networks [31]. These materials showed high BET surface area of 1094 m^2 g^{-1}

and the structure were predominantly microporous. BCMBP based networks showed storage capacity of 3.68 wt% at 15 bar and 77.3 K. The isosteric heat of hydrogen sorption for these materials is between 6 and 7.5 kJ·mol^{-1}. The physical properties like absolute density, bulk density, pore width, and pore volume are stimulated by p-DCX networks. This model also revealed the same isotherm shape and isosteric heat for H_2 sorption. Methane uptake was also investigated at 20 bar and 298 K. BCMBP and DCX copolymers showed adsorption up to 5.2 mmol g^{-1} for methane, which is high enough for the microporous but falls for high micropores volume. Martin and coworkers used high pressure test and atmospheric pressure tests for the CO_2 adsorption capacity of these polymers [32]. The results were found that CO_2 capture ability related with HCPs textural properties. The CO_2 adsorption performance of these materials was characterized by maximum uptake of CO_2 at atmospheric pressure. The polymer can show maximum uptake at high pressure.

Chaikittisilp synthesized hierarchically siloxane organic hybrid HCPs by Friedel-Crafts self poly-condensation of benzyl chloride and using its terminated double four ring cubic siloxane cages as molecular precursor (Fig. 1.3) [33]. Simultaneous polymerization of the organic functional groups and destruction of the siloxane cages during synthesis yielded PSN-5, which has an ultrahigh BET surface area (2500 m^2·g^{-1}) and large pore volume (3.3 cm^3·g^{-1}). PSN-5 also showed a high H_2 uptake of 1.25 wt % at 77 K and 760 Torr. Porous organic frame works and MOFs for H_2 storage application has the similar values. Yuan et al. studied Friedel-Crafts alkylation reactions on the basis of organic porous containing carborane (Fig. 1.4) [34]. In this study m-carborane was linked with two methyl functional groups in the design and synthesis of carborane monomer. The prepared monomer was used in the synthesis of PmCB3 and PmCB4 *via* copolymerization with BCMBP and self poly-condensation. The maximum uptake of hydrogen for PmCB3 and PmCB4 was 862 m^2·g^{-1} and 1037 m^2·g^{-1}, respectively.

Tan et al. synthesized hypercrosslinked polymer networks by the self-condensation of monohydroxymethyl compound, benzyl alcohol (BA) (Fig. 1.5) [35]. Network HCP-BA shows the BET surface areas of 742 m^{2}g^{-1}. Based on the reports in the synthesis of polymer, the pore formation mechanism can be explained. The benzayl carbocation generation is the primary step which attacks other monomers to form dimmers. In dimer the ortho-para groups are present in one phenyl ring which can attack by the cations significantly. Other molecules are attacked by dimmers and benzyl alcohol which can grow with its hydroxymethyl group. Many short

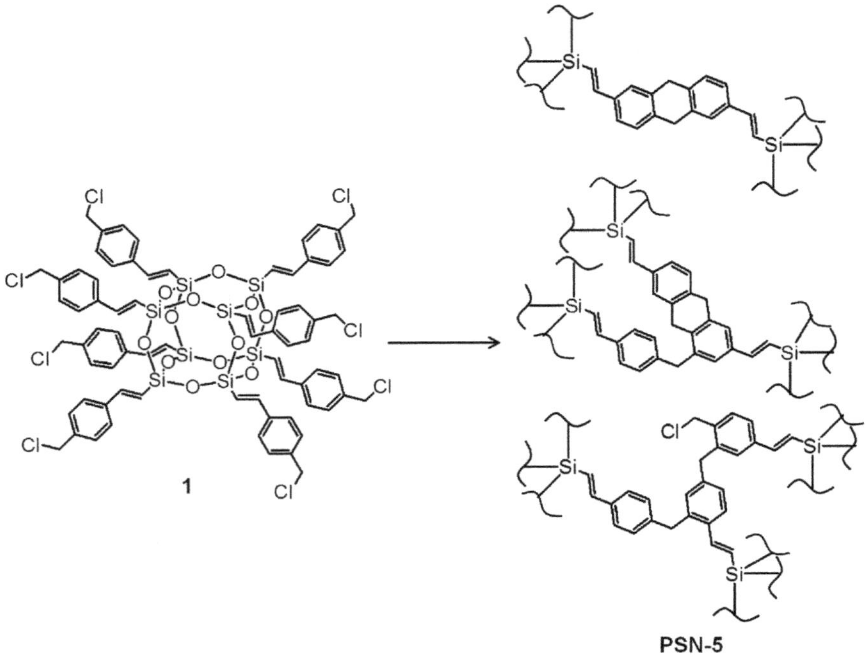

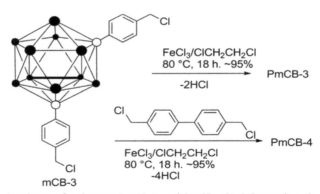

Figure 1.3 Synthesis of PSN-5 through Friedal-Crafts selfcondensation [33]. Reproduced with permission, copyright 2011, American Chemical Society.

Figure 1.4 Synthesis of polymers PmCB-3 and PmCB-4 [34]. Reproduced with permission, copyright 2012, John Wiley & Sons, Inc.

branches and substituted ring are produced by the addition of benzyl group to the dimmers. BA has the capability to store significant amount of H_2 (0.9 wt%) and CO_2 (8.46 wt%) at 0.1 bar at 77 K and 273 K. The study assumed that monofuntional monomers can be used to synthesize highly

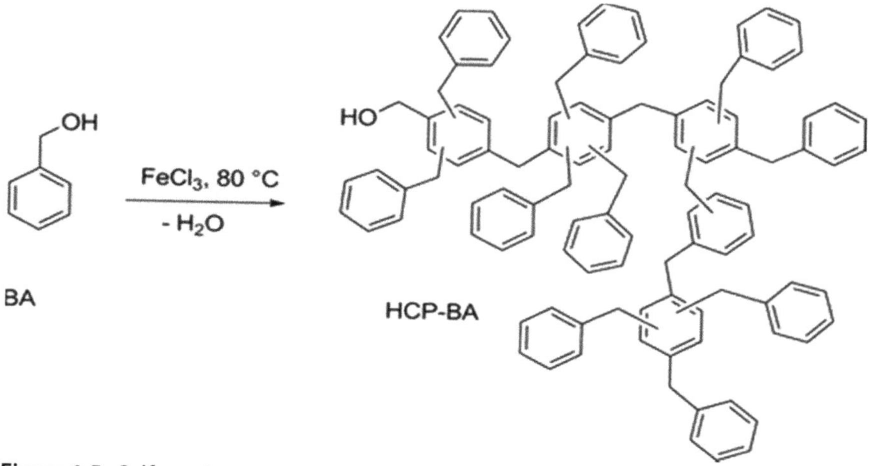

Figure 1.5 Self condensation synthesis of HCPs-BA [35]. Reproduced with permission, copyright 2013, Royal Society of Chemistry.

porous materials. These materials are the promising gas adsorbent candidates for clean energy and environmental applications.

1.4 Properties of the hypercrosslinked polymers

1.4.1 The effect of dilution on the initial system

Fig. 1.6 shows the swelling capacity of networks obtained by crosslinking linear polystyrene with xyelene dichloride and monocholrodimethyl (MCDE), which is the function of polystyrene concentration in ethylene dichloride (EDC). In the postcrosslinking, the swelling capacity increased with diluting the system. The swelling capacity of crosslinking styrene-DVB copolymers also vary with concentration of starting copolymer in EDC (Fig. 1.7). With higher DVB content, the swelling of the starting copolymer was smaller because the amount of EDC remaining in the final network was lower. Interestingly, replacing conventional MCDE with XDC or CMDP did not alter the effect of starting polymer concentration on swelling ability [36].

Figs. 1.6 and 1.7 showed that the polymer content up to 18% for linear polystyrene and ranged from 28% to 47% for swelling copolymers of styrene with 1%–4% DVB. In this case, the meeting of two styrene unit formed intermolecular loop will be regarded as negligible. More significantly, for hypercrosslinked networks with crosslinking degree ranging from 43% to 100% (plot lines 2 and 3 in Fig. 1.6), in which no looped may remained unbound towards the entire system, there is no need to distinguish between

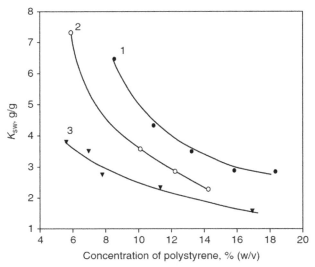

Figure 1.6 The swelling ability of networks in toluene produced by crosslinking linear polystyrene with XDC (line 1, 2) and MCDE (line 3) [36]. Reproduced with permission, copyright 2011, Elsevier.

intermolecular and intra molecular crosslinks. In order to explain the effect of dilution in the system under the postcrosslinking, the network was considered as collection of mutually condensed and interpenetrating meshes, which comprised polymeric chains and crosslinking bridges. It does not matter whether these meshes are belonging to the same initial chains, but each mesh can embrace chains belonging to the neighboring meshes. The interpenetration of meshes in the networks is strongly dependent on the polymeric chains concentration in the initial solution during the closure of meshes. It means that the swollen styrene-DVB copolymer or the linear styrene after postcrosslinking in solution has the higher swelling ability than gel-type styrene-DVB copolymer at equivalent degree of crosslinking. Because gel-type styrene-DVB copolymers are prepared without any solvent, therefore, each mesh of their network must be densely filled with alien polymeric chains and meshes.

1.4.2 The effect of initial copolymer network

Copolymers of the two different series with MCDE are prepared by additional crosslinking under the same conditions. Before the reaction of postcrosslinking, one copolymer of styrene partially swollen in 0.3% content of DVB with EDC while the other copolymer styrene swollen in 1%

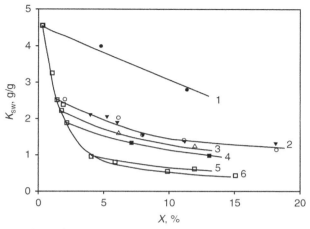

Figure 1.7 The effect of crosslinking degree on the equilibrium swelling in toluene for networks prepared by crosslinking gel-type styrene–DVB copolymers with 1.4-bis(chloromethyl)diphenyl (line 1, 2, 4, 5), monochlorodimethyl ether (line 3), and styrene–DVB copolymers (line 6) [36]. Reproduced with permission, copyright 2011, Elsevier.

content of DVB with EDC to the maximum, and both polymers absorb same amount of EDC before the postcrosslinking reaction. Due to the concentration of polymeric chains (0.485 mg mL^{-1}) in the two starting systems is the same, it can be expected that the swelling capacity of the respective final hypercrosslinked products is exactly equivalent for such two systems. Remarkably, the products derived from the partly swollen copolymer containing 0.3% DVB show a significantly higher absorption capacity for toluene (Fig. 1.8). After introducing more additional crosslinks, it appears that the small variation of 0.7% DVB between the two series of products is still noticeable.

The swelling capacity for two more series of samples with similar concentrations was analyzed to confirm the effect of polymeric chains concentration (Fig. 1.8, line 1 and 2). One sample was prepared by postcrosslinking of the styrene–0.3% DVB copolymer swollen to its maximum, the other one was obtained by crosslinking soluble linear polystyrene dissolved in EDC. According to the effects of concentration and crosslinking process, the swelling capacity of two components crosslinked at concentration of 0.485 mg mL^{-1} was noticeably lower than these two series. However, the swelling performance of completely swollen styrene–0.3% DVB copolymer appears to be significantly smaller than linear polystyrene.

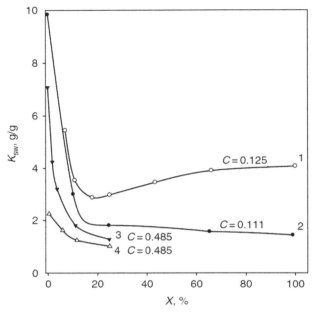

Figure 1.8 The effect of crosslinking degree on the equilibrium swelling capacity in toluene. Concentration of (1) linear polystyrene $C_0 = 0.125$ mg mL^{-1}; (2) Styrene 0.3% DVB copolymer swollen maximum, $C_0 = 0.111$ mg mL^{-1}; (3) Styrene 0.3% DVB partially swollen $C_0 = 0.485$ mg mL^{-1}; (4) Styrene 0.1% DVB swollen maximum at $C_0 = 0.485$ mg mL^{-1} [36]. Reproduced with permission, copyright 2011, Elsevier.

The structure of copolymer of styrene with 0.3% or 1% DVB prepared without solvent is unstressed in dry condition. On swelling these products with EDC, their network expands until further swelling is stopped when the stress of the elastically active chain increases to a maximum equilibrium level. In this maximally pre-strained state, the network receives many additional crosslinks upon reaction with bifunctional MCDE. The pre-strained structural fragments severely limit the solvent uptake of the final hypercrosslinked product. Materials prepared from dissolved linear polystyrene will not contain any prestrained fragments, therefore, these materials will always show the highest swelling capacity.

1.4.3 The effect of crosslinking distribution

When the initial reaction mixture of linear polystyrene with monochlorodimethyl is homogenous, the dispersion of crosslinking bridges in final network is almost homogeneous. However, when MCDE partly chloromethylated polystyrene links are used instead of unsubstituted

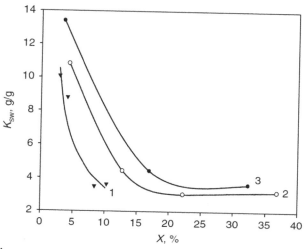

Figure 1.9 The network generated by crosslinking linear polystyrene with partly chloromethylated polystyrene containing (1) 3%; (2) 11.5%; and (3) 14.8 % of chlorine has an effect on equilibrium swelling in toluene [36]. Reproduced with permission, copyright 2011, Elsevier.

polystyrene links for such reaction, a certain divergence from statistical distribution of crosslinks must be predicted. The further chlorine atoms in the polymeric crosslinking agent links, and less of the latter is required to achieve the appropriate net crosslinking grade, but the final gel structure will be less uniform.

The relationship between toluene swelling and the degree of crosslinking of the resulting products was shown in Fig. 1.9. The networks formed by selfcrosslinking of the chloromethylated polystyrene containing 3% chlorine, with the crosslinking degree of 10% show a high toluene adsorption capacity (3.5g g^{-1}). By using chloromethylated polystyrene with more reactive chlorine, such as 11.5% or 14.8% Cl, as the crosslinking agent and combining it with a sufficient amount of unsubstituted polystyrene, products with the same nominal 10% or higher degree of crosslinking can be generated.

1.4.4 Role of crosslinking bridging structure and stresses of hypercrosslinked polymer

Linear polymer can be dissolved in the solvents having strong affinity toward the polymers and 3D polymers can also be dissolved in such solvents. Both thermodynamically good solvents (such as methylene dichloride and toluene solvent) and precipitating media for linear polystyrene (such as n-hexane)

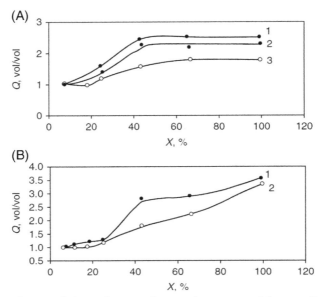

Figure 1.10 The crosslinking degree of networks generated by crosslinking linear polystyrene with (A) p-xylylene dichloride and (B) methylal is dependent on volume swelling in (1) n-hexane; (2) methanol; and (3) water [36]. Reproduced with permission, copyright 2011, Elsevier.

can be used for the swelling of hypercrosslinked polymer networks (Fig. 1.10). Gel type polystyrene copolymer swollen with toluene rapidly shrinks when placed it in the methanol, while hypercrosslinked gel swollen with toluene shows no change in their volume in any other organic solvent, including any precipitating media. The shrinkage of the swollen polymer only occurs during drying process by the removal of solvents. The dry polymer swells rapidly in organic solvents, thus the shrinkage is total reversible. The swelling values are independent on the sequence of solvents that touch the polymer, which confirm the equilibrium character of the swelling. Indeed, polymers absorb the same amount of n-heptane whether in direct contact with it or after replacing EDC with the solvent.

1.4.5 The effect of reaction medium

Cyclohexane, nitrobenzene, tetrachloroethane, and EDC are the appropriate solvents for reaction of Friedel-Crafts crosslinking. The swelling behavior of products prepared in EDC and tetrachloroethane are same while in nitrobenzene is different. The reaction polystyrene with XDC is accelerated

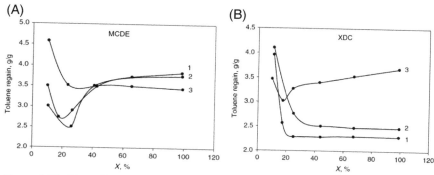

Figure 1.11 The effect of crosslinking degree on the evolution of toluene regain, hypercrosslinked polymer formed by crosslinking linear polystyrene with (A) monochlorodimethyl ether and (B) p-xylylene dichloride in (1) ethylene dichloride, (2) ethane tetrachloride, and (3) nitrobenzene [36]. Reproduced with permission, copyright 2011, Elsevier.

by nitrobenzene. The rate of hydrogen chloride generation increases to 4.26 in nitrobenzene as compared with 1.90 in EDC under identical conditions. Rigid HCPs prepared in the fast formation have the ability to enhance swelling capacity and trap large amount of nitrobenzene. MCDE can also accelerate the bridging of polystyrene, but it is unstable in nitrobenzene. The mixture of SnCl$_4$ and MCDE in nitrobenzene evolves HCl at 80–900 °C without adding any polystyrene mixture. Fig. 1.11 revealed the unusual swelling behavior of product obtained in the side reaction. Cyclohexane was used for the synthesis of hypercrosslinked polymer at 60 °C because the crosslinking reaction proceeds too slowly at low temperature. The affinity of cyclohexane to polystyrene increased with the increase of temperature, however, cyclohexane still remains not as good as EDC at 80 °C.

1.4.6 The effect of molecular weight

Molecular weight of the polymers also affects the swelling behavior of hypercrosslinked and macronet isoporous. As shown in Fig. 1.12, the resultant products show lower swelling capacity due to the higher molecular weight of polystyrene. It's tough to come up with a sensible answer for this discovery. We could only summarize that the microstructure of the initial solutions at 10% concentration in the postcrosslinking process is definitely influenced by the size of the polystyrene coils. When all polymeric coils achieve a capacious, unaffected, or even extended size, and totally interpenetrate one another to create a stable homogenous system, as the classic theories of

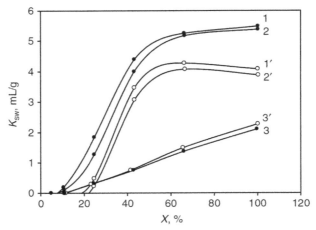

Figure 1.12 Swelling in (1, 1′) n-hexane, (2, 20′) methanol, and (3, 3′) water of the networks prepared by crosslinking with monochlorodimethyl ether in ethylene dichloride of polystyrene with molecular weight of (1-3) 8800 and (1′-3′) 300,000 Da [36]. Reproduced with permission, copyright 2011, Elsevier.

polymer solutions would imply, then chain length will have no discernible effect [37].

1.5 Morphology of hypercrosslinked polymers

1.5.1 Morphology of surface modification of polymer resins

Gel type precursors (GTP) were synthesized through free radical suspension polymerization, hypercrosslinked "Davankov Resins" (HCPs–DR) was prepared by crosslinking of GPT. Due to hydrophobic and low density, HCPs–DR shows weak aqueous wettability and less useful for the removal of metal ions from water. HCPs–DR does not disperse well in metal ion solution and remains floating on surface even after long time stirring (at least 2 h). Therefore, the materials could not sufficiently contact with the solution and show a poor adsorption performance. By introducing the –SO_3H hydrophilic group to HCPs–DR, the resulting sulfonic acid modified HCPs–DR (SAM-HCPs–DR) was well dispersed in a very short time under stirring. The modification does not make any change and the morphology after introducing –SO_3H was similar to the HCPs–DR but it was dark in color (Fig. 1.13). TEM images revealed that both materials have spherical morphology with diameter of 100 nm [38].

(A)

(B)

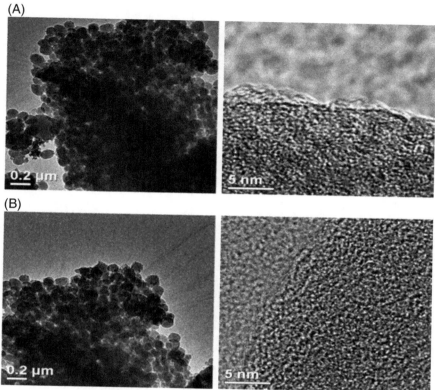

Figure 1.13 TEM images of (A) HCPs-Dr (B) SAM-HCPs-DR [38]. Reproduced with permission, copyright 2011, Elsevier.

1.5.2 Morphology of tubular hypercrosslinked porous polymer

$FeCl_3$ catalyzed Friedel–Crafts polymerization of benzene with formaldehyde dimethyl was used for the synthesis of benzene-based tubular hypercrosslinked porous polymer (HCPT-B). HCPT-B was directly carbonized at 700°C for 3 h under the nitrogen atmosphere to obtained benzene base porous carbon nanotubes (PCNT-B). The result revealed that monomer concentration has great influence on the morphology. Fig. 1.14 showed that the HCPT-B had uniform tubular morphology due to the monomer concentration of 0.5 M for benzene. Without any collapse of the tubes, the PCNT-B retained tubular morphology after carbonization. Such uniform and 1D tubular structure of carbon nanotubes with external diameter of 57 nm and inner diameter of 34 nm was synthesized at first time [39].

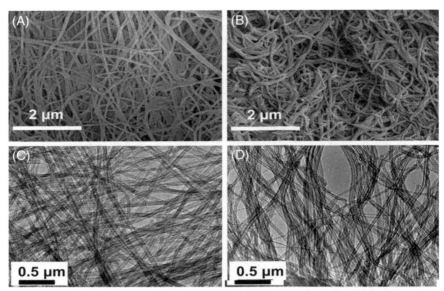

Figure 1.14 SEM images of (A) HCPT-B and (B) PCNT-B, TEM images of (C) HCPT-B and (D) PCNT-B [39]. Reproduced with permission, copyright 2017, American Chemical Society.

1.5.3 Morphology of mesoporous hypercrosslinked polymers

A mesoporous hypercrosslinked polymer (HCP) was constructed using heterocyclic-IL-containing precursors as building blocks and 1,4-bis(chloromethyl)benzene as crosslinker *via* the Friedel-Crafts reaction. The prepared HCP had a high specific surface area with abundant mesopore and specific porosity. FE-TEM and FE-SEM were used to examine the morphologies (Fig. 1.15) [40]. Amorphous structure and rough surface were stacked together loosely. TEM images indicated that disorders and abundant mesopores existed in the synthesized HCPs. The presence of mesopores and micropores described HCPs inherent nature.

1.5.4 Morphology of hollow structured hypercrosslinked polymer

Functionalized hollow structure HCPs were synthesized *via* Friedel-Crafts reaction, and harmonious coexistence of acid (sulfonic acid) and base (amine) sites peacefully coexist in a hollow microporous HCP catalyst (HCP–A–B). The introducing of functional groups to HCP did not affect the size and morphology. Fig. 1.16 also shows the spherical shape of HCP with flat surfaces, uniform size distribution and microsphere diameter of

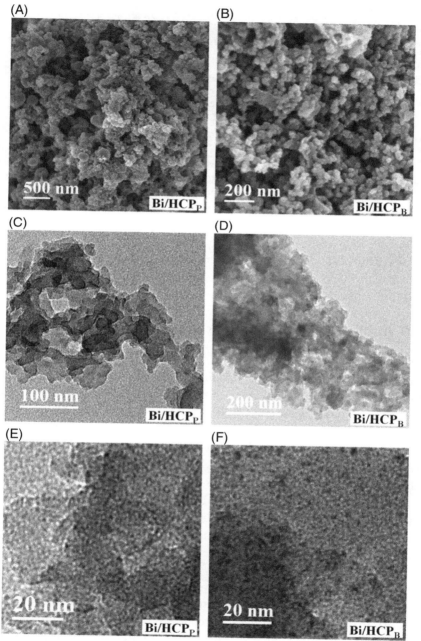

Figure 1.15 FE-SEM and FE-TEM images of hypercrosslinked polymers [40]. Reproduced with permission, copyright 2020, Elsevier.

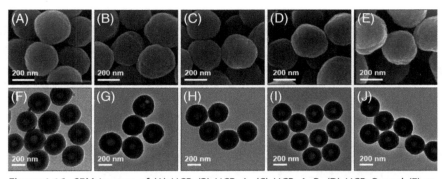

Figure 1.16 SEM images of (A) HCP, (B) HCP–A, (C) HCP–A–B, (D) HCP–B, and (E) recovered HCP–A–B. TEM images of (F) HCP, (G) HCP–A, (H) HCP–A–B, (I) HCP–B, and (J) recovered HCP–A–B [41]. Reproduced with permission, copyright 2017, American Chemical Society.

300-330 nm [41]. The hollow sphere's structure was also confirmed by TEM with thickness of 100–110 nm and inner diameter of approximately 100–120 nm.

1.6 Applications of hypercrosslinking polymers

1.6.1 Sensing

A novel porous polymer was synthesized by Friedel-Crafts alkylation polymerization, the porous polymers were used as a host to prepare electrochemical humidity sensor by loading LiCl, and the sensor showed good humidity sensing properties [42]. Compared with the pure polymer, LiCl-loaded composites improved humidity sensing properties. The hydrophilic properties of porous polymers were enhanced by using LiCl salt for the lithium modification, which enhances the interaction between materials and water molecules. The resulted modified sensors showed decrease with the increase of relative humidity. Lithium hydroxide instead of LiCl can also enhance the sensitivity of electrochemical humidity sensors.

1.6.2 Photocatalysis

Pure organic, heterogeneous, metal-free, and visible light-active photocatalysts offer a more sustainable and environmentally friendly alternative to traditional metal-based catalysts. Zhang et al. reported a series of microporous organic polymers containing photoactive conjugated organic semiconductor units as heterogeneous photocatalysts for a visible-light-promoted [43]. *Via* a simple Friedel-Crafts alkylation reaction, the microporous organic polymers were obtained by crosslinking of organic semiconductor compounds with defined valence and conduction band positions. The prepared microporous organic polymers showed high efficiency and reusability. Their results showed that the knitting strategy is the new opportunity to utilize it for the efficient heterogeneous photocatalysts.

1.6.3 Drug delivery

Hollow microporous organic capsules (HMOCs) have great potential to be used in storage materials and supplying path for the design of control uptake. Tan and coworkers investigated the drug loading and release in HMOCs by using ibuprofen (IBU) as model drug [44]. The drug loading efficiency of HMOCs was calculated from the data obtained from the thermogravimetric and UV-vis spectroscopic analysis. HMOCs can uptake 1.68–2.04 g of IBU g^{-1}, which is much higher than the solid HCP nanoparticles (0.80 g of IBU/g). The results indicated that the porous structure of the shell also affect the drug kinetics. Due to their tailored porous structure, these capsules possessed high drug loading efficiency, zero–order drug release kinetics are also demonstrated to be used as nanoscale reactors for the preparation of nanoparticles (NPs) without any external stabilizer.

1.6.4 Membrane gas separation

The gas separation technology is the prominent alternative technique for industrials process such as combustion carbon capture, hydrocarbon purification, hydrogen recovery and natural gas purification. It is a challenge to prepare polymeric sieve membrane with permanent microporosity. Tan and coworkers introduce formaldehyde dimethyl acetal (FDA) as an external crosslinker to form solution processable hypercrosslinked polymers (SHCPs) from commercial polystyrene in the presence of anhydrous ferric chloride at high dilution [45]. Despite being highly crosslinked, the resulting HCPs dissolve in a range of organic solvents to form thermodynamically stable homogenous liquids. By increasing the concentration of polystyrene and the amount of crosslinker, the BET surface area of SHCPs increased with the largest surface area of 724 m^2 g^{-1}. Moreover, they also show comparable gas uptake properties reaching 8.11 wt% CO_2 adsorption (1.13 bar and 273 K), 1.01 wt% H_2 adsorption (1.13 bar and 77 K), and 0.14 wt% CH_4 adsorption (1.13 bar and 273 K). Cooper and Bud investigated composite membrane permeability by adding HCP fillers onto PIM membrane. The CO_2 permeability showed increase of 250% as compared to the pristine [46].

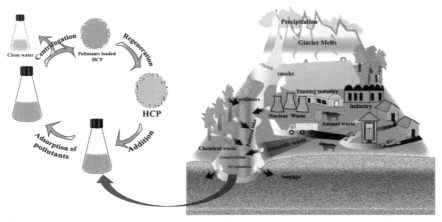

Figure 1.17 Hypercrosslinled polymer used for removal hazardous dyes [48]. Reproduced with permission, copyright 2021, American Chemical Society.

1.6.5 Removal of pollutants

HCPs are widely used as solid adsorbent for solid phase extraction, organic vapors adsorption, waste water treatment and chromatographic analysis. Due to the micropores, the adsorption is not only surface phenomena but also the adsorbate transfers from surface to inner of micropores through the channels. As a result, the adsorption rate as well as adsorption capacity also enhanced. Hydrophobic skeleton of HCPs materials showed good adsorption performance in waste water treatment as compared to the inorganic materials. Hydrophilic hypercrosslinked polymer was prepared in the form of microspheres from the precursors of 2-hydroxymethyl methacrylate for solid phase extraction [47]. Due to the high surface area of the hydrophilic hypercrosslinked polymer, they had the capability to adsorbed 90% polar compounds from water. HCPs materials were used as adsorbent for the removal of toxic metal ion from waste water. Atoms in the ligand with lone pair of electrons enhance the removal of metal ions.

Nowadays, the human health and the environment are adversely affected by the emission of volatile organic compounds. To search for effective materials for wastewater treatment, HCPs have emerged as an excellent class of porous materials for wastewater treatment due to their unique features, such as high surface area, tunability, biodegradability, and chemical versatility (Fig. 1.17) [48]. Wang and coworkers prepared microporous hypercrosslinked polymer adsorbents which can be used for adsorption and desorption process [49]. The microporous structure of HCPs used as adsorbents for the small organic compounds, it cannot be used for the large-scale removal of oil and organic solvents. Hierarchically monolith was synthesized by knitting polystyrene precursors and used for the oil sorption ability. Due to the hydrophobic properties, prepared monolith showed fast adsorption kinetics to different organic solvents and the adsorption capacities was in the range of 800%–1900% which depended on the organic solvent's density.

1.6.6 Capturing carbon dioxide gas

In recent years, the emission of greenhouse gases is increased at high level especially carbon dioxide gas which are affecting our climate and it has got attention of researchers. Due to the thermal stability and high surface area, porous materials were used as adsorbents in carbon capture and storage by reducing the concentration of CO_2 *via* physical adsorption in the air.

Cooper and coworkers had prepared number of polymer materials and they introduced various functional groups to polymer matrix [50]. Functional groups like carboxyl and amine enhanced the adsorption capability of CO_2 and it can increase the interaction between adsorbate and adsorbent. The microporous organic polymer was the product of the "click" reaction between tetrakis(4-azidophenyl)methane and tetrakis(4-ethynylphenyl)methane with BET surface areas of 1237 $m^2 \cdot g^{-1}$. The polymers showed a promising CO_2 adsorption capability with CO_2 uptake of 3.36 mmol g^{-1} at 1 bar and 298 K, possibly because of contributions from the electron–rich triazole unit in the network which enhanced the interaction between polymer matrix and CO_2 molecule.

FDA knitting method was used to get amino functional groups containing polymer network by copolymerization of benzene and aniline [51]. Due to the strong ability of donating effect, the surface area by incorporation of aniline onto benzene was decreased when the aniline concentration was increased. With the decrease of surface area, the CO_2 adsorption capability was also decreased, but the selectivity of CO_2 N_2^{-1} was increased which indicated the role of functional group of amines.

1.7 Future consideration and challenges

Nonetheless, many efforts have been focused toward the development of low-cost and effective smart materials for wastewater treatment. In order to continue these efforts, unique and smart materials that may efficiently and robustly serve to cleanse wastewater are still needed. The rational development of such smart and adaptable porous materials might lead to water purification without the need of severe efforts or the use of resources such as electrical and thermal energy, which have become an essential component of other water treatment systems. Still, there is a large opportunity to investigate the potential of new monomers with unique functionalities, which might present interesting chances to build functionalized HCPs for improving performance in the separation and purification of polluted water. It is crucial to note that in the synthesis of HCPs, noble metal-based catalysts are required, which makes the whole process costly and unsuitable for scale-up manufacturing. As a result, efforts must be made to create metal-catalyst-free approaches for the synthesis of smart porous materials in order to reduce the total cost of the water purification process. More emphasis should be placed on the production of hybrid adsorbents wherein the HCPs are developed on the surfaces of the nanostructures to increase the surface area and effectiveness for water treatment.

Nanostructures not just to give more surface area, but they may also be beneficial in removing pathogens from water. The consistency of the pore structure of HCPs is a fundamental barrier for them, since it affects their repeatability and difficultly makes scale-up manufacture. The establishment of everlasting permeability in the framework of hypercrosslinked porous polymers is required, which involves the development of new polymerization processes and the creation of innovative smart monomers with the appropriate functionalities.

Additionally, for wide range of applications, the management of heat released during in the exergonic crosslinking reaction, and the scale-up manufacturing of HCPs remains a difficulty. Moreover, the harmful effects of utilizing HCPs in wastewater treatment, particularly their influence on biological wastewater treatment, must do further investigation, so that HCPs may be changed to meet the fauna and flora of aquatic waste bodies. The characteristics of HCPs, particularly their enormous surface area of more than $4000 \ m^2 \ g^{-1}$, make them be smart water cleaning materials for investigation and eventual development to deal with the ever-increasing harmful pollutants in water.

References

[1] S.K. Das, X.B. Wang, Z.P. Lai, Facile synthesis of triazine-triphenylamine-based microporous covalent polymer adsorbent for flue gas CO_2 capture, Microporous Mesoporous Mater. 255 (2018) 76–83.
[2] J.Y. Jang, H.T.T. Duong, S.M. Lee, H.J. Kim, Y.J. Ko, J.H. Jeong, et al., Folate decorated hollow spheres of microporous organic networks as drug delivery materials, Chem. Comm. 54 (29) (2018) 3652–3655.

[3] H. Bildirir, V.G. Gregoriou, A. Avgeropoulos, U. Scherfd, CL. Chochos, Porous organic polymers as emerging new materials for organic photovoltaic applications: Current status and future challenges, Mater. Horiz. 4 (4) (2017) 546–556.

[4] S. Krishnan, CV. Suneesh, Fluorene - Triazine conjugated porous organic polymer framework for superamplified sensing of nitroaromatic explosives, J. Photoch. Photobio. A 371 (2019) 414–422.

[5] J. Huang, SR. Turner, Hypercrosslinked polymers: A review, Polym. Rev. 58 (1) (2018) 1–41.

[6] J. Germain, J. Hradil, J.M.J. Frechet, F. Svec, High surface area nanoporous polymers for reversible hydrogen storage, Chem. Mater. 18 (18) (2006) 4430–4435.

[7] M.P. Tsyurupa, VA. Davankov, Hypercrosslinked polymers: Basic principle of preparing the new class of polymeric materials, React. Funct. Polym. 53 (2-3) (2002) 193–203.

[8] N. Fontanals, R.M. Marce, F. Borrull, PAG. Cormack, Hypercrosslinked materials: Preparation, characterisation and applications, Polym. Chem. 6 (41) (2015) 7231–7244.

[9] I.F. Khirsanova, V.S. Soldatov, R.V. Martsinkevich, M.P. Tsyurupa, V.A. Davankov, Sorption properties of polystyrene gels cross-linked by para-xylylene dichloride, Colloid J. USSR 40 (5) (1978) 871–874.

[10] J. Germain, J.M.J. Frechet, F. Svec, Hypercrosslinked polyanilines with nanoporous structure and high surface area: Potential adsorbents for hydrogen storage, J. Mater. Chem. 17 (47) (2007) 4989–4997.

[11] H. Staudinger, W. Heuer, High polymer compounds, 93. Announcement. Decomposition of thread molecules of polystyrene, Ber. Dtsch. Chem. Ges. 67 (1934) 1159–1164.

[12] V. Davankov, M. Tsyurupa, M. Ilyin, L. Pavlova, Hypercross-linked polystyrene and its potentials for liquid chromatography: A mini-review, J. Chromatogr. A 965 (1-2) (2002) 65–73.

[13] I.M. Abrams, JR. Millar, A history of the origin and development of macroporous ion-exchange resins, React. Funct. Polym. 35 (1-2) (1997) 7–22.

[14] V.A. Davankov, MP. Tsyurupa, Hypercrosslinked polymeric networks and adsorbing materials: Synthesis, properties, structure, and applications, Comprehensive Analytical Chemistry, Elsevier, 2011.

[15] F. Svec, J. Germain, J.M.J. Frechet, Nanoporous polymers for hydrogen storage, Small 5 (10) (2009) 1098–1111.

[16] V.A. Davankov, S.V. Rogoshin, MP. Tsyurupa, Macronet isoporous gels through crosslinking of dissolved polystyrene, J. Polym. Sci. Pol. Sym. (47) (1974) 95–101.

[17] N. Grassie, J. Gilks, Friedel-Crafts crosslinking of polystyrene, J. Polym. Sci. Pol. Chem. 11 (7) (1973) 1531–1552.

[18] N. Grassie, J. Gilks, Thermal-analysis of polystyrenes crosslinked by para di(chloromethyl)benzene, J. Polym. Sci. Pol. Chem. 11 (8) (1973) 1985–1994.

[19] D.C. Wu, C.M. Hui, H.C. Dong, J. Pietrasik, H.J. Ryu, Z.H. Li, et al., Nanoporous polystyrene and carbon materials with core-shell nanosphere-interconnected network structure, Macromolecules 44 (15) (2011) 5846–5849.

[20] B. Gawdzik, J. Osypiuk, Modification of porous poly(styrene-divinylbenzene) beads by Friedel-Crafts reaction, Chromatographia 54 (5-6) (2001) 323–328.

[21] V.V. Azanova, J. Hradil, Sorption properties of macroporous and hypercrosslinked copolymers, React. Funct. Polym. 41 (1-3) (1999) 163–175.

[22] R.V. Law, D.C. Sherrington, C.E. Snape, I. Ando, H. Kurosu, Solid-state C-13 MAS NMR studies of hyper-cross-linked polystyrene resins, Macromolecules 29 (19) (1996) 6284–6293.

[23] R.H. Wiley, Crosslinked styrene-divinylbenzene network systems, Pure Appl. Chem. 43 (1-2) (1975) 57–75.

[24] B.T. Storey, Copolymerization of styrene and P-divinylbenzene. Initial rates and gel points, J. Polym. Sci. Part A 3 (1pa) (1965) 265–282.

[25] J. Huang, SR. Turner, Recent advances in alternating copolymers: The synthesis, modification, and applications of precision polymers, Polymer 116 (2017) 572–586.
[26] N. Ogawa, K. Honnyo, K. Harada, A. Sugii, Preparation of spherical polymer beads of maleic anhydride-styrene-divinylbenzene and metal sorption of its derivatives, J. Appl. Polym. Sci. 29 (9) (1984) 2851–2856.
[27] M. Maciejewska, L. Szajnecki, B. Gawdzik, Investigation of the surface area and polarity of porous copolymers of maleic anhydride and divinylbezene, J. Appl. Polym. Sci. 125 (1) (2012) 300–307.
[28] R.S. Frank, J.S. Downey, HDH. Stover, Synthesis of divinylbenzene-maleic anhydride microspheres using precipitation polymerization, J. Polym. Sci. Pol. Chem. 36 (13) (1998) 2223–2227.
[29] D. Donescu, V. Raditoiu, C.I. Spataru, R. Somoghi, M. Ghiurea, C. Radovici, et al., Superparamagnetic magnetite-divinylbenzene-maleic anhydride copolymer nanocomposites obtained by dispersion polymerization, Eur. Polym. J. 48 (10) (2012) 1709–1716.
[30] R. Gonte, K. Balasubramanian, P.C. Deb, P. Singh, Synthesis and characterization of mesoporous hypercrosslinked poly(styrene Co- maleic anhydride) microspheres, Int. J. Polym. Mater. 61 (12) (2012) 919–930.
[31] C.D. Wood, B. Tan, A. Trewin, H.J. Niu, D. Bradshaw, M.J. Rosseinsky, et al., Hydrogen storage in microporous hypercrosslinked organic polymer networks, Chem. Mater. 19 (8) (2007) 2034–2048.
[32] C.F. Martin, E. Stockel, R. Clowes, D.J. Adams, A.I. Cooper, J.J. Pis, et al., Hypercrosslinked organic polymer networks as potential adsorbents for pre-combustion CO_2 capture, J. Mater. Chem. 21 (14) (2011) 5475–5483.
[33] W. Chaikittisilp, M. Kubo, T. Moteki, A. Sugawara-Narutaki, A. Shimojima, T. Okubo, Porous siloxane-organic hybrid with ultrahigh surface area through simultaneous polymerization-destruction of functionalized cubic siloxane cages, J. Am. Chem. Soc. 133 (35) (2011) 13832–13835.
[34] S.W. Yuan, D. White, A. Mason, DJ. Liu, Porous organic polymers containing carborane for hydrogen storage, Int. J. Energy Res. 37 (7) (2013) 732–740.
[35] Y.L. Luo, S.C. Zhang, Y.X. Ma, W. Wang, B. Tan, Microporous organic polymers synthesized by self-condensation of aromatic hydroxymethyl monomers, Polym. Chem. 4 (4) (2013) 1126–1131.
[36] V.A. Davankov, M.P. Tsyurupa, Chapter 7 - Properties of hypercrosslinked polystyrene, Hypercrosslinked polymeric networks and adsorbing, Comprehensive Analytical Chemistry: Elsevier, 2011, 56, pp. 195–295.
[37] M. John, R.J.M. Chalmers, Chapter 6 Determination of molecular weights and their distributions, Molecular characterization and analysis of polymers, Comprehensive Analytical Chemistry: Elsevier, 2008, 53, pp. 205–251.
[38] B.Y. Li, F.B. Su, H.K. Luo, L.Y. Liang, B.E. Tan, Hypercrosslinked microporous polymer networks for effective removal of toxic metal ions from water, Microporous Mesoporous Mater. 138 (1-3) (2011) 207–214.
[39] X.Y. Wang, P. Mu, C. Zhang, Y. Chen, J.H. Zeng, F. Wang, et al., Control synthesis of tubular hyper-cross-linked polymers for highly porous carbon nanotubes, ACS Appl. Mater. Interf. 9 (24) (2017) 20779–20786.
[40] H.M. Zhou, Q.B. Chen, X.F. Tong, HL. Liu, Hypercrosslinked mesoporous polymers: Synthesis, characterization and its application for catalytic reduction of 4-nitrophenol, Colloid Interf. Sci. 37 (2020) 100285.
[41] Z.F. Jia, K.W. Wang, B. Tan, YL. Gu, Hollow hyper-cross-linked nanospheres with acid and base sites as efficient and water-stable catalysts for one-Pot tandem reactions, ACS Catal. 7 (5) (2017) 3693–3702.

[42] K. Jiang, T. Fei, T. Zhang, Humidity sensing properties of LiCl-loaded porous polymers with good stability and rapid response and recovery, Sensor Actuat. B-Chem. 199 (2014) 1–6.

[43] R. Li, Z.J. Wang, L. Wang, B.C. Ma, S. Ghasimi, H. Lu, et al., Photocatalytic selective bromination of electron-rich aromatic compounds using microporous organic polymers with visible light, ACS Catal. 6 (2) (2016) 1113–1121.

[44] B.Y. Li, X.J. Yang, L.L. Xia, M.I. Majeed, B. Tan, Hollow microporous organic capsules, Sci. Rep. 3 (2013) 2128.

[45] Y.W. Yang, B. Tan, CD. Wood, Solution-processable hypercrosslinked polymers by low cost strategies: A promising platform for gas storage and separation, J. Mater. Chem. A 4 (39) (2016) 15072–15080.

[46] T. Mitra, R.S. Bhavsar, D.J. Adams, P.M. Budd, AI. Cooper, PIM-1 mixed matrix membranes for gas separations using cost-effective hypercrosslinked nanoparticle fillers, Chem. Comm. 52 (32) (2016) 5581–5584.

[47] D. Bratkowska, N. Fontanals, F. Borrull, P.A.G. Cormack, D.C. Sherrington, R.M. Marce, Hydrophilic hypercrosslinked polymeric sorbents for the solid-phase extraction of polar contaminants from water, J. Chromatogr. A 1217 (19) (2010) 3238–3243.

[48] A. Waheed, N. Baig, N. Ullan, W. Falath, Removal of hazardous dyes, toxic metal ions and organic pollutants from wastewater by using porous hyper-cross-linked polymeric materials: A review of recent advances, J. Environ. Manage. 287 (2021) 112360.

[49] J. Wu, L.J. Jia, L.Y. Wu, C. Long, W.B. Deng, QX. Zhang, Prediction of the breakthrough curves of VOC isothermal adsorption on hypercrosslinked polymeric adsorbents in a fixed bed, RSC Adv. 6 (34) (2016) 28986–28993.

[50] R. Dawson, E. Stockel, J.R. Holst, D.J. Adams, AI. Cooper, Microporous organic polymers for carbon dioxide capture, Energy Environ. Sci. 4 (10) (2011) 4239–4245.

[51] R. Dawson, T. Ratvijitvech, M. Corker, A. Laybourn, Y.Z. Khimyak, A.I. Cooper, et al., Microporous copolymers for increased gas selectivity, Polym. Chem. 3 (8) (2012) 2034–2038.

CHAPTER 2

Novel synthetic method for tubular hypercrosslinked polymer nanofibers and its mechanism

Jiqi Wang[a], Yihao Fan[a] and Baoliang Zhang[a,b]
[a]School of Chemistry and Chemical Engineering, Northwestern Polytechnical University, Xi'an, China
[b]Xi'an Key Laboratory of Functional Organic Porous Materials, Northwestern Polytechnical University, Xi'an, China

2.1 Introduction

Hypercrosslinked polymer is a kind of porous material prepared based on Friedel-Crafts alkylation reaction [1–4]. The pore formation comes from the high crosslinking degree and rigidity of polymer network, which restrict the tight shrinkage of polymer chains. Therefore, the gaps between molecular chains become permanent pores. As an important member of the porous material family, hypercrosslinked polymers have attracted more and more attention of researchers due to their abundant nano-scale pores and ultra-high specific surface area. They have been widely used in the fields of gas adsorption, supercapacitor, water treatment, heterogeneous catalysis, and so on [5–8].

The preparation methods of hypercrosslinked porous polymer materials mainly include the following three methods: postcrosslinking method, self-polycondensation method, and external braiding method. Postcrosslinking method is generally to first prepare linear or crosslinked polymer precursors containing aromatic rings, and then realize hypercrosslinking through the Friedel-Crafts reaction between aromatic rings. The preparation process of this method is relatively cumbersome, but it can build micropores on materials with regular morphology, or make secondary pores on porous materials to obtain gradient pore structure. Self-polycondensation method utilizes a one-step Friedel-Crafts reaction between functional monomers (including benzyl halide, benzyl alcohol, etc.) to obtain hypercrosslinked polymers with a microporous structure. The reaction system and operation

Fabrication and Functionalization of Advanced Tubular Nanofibers and their Applications.
DOI: https://doi.org/10.1016/B978-0-323-99039-4.00006-1

process of this method are very simple, but there is certain requirements for monomer molecular structure. External braiding method realizes the hyper-crosslinking reaction between aromatic rings by using external crosslinking agent. The typical external crosslinking agent is formaldehyde dimethyl acetal (FDA). Compared with self-polycondensation method, this method has a wider application of monomers. In recent years, the research on hypercrosslinked polymers mainly focuses on the selection of precursors and monomers, the regulation of pore structure, and the application of materials, and related research has also made satisfactory progress [9–11]. With the deepening of the research on hypercrosslinked polymers, it is found that the urgent problems in this field includes: (1) developing new synthesis methods or reaction systems; (2) realizing one-step self-poly-condensation to prepare hyper-crosslinked polymer materials with regular morphology; (3) synthesizing functional composite hypercrosslinked polymer materials with novel structures. The solution of the above problems is of great significance for enriching the types of hypercrosslinked polymers and promoting the theoretical development and practical application of hypercrosslinked polymers.

2.2 The study on hypercrosslinked microporous materials with regular morphology

2.2.1 Research progress of hypercrosslinked microporous polymer microspheres

The preparation of hypercrosslinked microporous polymer microspheres or microcapsules is mainly based on the method of postcrosslinking of precursors. That is, the polymer particles precursors are prepared by conventional polymerization such as emulsion polymerization and suspension polymerization with styrene monomer and crosslinking agent. The main synthesis routes are shown in Fig. 2.1. If the precursor contains vinylbenzyl chloride (VBC) component, the hypercrosslinked structure can be obtained directly by Friedel-Crafts alkylation (Route 2). If it does not contain such functional group, it can be hypercrosslinked by adding external crosslinking agent (Route 1) or chloromethyl modification (Route 3); For the preparation of polymer microcapsules, the polymer must be coated on the template particles to form the core-shell morphology, and then the template can be removed after obtaining the hypercrosslinked structure (Route 4).

Bien Tan et al. first prepared monodisperse vinylbenzyl chloride-divinylbenzene (polyVBC-DVB) nanoparticles with diameter of 36–131 nm.

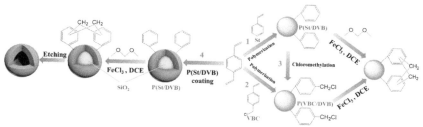

Figure 2.1 Schematic diagram of the preparation process of hypercrosslinked microporous polymer microspheres by postcrosslinking method.

After Friedel–Crafts reaction, uniform microporous polymer nanoparticles were prepared, surface area could up to 1500 m^2/g [12]. By using the postcrosslinking method, this team also performed postcrosslinking and sulfonation on the precursor microspheres prepared by suspension polymerization, and obtained functionalized porous microspheres with high adsorption capacity for Cu^{2+} [13].

Tan's team also carried out relevant work in the preparation of hypercrosslinked polymer microcapsules by postcrosslinking method. For example, the SiO_2@P(St/DVB) precursor was prepared by simple traditional emulsion polymerization, and hollow microporous organic capsules were obtained after hypercrosslinking reaction and chemical etching of SiO_2. The specific surface area of the resulting material increased accordingly, up to 1129 m^2 g^{-1} [14]. By the similar method, they also prepared an acid-base coexistence hypercrosslinked polymer with a hollow spherical structure, as shown in Fig. 2.2. Controlling the reaction conditions, 2-(hydroxymethyl)-isoindoline-1,3-dione and acetylsulfate were modified on the inner and outer surfaces of the microcapsules respectively. The obtained materials were used as catalysts for the conversion reactions of hydrolysis/Henry, hydrolysis/Knoevenagel and 2-ethoxy-3,4-dihydropyran and its derivatives, showing high catalytic efficiency, cycle stability and chemical stability [15].

Kun Huang et al. used polylactide-b-polystyrene diblock copolymers (PLA-b-PSt) in different proportions as precursors, and adopted a hypercrosslinking-mediated self-assembly strategy to crosslink PSt blocks to form micropores organic shell framework. Then, PLA blocks were degraded to generate hollow structures, and a series of hollow microporous organic nanospheres were obtained. The preparation process is shown in Fig. 2.3. After carbonization, these materials can be used in supercapacitors and nano reactors [16].

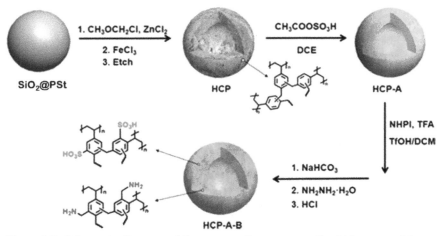

Figure 2.2 Schematic diagram of the preparation process of acid-base coexistence hypercrosslinked microcapsules. Reprinted (adapted) with permission from Copyright 2022 American Chemical Society [15].

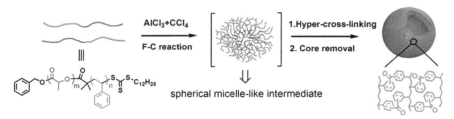

Figure 2.3 Schematic diagram of the process of preparing hollow microporous organic nanospheres by hypercrosslinking-mediated self-assembly strategy. Reprinted (adapted) with permission from Copyright 2022 American Chemical Society [16].

2.2.2 Research progress of hypercrosslinked microporous organic/inorganic composite microspheres

Hypercrosslinked microporous organic/inorganic composite microspheres are the hypercrosslinked microporous polymer microspheres or microcapsules functionalized by inorganic nanoparticles. The preparation method is usually to coat inorganic nanoparticles on polymer precursor, and then carry out hypercrosslinking reaction to build a microporous polymer shell. There are also reports of impregnating and loading inorganic nanoparticles on the basis of obtaining hypercrosslinked microporous polymer microcapsules.

Prof. Tan's team used oleic acid–modified Fe_3O_4 nanoparticles as magnetic functional components to synthesize magnetic microporous polymer microspheres with high specific surface area and superparamagnetic

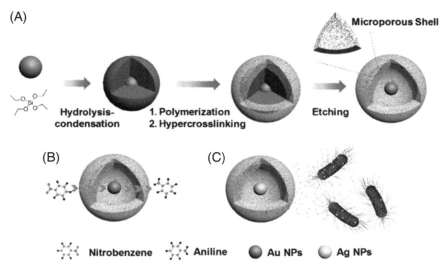

Figure 2.4 Schematic diagram of the preparation process of metal nanoparticles-loaded rattle-shaped ultra-microporous polymer microspheres. The copyright of the figure obtained from Elsevier under the license No: 5274651319554, Mar 23, 2022 [19].

properties through miniemulsion polymerization and Friedel–Crafts alkylation. The specific surface area of the obtained material was higher than $450 \ m^2 \ g^{-1}$. The characteristics of easy separation and high specific surface area make it show strong advantages in the removal of organic pollutants (ibuprofen, phenol, and Rhodamine B) in water [17,18].

Wu et al. used tetraethoxysilane hydrolysis–polycondensation-coated Au nanoparticles as the core, coated P(St/DVB) shell as the precursor by emulsion polymerization. After that, CCl_4 and formaldehyde dimethyl acetal (FDA) were used as external cross-linking agents to crosslink the polymer shell by using Friedel–Crafts alkylation reaction, resulting in a permanent pore structure, and then removing SiO_2 by hydrofluoric acid etching to form ringing compound particles. It was found that for shells with almost the same thickness, higher specific surface area, and larger porosity resulted easier mass transfer in the shell. This material was proved to be an excellent catalyst for the reduction of nitrobenzene. The team replaced Au nanoparticles with Ag nanoparticles and prepared Ag@MPNPs composite particles coated with hypercrosslinked polymer, which had inhibitory effects on *Escherichia coli* and *staphylococcus aureus* [19]. The preparation process was shown in Fig. 2.4.

Tan's team took pure SiO_2 nanoparticles as the template and used similar methods to obtain hypercrosslinked hollow microporous organic capsules m-HMOCs, and then embedded platinum (Pt) nanoparticles into

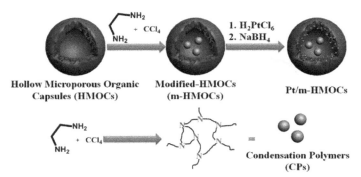

Figure 2.5 Schematic diagram of the preparation process of hollow microporous organic capsules loaded with Pt nanoparticles. The copyright of the figure obtained from the John Wiley and Sons under the license No: 5274690875569, Mar 23, 2022 [20].

m-HMOCs by means of deposition and reduction to obtain Pt/m-HMOCs composite microspheres. The preparation process was shown in Fig. 2.5. The microspheres could complete the hydrogenation of various nitroaromatic compounds with high yield within 2–5 hours [20].

2.2.3 Design and synthesis of hypercrosslinked microporous nanofibers

Compared with zero-dimensional and two-dimensional nanomaterials, one-dimensional (1D) nanomaterials have special shape anisotropy. Therefore, they have irreplaceable performance advantages in specific application fields such as optoelectronic devices, magnetic storage, sensors, and mechanical enhancement. At present, the development of synthetic methods and properties of 1D nanomaterials have become the focus of attention. Using hypercrosslinking method to construct 1D nanomaterials and then obtain 1D polymer nanofibers or nanotubes with microporous structure has become the goal pursued by researchers. and some progress has been made.

It is a relatively easy method to synthesize hypercrosslinked microporous nanofibers from the prepared polymer nanofiber precursors by postcrosslinking technology. Dai et al. used free radical addition reaction to graft VBC onto the polyethylene fiber matrix, and then the porous shell was constructed on the surface of polyethylene fibers by means of the hypercrosslinking reaction between benzyl chlorides. The porous fibers with a specific surface area of 529 m^2 g^{-1} were prepared, as shown in Fig. 2.6A [21]. This method can realize the porous functional modification of the surface of polymer matrix with complex morphology, and the obtained material shows strong adsorption capacity for methylene blue. Huang et al. grafted

(A)   (B)

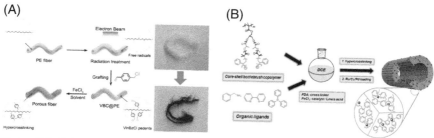

Figure 2.6 Schematic diagram of the preparation process of core-shell structured porous polymer fibers (A) and hypercrosslinked microporous polymer nanotubes (B). The copyright of the figure obtained from the John Wiley and Sons under the license No: 5274690875569 [21] and Elsevier under the license No: 5274651319554 [22], Mar 23, 2022.

small molecular organic ligands onto brush-like copolymer molecular templates, and obtained hypercrosslinked microporous polymers nanotubes with abundant pore structures by means of Friedel–Crafts alkylation reaction and template removal process between ligands (see Fig. 2.6B). The hierarchical porous structure can not only realize the uniform dispersion of metal catalyst particles in the microporous structure, but also the existence of hollow structure is conducive to the mass transfer of the reaction. Therefore, the nanotube significantly improves the performance of the catalyst when it is used as a catalyst carrier [22].

Jiang et al. used aromatic compounds as monomers and methylal as crosslinking agent to synthesize hypercrosslinked polymer nanotube precursors by an external braiding method. After carbonization, a novel porous carbon nanotubes were obtained. The specific surface areas could reach 1034 and 921 m^2 g^{-1} before and after carbonization, respectively. The preparation process was shown in Fig. 2.7. It was found that monomer concentration and mechanical stirring have important effects on the formation of polymer nanotube precursors. The size of carbon nanotubes was controlled by the molecular weight of monomers, and the formation mechanism was not clear. The material could be used as electrode material of supercapacitor, and its specific capacity was much higher than that of commercial multiwall carbon nanotubes [23].

2.3 Preparation and formation mechanism of hypercrosslinked tubular nanofibers

The synthesis of hypercrosslinked tubular nanofibers is carried out in a novel dual-oil phase system developed by our group [24]. The formation

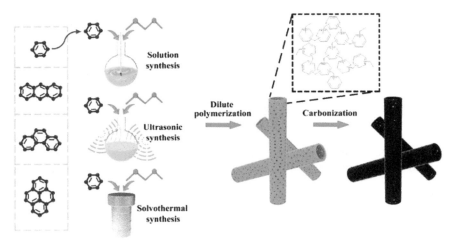

Figure 2.7 Schematic diagram of preparation process of porous carbon nanotubes by external braiding method. Reprinted (adapted) with permission from Copyright 2022 American Chemical Society [23].

schematic diagram and morphology of tubular nanofibers are displayed in Fig. 2.8. Polydimethylsiloxane is used as continuous phase, and 1,2-dichloroethane solution dissolved with monomer 1,4-p-dichlorobenzyl and catalyst $FeCl_3$ are used as dispersed phase. Under the action of mechanical stirring, the monomer droplets and catalyst droplets coalesce and disperse to form a homogeneous system. With the volatilization of 1,2-dichloroethane in the system, the catalyst anhydrous $FeCl_3$ first crystallizes and precipitates forming acicular crystal, which acts as a hard template. The monomer 1,4-p-dichlorobenzyl polymerizes on its surface forming hypercrosslinked polymer shell. After removing the template by Soxhlet extraction, tubular polymer nanofibers with axial through-holes are obtained, and the specific surface area is 17.19 m^2 g^{-1}. Compared with the hypercrosslinked polymers prepared by the traditional solution polymerization system, the specific surface area of the tubular polymer fibers decreased significantly, which was 996.21 m^2 g^{-1}. This might be explained that the catalyst was encapsulated in the interior of nanofibers as template, resulting in incomplete hypercrosslinked reaction on the tube wall and insufficient rigidity of the polymer framework. Tubular polymer nanofibers can be transformed into tubular carbon nanofibers by calcining at high temperature in a vacuum tubular furnace. During the calcination process, a large number of 4–5 nm small pores are formed on the tube wall due to the collapse and shrinkage of the polymer framework. Benefiting from the existence of secondary pore structure, the specific

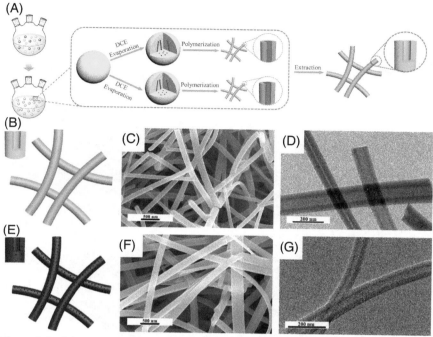

Figure 2.8 Schematic diagram of the proposed formation mechanism of tubular polymer nanofibers (A); Schematic illustration (B), SEM (C) and TEM (D) images of tubular polymer nanofibers; schematic illustration (E), SEM (F) and TEM (G) images of tubular carbon nanofibers. The copyright of the figure obtained from Royal Society of Chemistry under the license No: 1203116-1, Mar 24, 2022 [24].

surface area of tubular carbon nanofibers increases significantly, reaching 563.09 m^2 g^{-1}.

In order to verify the universality of the developed method, several difunctional and trifunctional benzyl halides were selected as polymeric monomers. The SEM and TEM images of the products were shown in Fig. 2.9A–F. This method possessed universality for different benzyl halides, and tubular nanofibers could be obtained similarly. The nanofibers prepared with alpha,alpha'-dichloro-o-xylene (Fig. 2.9B) had poor morphology and exhibited adherent flat structure. The yields of the products synthesized from 1,3–bis(bromomethyl)–5–methylbenzene (Fig. 2.9E) and 1,3,5–tris(bromomethyl)benzene (Fig. 2.9F) were lower at same inventory rating, which might be attributed to the reduce of reactive sites on the benzene ring. Interestingly, the tubular nanofibers prepared by alpha,alpha'-dibromo-p-xylene and alpha,alpha'-dibromo-m-xylene exhibited similar morphology to candied haws, seen Fig. 2.9C and D. The obtained nanofibers

Figure 2.9 SEM images of the obtained samples: the tubular polymer nanofibers prepared by alpha,alpha'-dichloro-m-xylene (A), alpha,alpha'-dichloro-o-xylene (B), alpha,alpha'-dibromo-p-xylene (C), alpha,alpha'-dibromo-m-xylene (D), 1,3-bis(bromomethyl)-5-methylbenzene (E), 1,3,5-tris(bromomethyl)benzene (F); the samples prepared by different solvents instead of polydimethylsiloxane (G–L), n-hexane (G), cyclohexane (H); n-Heptane (I), kerosene (J), mineral oil (K), paraffin (L). The scale bar is 0.5 μm. The copyright of the figure obtained from Royal Society of Chemistry under the license No: 1203116-1, Mar 24, 2022 [24].

had thicker tube wall and smaller inner diameter. The reasons for the above morphological differences are not yet clear, which may be related to the monomer activity and the catalytic ability of the catalyst for the external crosslinking reaction. The effect of solvent type on the product morphology was investigated by replacing the continuous phase polydimethylsiloxane. Nanoparticles were obtained in the n-hexane (Fig. 2.9G) and cyclohexane (Fig. 2.9H) systems. It was similar to the product prepared by conventional solvent DCE (Fig. 2.9G). When n-heptane and kerosene were used as solvent, large amounts of vesicle-like particles were present in the products, seen Fig. 2.9I and J. Using liquid paraffin as solvent, the product was viscous with yield. The product obtained from the solid paraffin system contained fibrous material but encapsulated in viscous substance (Fig. 2.9L). The above results suggested that the morphology of products was significantly affected by solvents. The compatibility with DCE and the solubility of DCX and $FeCl_3$ were the core causes of this difference in morphology. A continuous phase, which could form an obvious dual phase system with DCE and possess no solubility for DCX and $FeCl_3$, was conducive to the formation of regular morphology.

2.4 Application of hypercrosslinked tubular nanofibers

Due to the special secondary pore structure, ultra-high specific surface area, easy modification property, and adjustable structural properties, hypercrosslinked tubular nanofibers can be used as microwave absorbers, lithium-ion battery anode materials, surface protein imprinting carriers, sewage treatment adsorbents, etc. Large-scale applications have made a lot of progress in related fields in recent years.

(1) Application in microwave absorption

With the large-scale application of electromagnetic technology, the problems of electromagnetic radiation and electromagnetic pollution have become increasingly prominent. Due to the strong penetrability, electromagnetic wave can penetrate human body, causing physical and psychological damage and threatening people's health. Furthermore, the problem of signal interference and the risk of information leakage in microwave propagation are also extremely challenging. Therefore, the development of electromagnetic radiation and electromagnetic pollution response technologies has become a topic worthy of discussion. Microwave absorbing materials can completely absorb incident electromagnetic wave and convert it into heat, and there are almost no problems of transmission and reflection. It is a technical means to completely solve electromagnetic radiation and pollution.

The tubular carbon nanofibers calcined at high temperature have good dielectric loss ability, exhibit excellent electromagnetic wave attenuation performance, and can be directly used as wave absorbers [25]. Compared with granular or layered carbon materials, tubular carbon nanofibers can be dispersed in matrix to form an effective three-dimensional conductive network due to their one-dimensional structure, which can dissipate electromagnetic wave in the form of microcurrent. Compared with other one-dimensional carbon materials, tubular carbon nanofibers not only have through pores in the axis, but also have abundant small pores on the tube wall. The existence of these channels can not only reflect and scatter electromagnetic waves multiple times, but also serve as transmission channels for electromagnetic waves and improve impedance matching characteristics. At the same time, the abundant pores also provide a high specific surface area, which makes it have strong interfacial polarization ability. Therefore, the prepared tubular carbon nanofibers exhibit excellent microwave absorbing properties. In addition, tubular carbon nanofibers can also be easily combined with magnetic components and other materials with different dielectric properties to optimize electromagnetic parameters and obtain better electromagnetic wave attenuation ability and impedance matching.

(2) Application in oil absorption

Oil is one of the important energy sources for human survival. However, the leakage problem in the process of exploitation and transportation not only causes huge waste of oil resources, but also causes serious pollution to ocean and fresh water resources. Traditionally, there are three main methods of oil spill treatment methods: biological, chemical, and mechanical [26]. Biological methods use microorganisms to degrade oil spills. Chemical methods include in-situ burning and dispersion. Mechanical methods rely on manpower or machines to skim off oil spills or use oil-absorbing materials to absorb them. Compared with other oil removal methods, the use of adsorbent material for recovery is currently the most economical oil spill treatment method, with the characteristics of high efficiency and low consumption.

Many materials are used for oil adsorption, mainly including inorganic and organic [27]. Inorganic oil adsorption materials mainly include activated carbon, zeolite, and so on, which are characterized by low cost, less oil adsorption, and high transportation cost. Due to their light weight and high specific surface area, carbon nanotubes, graphene, and other new carbon materials are also used for oil adsorption. Compared with inorganic materials, synthetic polymers (polystyrene fiber, polymethacrylate fiber, polypropylene fiber, etc.) and natural polymers (cotton, straw, wheat straw, milkweed, etc.) show high oil adsorption rate and have obvious advantages in use.

It can be found that one-dimensional porous materials show advantages of high adsorption capacity, good buoyancy in water and good circulation performance in oil adsorption due to its large aspect ratio, high specific surface area and abundant pore structure. The commonly reported one-dimensional oil-absorbing materials are mainly various polymer fibers prepared by electrospinning, carbon nanotubes and various fiber materials. Compared with other oil-absorbing materials, fiber oil-absorbing materials show the following characteristics: (a) Fine, soft, large specific surface area, fast adsorption rate, convenient recycling treatment, can be processed into various forms of products through textile or nonwoven according to needs. (b) By controlling the spinning process, the fibers with different supramolecular structure and morphological structure can be prepared. On the basis of guaranteeing the good mechanical properties of the fibers, the adsorption area and the adsorption rate of the fibers can be effectively increased, and the application range is greatly extended.

(3)   Application in lithium-ion batteries

In recent years, many researches on anode materials for alkali metal ion batteries (AMIBs) have been deepening. Generally, anode materials of AMIBs can be divided into intercalation, alloy and conversion type materials according to the energy storage mechanism. As a typical intercalation material, carbon materials have the advantages of abundant reserves, easy availability, high energy, and power density. When alkali metal ions are inserted into the lattices or interlayers of carbon materials, the volume change hardly occurs due to the mechanism of adsorption/desorption, the excellent structural stability is thus ensured. One-dimensional carbon materials, especially hollow porous tubular carbon nanofibers (TCNFs), the hollow structure is conducive to the immersion of electrolyte, porous structure can provide more active sites for the adsorption of alkali metal ions, the larger length-diameter ratio, internal and external area can load a variety of active substances, TCNFs exhibit the superior programmability.

Currently, TCNFs have not been directly used in AMIBs due to the high cost. TCNFs are usually employed to combine with the conversion or alloy-type materials to prepare bilayer or sandwich structures, which exhibit excellent cycling and rate capability. Our group adopted a multistep reaction strategy to prepare sandwich structure hierarchical nanofibers composed of hollow carbon fibers as the substrate, MoS_2 as the interlayer with Co and/or ZnS nanoparticles anchoring in carbon skeletons as the outer shell (CNF/MoS_2/Co-ZnS⊂NC), see Fig. 2.10 [28]. As anode materials for sodium ion batteries, the unique

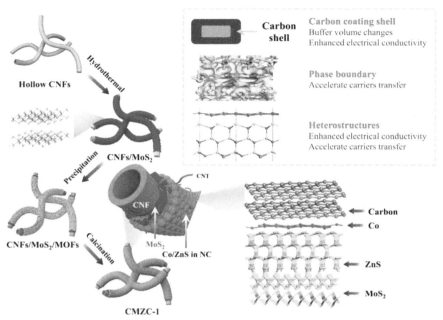

Figure 2.10 Schematic illustration of the formation process of the CNF/MoS$_2$/Co-doped ZnS hierarchical nanofibers. The copyright of the figure obtained from John Wiley and Sons under the license No: 5274710665358, Mar 23, 2022 [28].

hierarchical structure and abundant migration channels of Na+ ensured the rapid diffusion kinetics. The coaxial structure could evenly disperse volumetric strain, structural stability was guaranteed. Therefore, TCNFs show the broad application potential as an important component of anode materials for AMIBs.

(4) Application in surface protein imprinting

Biomacromolecules such as proteins are the material basis for maintaining normal life activities of organisms, and are also an important part of cells. The effective separation and analysis of proteins are of great significance to proteomics and biomedicine [29]. However, proteins are usually present in biological samples with complex matrices, making it difficult to analyze and detect them directly. Therefore, the establishment of separation and analysis methods with good selectivity, high stability, simple operation, and low cost is an important prerequisite for protein detection. The protein imprinted materials prepared by molecular imprinting technique (MIT) have three-dimensional imprinting sites, which are consistent with the shape, size, functional groups, and their arrangement of target proteins [30]. Accordingly, the protein imprinted materials possess high selectivity, affinity, and stability, which could achieve the selective identification and separation of targets in complex biological samples. In recent years, surface imprinting technique is currently the most widely used and most basic method for protein molecular imprinting. Surface protein imprinting refers to the coating of polymers containing template molecules on the surface of a matrix material, so the nature of the carrier is crucial [31]. The unique feature of tubular carbon nanofibers (TCFs) is that their surfaces have abundant pores and hollow internal structures. The protein imprinted materials prepared with TCFs as carriers could be endowed with self-driven adsorption properties. When the material is dispersed in high concentration solution of protein, a concentration gradient is formed between the inside and outside space of the tube. The generated osmotic pressure difference driven the protein molecules to spontaneously bind to the imprinting sites on the surface and further migrate to the internal cavity through the pores inside the tube wall. This characteristic enables the protein imprinted materials to exhibit high adsorption capacity and fast adsorption rate for the template protein, which provides a new idea for the development of high-performance protein imprinted materials.

(5) Application in water treatment

In recent years, water pollution has become one of the most serious problems. As a significant part of environmental governance, removing heavy metal ions and organic pollutants from wastewater has attracted extensive attention. Therefore, exploring rapid and effective methods to dislodge heavy metal ions and organic pollutants from wastewater has become a research hotspot [32]. Carbon materials possess large specific surface area, high porosity, and nanoscale pores. Therefore, they show high adsorption activity on heavy metal ions and organic pollutants. Compared with other one-dimensional carbon materials, tubular carbon nanofibers not only have through pores in the axis, but also have abundant small pores on the tube wall. These axial through channels and a large number of mesopores in the pipe wall are important containers for containing and absorbing pollutants. Studies have shown that the tubular carbon nanofibers have excellent removal effects for heavy metal ions such as Cu^{2+} and $Hg2+$, and organic dyes such as rhodamine B, methylene blue and methyl orange.

2.5 Summary and prospect

Our group have developed a novel dual-oil phase confined self-condensation system, which overcomes the problem that traditional solution polymerization systems can only prepare hypercrosslinked porous polymers with random morphology. This system can achieve the rapid preparation of hypercrosslinked polymer nanofibers with hollow tubular structure. The obtained tubular polymer nanofibers, as well as derived carbon nanofibers and their composites show excellent application properties in the fields of microwave absorption, oil absorption, lithium-ion battery cathode materials, surface protein imprinting, and water pollutants absorption due to their one-dimensional tubular structure, ultra-high specific surface area, and high porosity. Furthermore, the tubular hypercrosslinked polymer nanofibers also have potential applications in the fields of gas storage/separation/adsorption, heterogeneous catalysis, optoelectronics, sensors, and semiconductor devices.

References

[1] V. Davankov, V. Rogozhin, M. Tsjurupa, Macronet polystyrene structures for ionites and method of producing same, US 3729457, 1973.

[2] L. Tan, B. Tan, Hypercrosslinked porous polymer materials: design, synthesis, and applications, Chem. Soc. Rev. 46 (11) (2017) 3322–3356.

[3] L. Tan, B. Tan, Research progress in hypercrosslinked microporous organic polymers, Acta. Chim. Sinica. 73 (6) (2015) 530–540.

[4] J. Huang, S.R. Turner, Hypercrosslinked polymers: A review, Polym. Rev. 58 (1) (2017) 1–41.

[5] X. Liang, J. Wang, Q. Wu, C. Wang, Z. Wang, Use of a hypercrosslinked triphenylamine polymer as an efficient adsorbent for the enrichment of phenylurea herbicides, J. Chromatogr. A 1538 (2018) 1–7.

[6] R. Castaldo, G. Gentile, M. Avella, C. Carfagna, V. Ambrogi, Microporous hyper-crosslinked polystyrenes and nanocomposites with high adsorption properties: A review, Polymers 9 (12) (2017) 651.

[7] R.M.N. Kalla, A. Varyambath, M.R. Kim, I. Kim, Amine-functionalized hyper-crosslinked polyphenanthrene as a metal-free catalyst for the synthesis of 2-amino-tetrahydro-4H-chromene and pyran derivatives, Appl. Catal. A: Gen. 538 (2017) 9–18.

[8] D. Mukherjee, G. Gowda Y. K, H. Makri Nimbegondi Kotresh, S. Sampath, Porous, hyper-cross-linked, three-dimensional polymer as stable, high rate capability electrode for lithium-ion battery, ACS Appl. Mater. Interfaces 9 (23) (2017) 19446–19454.

[9] R. Castaldo, R. Avolio, M. Cocca, G. Gentile, M.E. Errico, M. Avella, et al., A versatile synthetic approach toward hyper-cross-linked styrene-based polymers and nanocomposites, Macromolecules 50 (11) (2017) 4132–4143.

[10] Y. Liu, X. Fan, X. Jia, X. Chen, A. Zhang, B. Zhang, et al., Preparation of magnetic hyper-cross-linked polymers for the efficient removal of antibiotics from water, ACS Sustain. Chem. Eng. 6 (1) (2017) 210–222.

[11] C. Cai, Z. Hou, T. Huang, K. Li, Y. Liu, N. Fu, et al., Preparation of monodisperse hyper-crosslinking polymer nanoparticles for highly efficient CO_2 adsorption, Macromol. Chem. Phys. 218 (7) (2017) 1700001.

[12] B. Li, X. Huang, L. Liang, B. Tan, Synthesis of uniform microporous polymer nanoparticles and their applications for hydrogen storage, J. Mater. Chem. 20 (35) (2010) 7444–7450.

[13] B. Li, F. Su, H.-K. Luo, L. Liang, B. Tan, Hypercrosslinked microporous polymer networks for effective removal of toxic metal ions from water, Micropor. Mesopor. Mater. 138 (1-3) (2011) 207–214.

[14] B. Li, X. Yang, L. Xia, M.I. Majeed, B. Tan, Hollow microporous organic capsules, Sci. Rep. 3 (2013) 2128.

[15] Z. Jia, K. Wang, B. Tan, Y. Gu, Hollow hyper-cross-linked nanospheres with acid and base sites as efficient and water-stable catalysts for one-pot tandem reactions, ACS Catal. 7 (5) (2017) 3693–3702.

[16] Z. He, M. Zhou, T. Wang, Y. Xu, W. Yu, B. Shi, et al., Hyper-cross-linking mediated self-assembly strategy to synthesize hollow microporous organic nanospheres, ACS Appl. Mater. Interfaces 9 (40) (2017) 35209–35217.

[17] Q. Li, Z. Zhan, S. Jin, B. Tan, Wettable magnetic hypercrosslinked microporous nanoparticle as an efficient adsorbent for water treatment, Chem. Eng. J. 326 (2017) 109–116.

[18] X. Yang, B. Li, I. Majeed, L. Liang, X. Long, B. Tan, Magnetic microporous polymer nanoparticles, Polym. Chem. 4 (5) (2013) 1425–1429.

[19] Y. Du, Z. Huang, S. Wu, K. Xiong, X. Zhang, B. Zheng, et al., Preparation of versatile yolk-shell nanoparticles with a precious metal yolk and a microporous polymer shell for high-performance catalysts and antibacterial agents, Polymer 137 (2018) 195–200.

[20] X. Yang, K. Song, L. Tan, I. Hussain, T. Li, B. Tan, Hollow microporous organic capsules loaded with highly dispersed Pt nanoparticles for catalytic applications, Macromol. Chem. Phys. 215 (12) (2014) 1257–1263.

[21] Y. Sheng, Q. Chen, S.M. Mahurin, R.T. Mayes, W. Zhan, J. Zhang, et al., Fibers with hyper-crosslinked functional porous frameworks, Macromol. Rapid Comm. 39 (8) (2018) 1700767.

[22] Y. Xu, T. Wang, Z. He, M. Zhou, W. Yu, B. Shi, et al., Organic ligands incorporated hypercrosslinked microporous organic nanotube frameworks for accelerating mass transfer in efficient heterogeneous catalysis, Appl. Catal. A: Gen. 541 (2017) 112–119.

[23] X. Wang, P. Mu, C. Zhang, Y. Chen, J. Zeng, F. Wang, et al., Control synthesis of tubular hyper-cross-linked polymers for highly porous carbon nanotubes, ACS Appl. Mater. Interfaces 9 (24) (2017) 20779–20786.

[24] J. Wang, Z. Yang, M. Ahmad, H. Zhang, Q. Zhang, B. Zhang, A novel synthetic method for tubular nanofibers, Polym. Chem. 10 (31) (2019) 4239–4245.

[25] J. Wang, Y. Huyan, Z. Yang, A. Zhang, Q. Zhang, B. Zhang, Tubular carbon nanofibers: Synthesis, characterization and applications in microwave absorption, Carbon 152 (2019) 255–266.

[26] M.Z. Iqbal, A.A. Abdala, Oil spill cleanup using graphene, Environ Sci Pollut. Res. 20 (5) (2013) 3271–3279.

[27] G. Thilagavathi, D. Das, Oil sorption and retention capacities of thermally-bonded hybrid nonwovens prepared from cotton, kapok, milkweed and polypropylene fibers, J. environ. manage. 219 (2018) 340–349.

[28] J. Chen, T. Wang, F. Zhang, N. Tian, Q. Zhang, B. Zhang, The multicomponent synergistic effect of sandwich structure hierarchical nanofibers for enhanced sodium storage, Small 18 (14) (2022) 2107370.

[29] H. Adam, S.C.B. Gopinath, M.K.M. Arshad, T. Adam, U. Hashim, Perspectives of nanobiotechnology and biomacromolecules in parkinson's disease, Process Biochem. 86 (2019) 32–39.

[30] J. Ashley, M.A. Shahbazi, K. Kant, V.A. Chidambara, A. Wolff, D.D. Bang, et al., Molecularly imprinted polymers for sample preparation and biosensing in food analysis: Progress and perspectives, Biosens. Bioelectron 91 (2017) 606–615.

[31] Y. Tang, Y. Yao, X. yang, T. Zhu, Y. Huang, H. Chen, et al., Well-defined nanostructured surface-imprinted polymers for the highly selective enrichment of low-abundance protein in mammalian cell extract, New J. Chem. 40 (12) (2016) 10545–10553.

[32] M. Ahmad, J. Wang, J. Xu, Z. Yang, Q. Zhang, B. Zhang, Novel synthetic method for magnetic sulphonated tubular trap for efficient mercury removal from wastewater, J. Colloid Interf. Sci. 565 (2020) 523–535.

CHAPTER 3

Design and preparation of self-driven BSA surface imprinted tubular nanofibers and their specific adsorption performance

Zuoting Yang[a] and Baoliang Zhang[a,b]
[a]School of Chemistry and Chemical Engineering, Northwestern Polytechnical University, Xi'an, China
[b]Xi'an Key Laboratory of Functional Organic Porous Materials, Northwestern Polytechnical University, Xi'an, China

3.1 Introduction

Accurate analysis of protein structure is one of the research hotspots in the world today. This research can not only provide a material basis for exploring the rules of life activities but also provide a theoretical basis and effective solutions to overcome many diseases. The highly selective recognition of proteins can provide an important technical basis for proteomics research and is an important research content in the field of life sciences. In recent years, to achieve high-efficiency recognition of molecular, scientists have carried out a lot of work in the research and development of novel adsorption materials with high selectivity [1,2]. To realize selective adsorption of proteins, researchers modify specific ligands (such as enzymes, antibodies, nucleic acid aptamers, etc.) on the surface of nanomaterials. However, there are many problems such as many steps, time-consuming, high precision, and relatively expensive or difficult to obtain biological ligands, which limit their stability and adaptability, thus affecting their application expansion. Based on simulating the interaction between enzymes and substrates and receptors and antibodies in nature, molecularly imprinting technique (MIT) is proposed and developed. Compared with natural biological ligands, the MIT is based on polymer materials and has significant advantages in various

Fabrication and Functionalization of Advanced Tubular Nanofibers and their Applications.
DOI: https://doi.org/10.1016/B978-0-323-99039-4.00009-7
47

aspects. Therefore, it has attracted attention in the field of protein separation [3–5].

In 1972, MIT was first proposed by Wulff, a German scientist [6]. This technique means that multiple action sites were formed between the template molecules and the functional monomer through covalent or noncovalent bonds during the imprinting process. And a cross-linking agent is added to fix the complex of the functional monomer and template molecules on the imprinted polymers (MIPs). After elution of template molecules, a cavity with multiple binding sites matching the special structure of templates is formed, which can specifically recognize template molecules [7]. The specific selectivity of MIPs for target molecules is graphically described as a technique to create an "artificial lock" that identifies "molecular keys." MIPs is widely used in purification separation, solid-phase extraction [8–11], membrane separation [12,13], chromatographic separation and analysis [14–16], drug delivery [17–19], biosensing [20–22], simulation enzyme catalysis, and chemical sensors due to their obvious advantages of good chemomechanical stability, strong selectivity, repeatability, and easy preparation. At present, MIT has made great progress in the aspect of separation and recognition for small molecules [23–27]. Nevertheless, it still faces great challenges in the preparation of biological macromolecular imprinted polymers such as proteins. The major reason is that proteins exhibit the characteristics of large molecular weight, complicated structure, changeable conformation, and sensitivity to the environment [28–30]. Therefore, the preparation of protein imprinted polymers is faced with some problems such as low effective imprinting capacity, slow mass transfer, difficult elution and rebinding, and poor recognition ability.

To solve the above problems, researchers have made plentiful fruitful attempts and explorations. Aiming at the problem of low effective imprinting amount, it can be solved by improving the surface area of imprinted materials and increasing the strength of interaction between the surface of carrier materials and template molecules. Surface protein imprinting is an effective way to improve mass transfer [31–33]. At the same time, the introduction of responsive components into the polymer imprinted layer can improve the mass transfer and elution efficiency by intelligently adjusting the swelling degree of polymers [34]. To solve the poor selectivity, antiprotein components can be introduced in the imprinting layer or the cross-linking structure of the polymer can be adjusted [35–37]. The milder conditions of the imprinting process can keep the template protein conformation stable and thus improve selectivity [38].

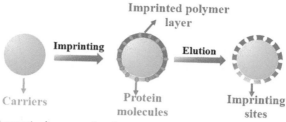

Figure 3.1 Schematic diagram of surface protein imprinted materials.

3.2 Surface protein imprinting technique

The surface imprinting technique refers to coating an ultrathin polymer layer on the carrier materials. The imprinting sites are loaded near the surface of the shell layer, which can significantly improve the binding capacity, kinetics, and location accessibility of the imprinted materials. Strictly speaking, the materials whose thickness of the imprinted layer is less than or equal to the largest dimension of protein molecules are called surface protein imprinted materials, as shown in Fig. 3.1. The most prominent feature of surface imprinted materials is that proteins, including target or template molecules, are not required to pass through the polymer layer with an obstructive effect and the imprinting cavity is bare. The core advantages of this kind of material are to reduce the mass transfer resistance and improve the separation efficiency. Therefore, surface imprinting has become the focus in the field of protein imprinting.

Imprinting factor and selectivity factor are two core indicators for evaluating imprinted materials. Imprinting factor (IF) is applied to measure the recognition ability of imprinted materials. The calculation formula is as follows:

$$IF = Q_{SIP} Q_{NIP}^{-1}$$

Q_{SIP} (mg g^{-1}) is the adsorption capacity of imprinted materials toward templates, Q_{NIP} (mg g^{-1}) refers to the adsorption capacity of non–imprinted materials toward templates.

Selectivity factor (β) is employed to assess the selective ability of imprinted materials. The calculation formula is as follows:

$$\beta = IF_{tem} IF_{com}^{-1}$$

IF_{tem} means the imprinting factor of imprinted materials to template proteins, IF_{com} represents the imprinting factor of imprinted materials to competitive proteins.

3.2.1 Functional carrier

For surface protein imprinted materials, the choice of carrier is very important. Compared with other solid carriers used for surface imprinting such as silica, nano-porous alumina membranes, and zinc oxide, the functional carrier is easier to satisfy the needs of protein imprinted materials with high performance, thus attracting extensive attention from researchers. Common functional carriers include magnetic nanomaterials, fluorescent materials, and electrodes.

(1) Magnetic nanomaterials

As we all know, magnetic materials can move directionally in a controlled manner under the action of an external magnetic field. Based on this characteristic, they have been applied in many fields. Magnetic protein imprinted polymers obtained by combining the magnetic nanomaterials and surface protein imprinting technique can realize rapid solid–liquid separation under an external magnet, avoiding tedious sample handling and separation processes. Compared with centrifugation, filtration, membrane separation, and other methods, magnetic MIPs have the advantages of simple operation, rapid, and efficient separation. Common magnetic materials can be divided into zero-dimensional magnetic nanomaterials, one-dimensional magnetic nanomaterials, and two-dimensional magnetic nanomaterials.

Zero-dimensional magnetic nanomaterials generally refer to magnetic nanoparticles or microspheres. Several hundred nanometers (200–500 nm) magnetic particles in the submicron scale have also been reported as zero-dimensional magnetic nanomaterials by some reports. Traditional magnetic nanoparticles include Fe_3O_4 nanoparticles (NPs), $CoFe_2O_4$, $CuFe_2O_4$, Fe_3GeTe_2, $ZnFe_2O_4$, Nb-Fe-B/Fe, magnetic nickel-based nanoparticles and bimetallic particles (BMNPs) composed of Fe (0) and other zero-valent metals such as Cu (0) or Ag (0). Among them, Fe_3O_4 NPs have been widely used as the carrier of protein surface imprinting due to their simple preparation, low cost, good dispersibility, and biocompatibility, excellent specific saturation magnetization, paramagnetism, and other advantages [31,39–41]. As solid-liquid adsorption and separation material, it is important to have a suitable density so that it can be suspended in a liquid system, which can increase the contact between the adsorbent and the adsorbate. The density of Fe_3O_4 is $5.18 \text{ g cm}^{3 \ -1}$, and the deposition rate is very fast in the system. Preparation of hollow structure is an effective method to reduce its density in disguised form. Simultaneously, the low density of carriers is also conductive to the improvement of adsorption capacity per unit mass. In previous work,

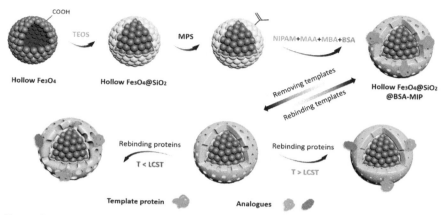

Figure 3.2 Schematic diagram of the preparation process and thermal recognition mechanism of HMS@MIPs. The copyright of the figure obtained from Elsevier under the license No: 5277521486463, Mar 28, 2022 [42].

hollow $Fe_3O_4@SiO_2$ was used as the carrier to realize the preparation of thermo-sensitive BSA surface imprinted microspheres HMS@MIPs by surface graft copolymerization [42]. The process is shown in Fig. 3.2. The microspheres exhibited high adsorption capacity and binding ability toward BSA. The adsorption capacity and imprinting factors were 430.67 mg g^{-1} and 2.01.

One-dimensional magnetic nanomaterials refer to magnetic materials with large aspect ratios and small-scale dimensions in nano-size, mainly including magnetic nanowires, magnetic nanobelts, magnetic nanotubes, and magnetic nanofibers. In recent years, the application of one-dimensional magnetic nanomaterials as a carrier for protein surface imprinting has been widely reported by researchers. Our group prepared nitrogen-doped magnetic carbon nanotube N-MCNTs with high specific surface area, good hydrophilicity, abundant nitrogen functional groups, and excellent magnetic properties through high-temperature pyrolysis and vapor deposition [43]. The high specific surface area is conducive to the formation of more imprinting sites. Good water dispersibility enables protein imprinting to be performed in an aqueous environment, ensuring the stability of the template protein conformation. Abundant nitrogen functional groups can form noncovalent interactions with proteins such as hydrogen bonds and electrostatic attraction, which is conducive to the immobilization of template proteins. High magnetic responsiveness simplifies the separation process and facilitates rapid separation. A novel BSA surface imprinted nitrogen-doped

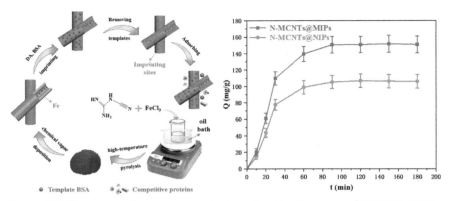

Figure 3.3 The preparation process and adsorption kinetics curve of N-MCNTs@MIPs. The copyright of the figure obtained from Elsevier under the license No: 5277560996067, Mar 28, 2022 [43].

magnetic carbon nanotubes (N-MCNTs@MIPs) were prepared by deposition of polydopamine (PDA) on the N-MCNTs surface. The adsorption capacity of imprinted materials for BSA reached 150.86 mg g^{-1} within 90 min. The imprinting factor was 1.43. The results of competitive adsorption and separation of fetal bovine serum indicated that N-MCNTs@MIPs can specifically recognize BSA. The preparation process and adsorption kinetics curve of N-MCNTs@MIPs are shown in Fig. 3.3.

Two-dimensional magnetic nanomaterials refer to magnetic materials that exist in a lamellar morphology and whose thickness is less than 100 nm. Currently reported two-dimensional magnetic nanomaterials mainly include magnetic graphene nanosheets and magnetic Mxene nanosheets. Due to the advantages of magnetism and graphene, surface protein imprinted magnetic graphene materials have been developed. Chen et al. [44] prepared two-dimensional BSA imprinted polymer $Fe_3O_4/GO/PDA$ (2D-MIPS) on the surface of magnetic graphene by self-polymerization of dopamine. Its adsorption capacity for BSA was 117.1 mg g^{-1}, and the imprinting factor was 2.7.

(2) Fluorescent materials

Fluorescence detection has become one of the most widely used detection methods owing to its characteristics of high sensitivity, quick response, and easy operation. The fluorescent molecularly imprinted polymer prepared by combining the fluorescent material with the surface protein imprinting has both a high sensitivity of fluorescence and the specific recognition performance of the imprinted polymers. Fluorescent molecularly imprinted polymers realize the detection of target proteins by converting the binding of template molecules and imprinting sites into changes in fluorescent signals. In recent years, the widely studied fluorescent materials mainly include quantum dots (QDs) [45–47], carbon dots (CDs) [48,49], and up-conversion nanoparticles (UCNPs) [50]. Quantum dots are semiconductor crystal materials composed of inorganic compounds such as cadmium selenide, cadmium sulfide, zinc sulfide, and indium arsenide. It has a regular crystal structure and strong fluorescence emission. Due to different sizes, QDs have fluorescence emission of different wavelengths, which possess huge advantages in the fields of biomedical imaging, fluorescent labeling, and biosensing. So far, molecularly imprinted polymers based on quantum dots have been widely reported. Additionally, such polymers have been successfully applied in the field of biological macromolecules. There are two main reasons for the changes in the fluorescence signal of QDs in protein imprinted materials. One is that the target protein binds to the imprinting sites to quench the fluorescence, resulting in a decrease in fluorescence intensity. Selecting acrylamide as a functional monomer and N, N-methylene diacrylamide as a crosslinking agent, Zhang et al. [45] successfully prepared fluorescent MIPs (MIP-QDs) on the surface of L-cysteine modified Mn^{2+} doped ZnS quantum dots by surface molecular imprinting technique for the recognition of lysozyme. The results suggested that the fluorescence quenching of the MIP-QDs complex was directly proportional to the concentration of lysozyme. Thus, the fluorescent acceptor based on MIP was successfully applied to the direct fluorescence quantitative analysis of lysozyme. Under optimal conditions, MIP-QDS could detect lysozyme in the range of 0.1 μM–2.0 μM, and the detection limit was 25.2 nM.

Another reason for the change in fluorescence signal is that the protein recognition process leads to increased fluorescence of quantum dots. For example, the combination of cis-diol groups in template glycoprotein and boric acid groups on the surface of quantum dots can block electron transfer from quantum dots to boric acid groups, resulting in enhanced fluorescence intensity. Chang et al. [51] designed a high-sensitivity fluorescent biosensor based on boric acid functional polymer-terminated Mn-doped ZnS quantum dots (QDs@MPS@AAPBA). The preparation process is given in Fig. 3.4. In the absence of glycoproteins, the fluorescence emission intensity of QDs@MPS@AAPBA is relatively weak due to the effective electron transfer of QDs to the boronic acid groups on its surface. When glycoproteins are introduced into the system, the electron transfer process is blocked and an increase in fluorescence is observed (see Fig. 3.5). The sensor was used to detect horseradish peroxidase (HRP) and transferrin (TRF). The

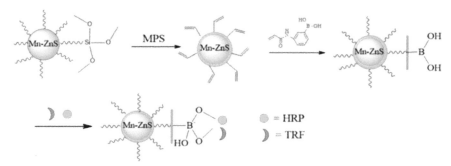

Figure 3.4 Schematic diagram for the preparation of QDs@MPS@AAPBA. The copyright of the figure was obtained from Elsevier under license No: 5277580280167, Mar 28, 2022 [51].

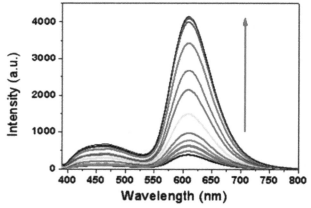

Figure 3.5 Changes in fluorescence intensity of QDs@MPS@AAPBA with the concentration of TRF (concentration increases gradually from bottom to top). The copyright of the figure was obtained from Elsevier under license No: 5277580280167, Mar 28, 2022 [51].

binding constant and detection limit were $7.23 \times 10^6 \ M^{-1}$, $1.53 \times 10^7 \ M^{-1}$ and $1.44 \times 10^{-10} \ M$, $3.36 \times 10^{-10} \ M$. Furthermore, the detection of TRF in serum by this detector has a recovery rate in the range of 95.7%–103.0%.

Compared with QDs, CDs and UCNPs are more and more favored by researchers due to their advantages of simple synthetic route, good biocompatibility, and environmental friendliness. The fluorescent molecularly imprinted polymer constructed by combining CDs and UCNPs with the highly selective surface protein imprinting technique has a promising application prospect in the detection and separation of biological macro-molecules. As shown in Fig. 3.6, selecting magnetic carbon dots (M-CDs)

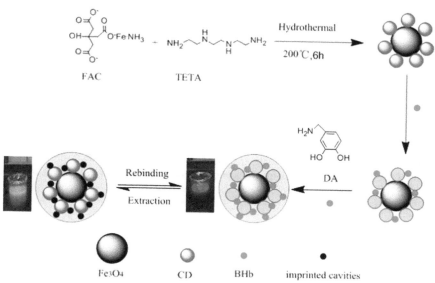

Figure 3.6 The preparation of M-CDs@MIPs. The copyright of the figure obtained from Elsevier under license No: 5277541367225, Mar 28, 2022 [48].

as the fluorescent source and carrier, Lv et al. [48] prepared fluorescent molecularly imprinted polymers M-CDs@MIPs on its surface through self-polymerization of dopamine for selective and sensitive fluorescence recognition of bovine hemoglobin. Under optimal conditions, the detection range of this imprinted polymer for bovine hemoglobin was 0.05–16.0 μM, and the detection limit was 17.3 nM. The magnetic fluorescent imprinted polymers were used to detect bovine hemoglobin in real samples. The recovery rate was 99.0%–104.0%. Guo et al. [52] prepared UCNPs@MIP by in-situ coating cytochrome C (Cyt C) imprinting material on the surface of carboxy-modified UCNPs using sol-gel technology, as observed in Fig. 3.7. The fluorescence intensity of UCNPs@MIP gradually weakened with the increase of Cyt C concentration. The imprinting factor under optimal conditions was 3.19. In summary, fluorescent molecularly imprinted polymers exhibit broad application prospects in protein detection.

(3) Electrode

In addition to nanomaterials, electrodes can also be chosen as carriers for surface protein imprinting. The immobilization of the protein imprinted polymers on the electrode surface by surface coating, in-situ polymerization, electrochemical polymerization, or self-assembly method can be utilized for the preparation of biosensors. Imprinted polymers exhibit highly selective to

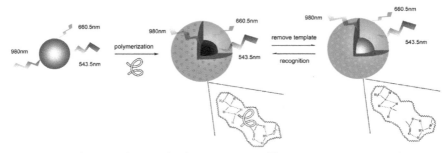

Figure 3.7 Schematic diagram for the preparation of fluorescent UCNPs@MIP. The copyright of the figure was obtained from Elsevier under license No: 5277550992848, Mar 28, 2022 [52].

the substance being isolated or analyzed. Fix it on the surface of the transfer electrode, the physical or chemical signal generated by the specific binding of the target protein to the imprinted sites of the polymers is amplified through a signal converter to generate an output signal. According to the difference between the output signal and the concentration of the analyte, the relationship can be analyzed quantitatively or qualitatively. As a sensor recognition component, the protein imprinted polymers not only possess good selectivity and high sensitivity but also carry the advantages of good stability, high reliability, and reusable. Therefore, protein imprinted sensors have become a research hotspot in the field of a sensor in recent years [53–55].

According to the different working principles of the sensor signal conversion element, molecularly imprinted sensors can be divided into three types: molecular imprinted mass–sensitive sensors [56], molecular imprinted optical sensors [57–59], and molecular imprinted electrochemical sensors [60–62]. Molecular imprinting electrochemical sensor has the advantages of simple electrochemical sensor equipment, high sensitivity, fast analysis speed, and simultaneous determination of multiple components. At the same time, it has a specific recognition ability deriving from the molecular imprinting technique, showing good stability and wide practicability. Ana et al. [60] took chrome black T (EBT) as the functional monomer and obtained BSA molecular-imprinted sensor on the surface of the electrode by electrochemical polymerization. The preparation process is shown in Fig. 3.8. The detection range for BSA was $5.00 \times 10^{-7} - 1.00 \times 10^{-5}$ mol L^{-1}, and the detection limit was 4.95×10^{-7} mol L^{-1}.

The molecular imprinted optical sensor not only has the advantages of the high sensitivity of the optical sensor but also has the characteristics

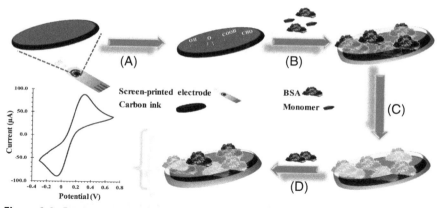

Figure 3.8 Design of BSA sensor: (A) oxidation of carbon electrode; (B) Electropolymerization of monomers in the presence of proteins; (C) template removal and (D) protein recombination. The copyright of the figure was obtained from Elsevier under license No: 5277590014387, Mar 28, 2022 [60].

of good selectivity of the molecular imprinting technique. Palladino et al. [57] synthesized TnT photo-biosensor by combining molecularly imprinted polymer (MIP) of polydopamine (PDA) with surface plasmon resonance (SPR) transduction using partial peptide of troponin T (TnT) as a template, as depicted in Fig. 3.9. The sensor can quickly and specifically recognize TnT and exhibited excellent sensitivity to the linear response of TnT concentration in the nanomolar range.

The mass-sensitive sensor mainly quantitatively detects the target substance through the slight changes in the surface quality of the sensor or the changes in the acoustic wave parameters such as amplitude, frequency, and wave speed caused by the change. As shown in Fig. 3.10, Orihara et al. [56] formed MIP film on a gold electrode by plasma-induced graft polymerization (PIP) technique and used it as an artificial recognition element to produce an unlabeled quartz crystal microbalance (QCM) biosensor for the detection of heparin. The sensor can specifically recognize heparin, and the selectivity factor for heparin analog chondroitin sulfate C was 4.0. The detection range for heparin was 0.001-0.1 wt% in the presence of CS (0.1 wt%). Fig. 3.11 shows the relationship between the frequency and time of heparin-PIP–MIP and PIP–NIP QCM to detect heparin, CS, and HSA at different concentrations.

With the aid of the surface protein imprinting technique, most sensors can show high selectivity and low detection limits for analytes, and play an irreplaceable role in biological separation and analysis.

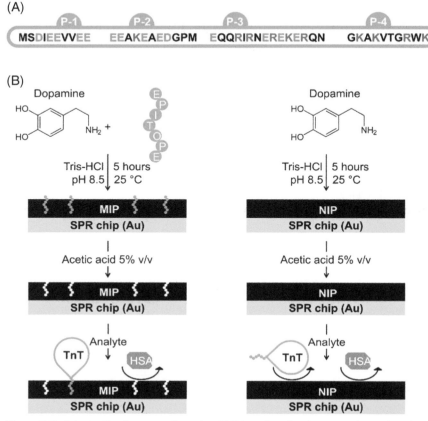

Figure 3.9 Schematic diagram of coating TnT imprinted polymer and non-imprinted polymer on the surface of gold SPR chip. The copyright of the figure was obtained from Elsevier under license No: 5277590452515, Mar 28, 2022 [57].

3.2.2 Surface design and construction of carrier materials

The types of surface functional groups, graft density, and surface roughness of carrier materials have important effects on the adsorption capacity and selective recognition ability of surface imprinted materials. Therefore, it is very important to design and construct the surface of the carrier before the imprinting process.

(1) Surface functional groups

Generally speaking, the characteristics of imprinted polymers are determined by their structure. For surface imprinted materials, the surface functional groups of the carrier material play a more important role in the adsorption and recognition of proteins. The hydrogen bond and electrostatic

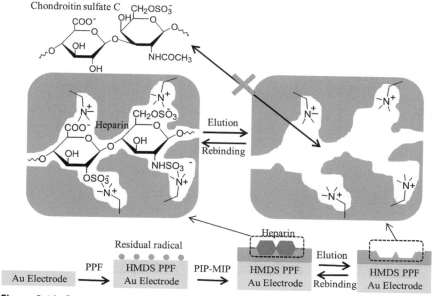

Figure 3.10 Preparation of heparin-PIP-MIP QCM. The copyright of the figure was obtained from Elsevier under license No: 5277591130016, Mar 28, 2022 [56].

attraction between different functional groups and the target molecules are different in strength, which means that the imprinting amount and binding amount of the imprinted materials prepared under similar conditions to the target molecules are different. In addition, the same functional groups with different grafting densities have a significant impact on the performance of the imprinted materials. Increasing the density of surface functional groups can improve the adsorption rate, adsorption capacity, and recognition performance of the imprinting sites toward the template proteins. At the same time, considering the different charges of various proteins in $pH = 7.0$ solution, the density of functional groups has different effects on the imprinting efficiency of proteins under the influence of electrostatic attraction. For example, the increase of surface carboxyl density is more beneficial to the improvement of the imprinting efficiency of proteins with a high isoelectric point, while the increase of surface amino density is more conductive to the improvement of the imprinting efficiency of proteins with a low isoelectric point. In addition, the effect of functional groups on the efficiency of imprinting is limited by the thickness of the imprinting layer, and the critical value of the thickness is related to the size of the template proteins. In previous studies, Fe_3O_4@HEA@Protein–MIPs were

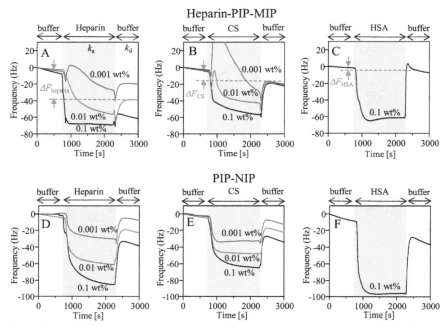

Figure 3.11 The relationship between the frequency and time of heparin-PIP-MIP and PIP-NIP QCM to detect heparin, CS and HSA at different concentrations. The copyright of the figure was obtained from Elsevier under license No: 5277591130016, Mar 28, 2022 [56].

obtained by imprinting on the surface of maleic anhydride (MAH) modified Fe_3O_4@hydroxyethyl acrylate (Fe_3O_4@HEA) microspheres, as shown in Fig. 3.12 [63]. The grafting density of carboxyl groups on the Fe_3O_4 surface was regulated by changing the concentration of HEA, and its effect on the recognition performance of imprinted materials was investigated. The results demonstrated that the increase in surface carboxyl density is beneficial to improving the adsorption rate, adsorption capacity, and recognition performance of imprinted materials. Meanwhile, as can be seen from Fig. 3.13, the increase in the density of carboxyl groups at the interface in pH 7.0 solution was more conducive to improving the imprinting efficiency of proteins with high electric points.

(2) Surface roughness

For surface imprinting, the surface area of the carrier directly determines the surface area of the imprinting layer, thereby affecting the number of effective imprinting sites. The increase of the external surface area of the carrier can simultaneously increase the mass transfer rate and the adsorption capacity, which is of great significance to the preparation of

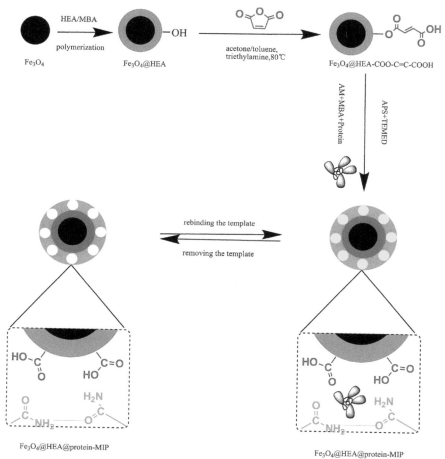

Figure 3.12 Schematic diagram of the synthesis process of Fe$_3$O$_4$@HEA@protein-MIPs. The copyright of the figure was obtained from Elsevier under license No: 5277650712805, Mar 28, 2022 [63].

protein imprinted polymers. Therefore, the preparation of carriers with a high external surface area has become the goal pursued by many researchers. By increasing the surface roughness, the external surface area of the material can be effectively improved. There are two main ways to improve the surface roughness of materials: one is to control the synthesis process to directly obtain the rough surface. As given in Fig. 3.14, SiO$_2$@GSH-MIP was prepared by imprinting glutathione (GSH) on the surface of fibrous silicon with a high specific surface [64]. The adsorption capacity of the microspheres for GSH was 85 mg g^{-1}, and the imprinting factor was 3.22.

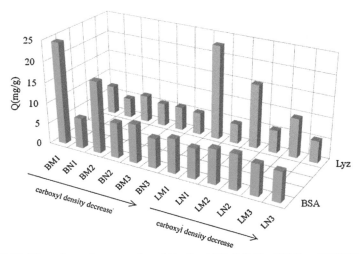

Figure 3.13 The adsorption capacity of Fe_3O_4@HEA (1/2/3)@BSA/Lyz-MIP/NIP toward BSA and Lyz. The copyright of the figure was obtained from Elsevier under license No: 5277650712805, Mar 28, 2022 [63].

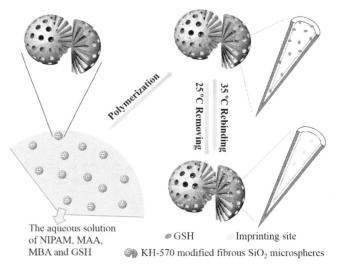

Figure 3.14 Schematic diagram for the preparation of GSH imprinted SiO_2. The copyright of the figure was obtained from Elsevier under license No: 5277680808732, Mar 28, 2022 [64].

Another method is to etch or grow unevenly on a smooth surface of the precursor. The former approach is relatively simple, but there are fewer systems that can be implemented.

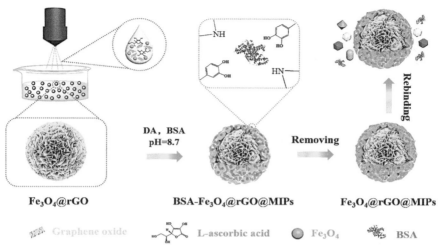

Figure 3.15 Preparation of BSA surface-imprinted magnetic graphene microspheres (Fe_3O_4@rGO@MIPs). The copyright of the figure was obtained from Elsevier under license No: 5277690752379, Mar 28, 2022 [33].

For polymer materials, raspberry-like microspheres can be obtained by controlling phase separation and volume shrinkage in suspension polymerization. Employing it for protein surface imprinting can solve the problems of limited diffusion, difficulty in elution, and recombination of target molecules to a certain extent. In our previous work, raspberry globular polyglycidyl methacrylate/polystyrene (PGMA/PS) microspheres were prepared by seed polymerization [38]. Selecting it as a carrier, protein imprinted microspheres PGMA/PS@BSA-MIPs were prepared by dopamine polymerization. The adsorption capacity and imprinting factor of the imprinted microspheres were 53.2 mg g^{-1} and 4.52, respectively. In addition, the preparation of raspberry-like nanoparticles by one-step soap-free emulsion polymerization by adjusting the compatibility of monomers was also reported [65].

Ultrasonic atomization or airflow drying technology can directly convert two-dimensional materials into composite microspheres with wrinkled surfaces. The wrinkled morphology brings more external surface area and provides more imprinting sites and low mass transfer resistance. While increasing the adsorption capacity and effective imprinting capacity, the elution of the template and the adsorption of the target are beneficial, thus reducing the non-specific adsorption and increasing the imprinting factor. As shown in Fig. 3.15, a novel magnetic graphene composite microsphere

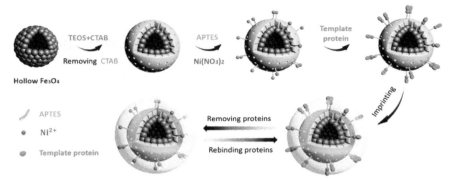

Figure 3.16 Preparation of HMMS@Ni^{2+}-MIPs. Reprinted (adapted) with permission from [66]. Copyright 2022 American Chemical Society.

(Fe_3O_4@rGO) was prepared by ultrasonic spray drying technique [33]. The polydopamine containing template protein BSA was deposited on the surface of Fe_3O_4@rGO, and the successful preparation of BSA imprinted magnetic graphene composite microspheres (Fe_3O_4@rGO@MIPs) was realized. The adsorption capacity of Fe_3O_4@rGO@MIPs on BSA reached 317.58 mg g^{-1} within 60 min. The imprinting factor was 4.24. More importantly, Fe_3O_4@rGO@MIPS can specifically recognize template BSA from mixed proteins and actual samples. There is no significant decrease in adsorption capacity and imprinting factor after repeated use eight times, which is expected to be used for protein separation in actual biological samples.

(3) Surface metal ion coordination

Different from the noncovalent interactions such as hydrogen bonding, electrostatic attraction, and Van Der Waals force, the five- or six-membered rings formed by the chelation of metal ions with the imidazole ring in the histidine residue of the protein have relatively high stability. This is beneficial to enhancing the interaction between the target proteins and the imprinted materials, thus increasing the number of effective imprinting sites. Therefore, protein imprinted materials that combine metal ion coordination with surface imprinting techniques have gradually emerged in recent years. As shown in Fig. 3.16, with the help of Ni^{2+}-BSA directional chelation, we achieved BSA imprinting on the surface of hollow Fe_3O_4@mSiO$_2$ through the self-polymerization of dopamine [66]. The prepared imprinted material HMMS@Ni^{2+}-MIPs exhibited a high adsorption capacity (266.99 mg g^{-1}) for BSA. The imprinting factor was 5.45. Later, to further study the effect of metal ion chelation, a series of metal ions (Mn^{2+}, Fe^{3+}, Ni^{2+}, Cu^{2+}, Zn^{2+})

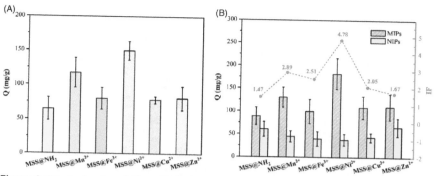

Figure 3.17 Anchoring ability of different microspheres to bovine serum albumin (BSA) (A) and adsorption performance of imprinted materials prepared from these substrates (B). The copyright of the figure was obtained from Elsevier under license No: 5277701058262, Mar 28, 2022 [67].

were chelated to the surface of $Fe_3O_4@mSiO_2$ to prepare BSA surface imprinted polymers [67]. As can be seen from Fig. 3.17, Ni^{2+} showed the best anchorage ability for BSA, and the prepared MSS@Ni^{2+}-MIPs showed the highest binding ability and rapid adsorption rate. The adsorption capacity of BSA was 182.18 mg g^{-1} within 30 min. The imprinting factor was 4.78.

3.2.3 Composition of an imprinted polymer layer

The imprinted polymer layer of surface protein imprinting is formed by the self-assembly of template proteins and functional monomers and cross-linking polymerization under the action of cross-linking agent. MIPs prepared by traditional methods are highly cross-linked polymers. Although they have the advantages of stable structure, their molecular recognition is simple and mechanical. This makes it difficult to balance the desorption rate, selectivity, and reusability in the separation and purification process, thus limiting its application in industrial separation. In recent years, to solve the above problems, researchers are increasingly interested in introducing some special components, such as smart response components, anti-protein components, and modified biomolecules, into imprinted polymer layers as functional monomers or cross-linking agents.

(1) Smart response imprinted polymers

Smart responsive imprinted polymers (S-MIPs) are introducing substances that can respond to external environmental stimuli, such as temperature, gas, pH, light, into the imprinted polymer layer as functional monomers or cross-linking agents. At present, research on S-MIPs has received a series of achievements. Thermo-sensitive MIPs (T-MIPs) [42,68],

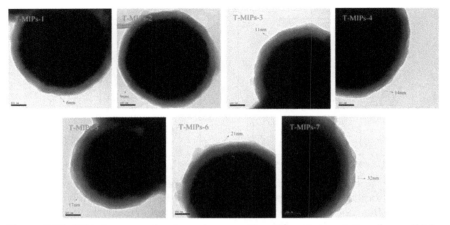

Figure 3.18 TEM images of microspheres with different imprinting layer thicknesses. The copyright of the figure was obtained from Elsevier under license No: 5277950043263, Mar 28, 2022 [68].

pH-sensitive MIPs (pH-MIPs) [69], and light-responsive MIPs (P-MIPs) [30] have been successfully prepared and shown a good application prospect in many fields such as drug delivery, biotechnology, separation science, sensors and so on. As shown in Fig. 3.18, a series of thermo-sensitive magnetic imprinted microspheres (T-MIPs) with different cross-linking degrees and imprinting layer thicknesses were prepared by precipitation polymerization [68]. The adsorption results showed that 17 nm was the best thickness for BSA imprinting. Considering both response performance and recognition capability, the crosslinking degree was determined to be 20% as the balance point. Under these conditions, the elution efficiency can reach 78.60% after one washing, ensuring good regeneration ability. The adsorption capacity and imprinting factors were 42.01 mg g^{-1} and 3.41, respectively. Xie et al. [30] prepared a novel photostimulation-responsive bovine hemoglobin (Bhb) imprinted polymer (Bhb-MIPs) on the surface of $Fe_3O_4@SiO_2$ nanoparticles modified by double bond using water-soluble azobenzene containing 4- [(4-methylacryloxy) phenyl azobenzene sulfonic acid (MAPASA) as photoswitching functional monomer. The preparation and application principles are shown in Fig. 3.19. The adsorption capacity of Bhb by Bhb-MIPs reached 81.1 mg g^{-1} within 60 min, and the imprinting factor was 3.95.

Subsequently, some dual-sensitivity MIPs and multi-sensitivity smart imprinted polymers have appeared successively. Using chitosan grafted N-isopropyl acrylamide (CS-G-NIPAM) as a pH- and temperature-sensitive

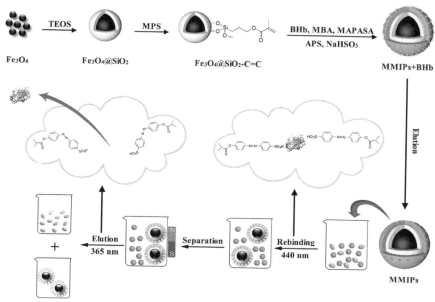

Figure 3.19 Preparation and application principle of BHb-MIPs. The copyright of the figure was obtained from Elsevier under license No: 5277950269531, Mar 28, 2022 [30].

monomer, acrylamide as a copolymer, N, N '-methylene bisacrylamide as a crosslinking agent, and BSA as a template protein, Dong et al. successfully synthesized pH- and temperature-sensitive protein imprinted microspheres with high adsorption capacity on the surface of SiO_2 [69]. The adsorption capacity of the prepared SiO_2@BSA molecularly imprinted polymer reached 119.88 mg g^{-1} within 2 h, and the imprinting factor was 2.25.

(2) Antiprotein imprinted polymer layer

To reduce the nonspecific adsorption of protein imprinted polymers, the researchers have introduced anti-protein adsorption chain segments during its preparation. Anti-protein adsorption segment refers to polymers with zero net charges, which contains neutral polymers and amphoteric polymers. These polymers with a total neutral charge can effectively reduce the electrostatic interaction with the charged protein patch part, and their hydrophilic nature can form an additional hydration layer on the surface of materials to prevent the protein from adsorbing on the surface materials. At the same time, the large space steric hindrance of the polymer itself can also effectively reduce the protein adsorption on the surface of materials. Oligo-polyethylene glycol/polyethylene glycol (OEG/PEG) is one of the most widely used neutral polymers in the field of anti-adsorption. Yang et al. [70] proposed a novel strategy for controlled synthesis of pegylated surface imprinting nanoparticles with reduced non-specific binding. This strategy was based on two consecutive steps of surface-initiated reversible addition-fragmentation chain transfer aqueous precipitation polymerization (SI-RAFT APP). Firstly, lysozyme was used as a template to form protein imprinted nanoshells on the surface of the nanocore through the first step of SI-RAFT-APP. Then, nonlinear PEG chains were grafted onto core-shell imprinted nanoparticles before template removal by the second step SI-RAFT-APP. The thickness of imprinted nanoshells and the length of the graft chain can be easily controlled by polymerization time. Compared with the unpegylated control group, the obtained pegylated core-shell particles showed significantly improved template binding selectivity. The adsorption capacity was 50.03 mg g^{-1} and the imprinting factor increased from 2.1 to 9.1.

Although PEG as an anti-protein adsorption chain segment has made remarkable achievements in the field of protein imprinting, the hydrophilic and hydrophobic properties of PEG itself lead to a significant decrease in protein activity. And the negative effect brought by PEG with larger molecular weight is more obvious. On the contrary, amphoteric polymers have charged positive and negative charge centers, which will not change the spatial conformation of the protein. Moreover, its super-hydrophilic characteristics determine that the protein can still maintain its activity during the imprinting process. At present, amphoteric anti-protein adsorption polymers extensively studied by researchers mainly include methylacryloxy phosphorylcholine (MPC) [35,71] and amphoteric betaine structure polymers [37,72]. In the early stage, we prepared novel BSA surface imprinted magnetic microspheres by copolymerizing anti-protein adsorption chain segment MPC, functional monomer acrylamide, and crosslinking agent MBA on the surface of $Fe_3O_4@SiO_2$ microspheres [71]. The preparation process is shown in Fig. 3.20. Binding specificity experiments revealed that the introduction of MPC can greatly improve the recognition specificity of BSA imprinted microspheres. The adsorption capacity and imprinting factors are 21.79 mg g^{-1} and 8.32, respectively. More importantly, the selectivity coefficients of the imprinted microspheres for human serum albumin, ovalbumin, lysozyme, cytochrome C, and ribonuclease A can reach 1.63, 5.23, 9.14, 7.43, and 7.23, respectively.

(3) Biomolecular imprinted polymer layer

Many commonly used functional monomers and cross-linking agents have defects in maintaining protein conformation, which makes proteins

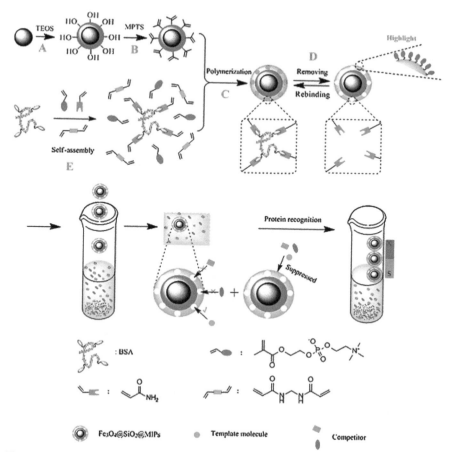

Figure 3.20 Schematic diagram for the preparation of $Fe_3O_4@SiO_2@BSA$-MIPs. The copyright of the figure was obtained from Elsevier under the license No: 5277951408540, Mar 28, 2022 [71].

prone to structural changes during the imprinting process. Therefore, the identifiable imprinted cavity of the imprinted polymers cannot be precisely complementary to the template protein in space and functional group distribution, thereby reducing the recognition performance of the materials. According to reports, selecting biological macromolecules as functional monomers with good biocompatibility, such as dopamine, cyclodextrin, and chitosan, can solve this problem to a certain extent. Among them, dopamine is a functional monomer that can undergo oxidation and self-polymerization in an alkaline solution to form an adhesive polydopamine membrane. Because dopamine is rich in amino groups and phenolic hydroxyl groups

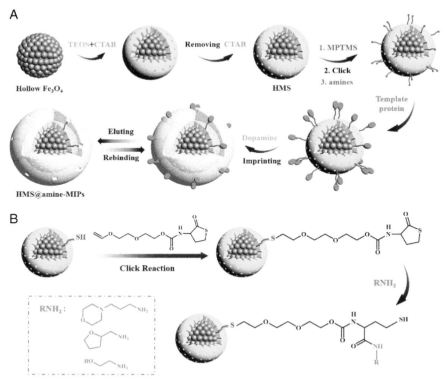

Figure 3.21 Preparation of HMS@amine-MIPs (a) and free radical mercapto-ene-amine coupling strategy. The copyright of the figure was obtained from Elsevier under license No: 5277961142407, Mar 28, 2022 [73].

and has a benzene ring structure, it can endow the polymer shell with considerable biocompatibility and mechanical strength. The protein imprinted polymers prepared using dopamine as a functional monomer has the characteristics of strong rigidity, low mass transfer resistance, and fast adsorption rate. Accordingly, dopamine is widely used as a functional monomer for protein imprinting [33,38,43,73]. For example, we introduced a series of mercapto–ene–amine chains as protein anchors into the imprinting system by click chemistry [73]. Using BSA as the template, imprinted microspheres HMS@amine-MIPs (amine refers to MA, THA, and EA) were prepared by self-polymerization of dopamine on the surface of hollow magnetic microspheres. The preparation process is shown in Fig. 3.21. Among them, the adsorption capacity of BSA imprinted microspheres HMS@MA-MIPs based on mercaptene-MA conjugate (MA is 3-morpholine propyl amine) reached 209.22 mg g^{-1} within 30 min, and the imprinting factor was 4.59.

Chitosan is a natural alkaline polysaccharide. Its chemical structure contains a large number of hydroxyl and amino groups, which can provide abundant reaction sites and is easy to chemically modify. Moreover, chitosan has excellent biocompatibility, biodegradability, non-cytotoxicity, and non-antigenicity. Based on the above advantages, chitosan can be introduced as a functional monomer and cross-linking agent in the preparation of imprinted polymers. Selecting $Fe_3O_4@SiO_2$ as the carrier, maleic anhydride modified chitosan as the functional monomer and crosslinking agent, and 2-hydroxyethyl acrylate (HEA) without damaging the protein structure as the functional monomer, BSA-imprinted magnetic microspheres $Fe_3O_4@SiO_2@MIPs$ were successfully prepared by precipitation copolymerization [37]. Additionally, methyl sulfobetaine methacrylate (SBMA) was introduced as an anti-protein segment into the imprinted polymer to reduce nonspecific adsorption and improve selectivity. The adsorption capacity of the imprinted microspheres on BSA was 116.39 mg g^{-1}, and the imprinting factor was 4.73.

3.3 Concept of self-driven protein imprinting

Self-driven protein imprinting refers to the imprinted materials that drive the target proteins to spontaneously adsorb to the imprinting sites on the surface of the imprinted materials under its conditions without external force. Carrier materials with self-driven properties need to meet the following characteristics: First, the surface of the materials should have abundant pores and the pore size should match the size of the BSA (14 × 4 × 4 nm), to ensure the smooth passage of BSA; secondly, the materials need to have a hollow internal structure, so that when it is dispersed in a high-concentration protein solution, a concentration gradient is formed between the cavity inside the tube and outside of the tube, resulting in an osmotic pressure difference. The osmotic pressure difference drives BSA molecules to spontaneously bind to the imprinting sites on the surface, and further migrate into the internal cavity through the pores of the tube wall. As a result, the imprinted materials exhibit a high adsorption capacity and a fast adsorption speed for the target proteins [74]. The specific structure of the carrier materials is shown in Fig. 3.22.

3.4 Self-driven surface imprinted tubular carbon nanofibers

Protein imprinted material SIPTCFs with self-driven function were successfully prepared by using tubular nanofibers with abundant pores in the

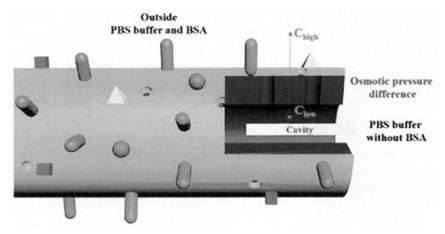

Figure 3.22 The specific structure of the carrier materials with self-driven properties. The copyright of the figure was obtained from Elsevier under license No: 5277971347796, Mar 28, 2022 [74].

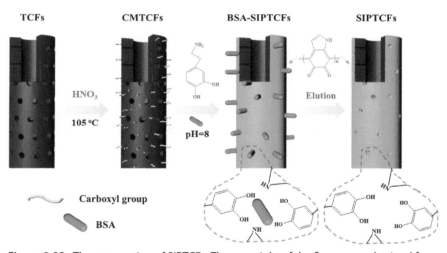

Figure 3.23 The preparation of SIPTCFs. The copyright of the figure was obtained from Elsevier under license No: 5277971347796, Mar 28, 2022 [74].

carboxyl modified tube wall as the carrier, dopamine as the imprinting polymer monomer, and BSA as the templates [74]. The preparation process is shown in Fig. 3.23. The carboxyl groups on the interface play an effective role in improving the hydrophilicity of tubular carbon nanofibers (TCFs), adsorbing proteins to the surface through weak interaction with proteins, and guiding the polymerization of dopamine to facilitate the coating of imprinted layers. Due to the abundant pores on the tube wall of the tubular

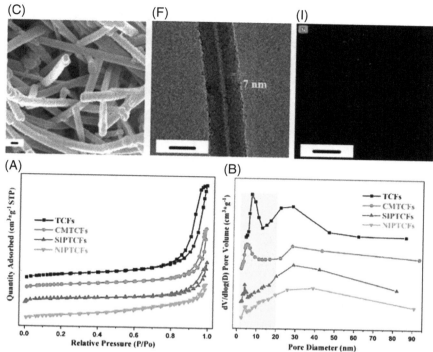

Figure 3.24 SEM (A) and TEM (B, C) electron images of SIPTCFs, as well as nitrogen adsorption-desorption curves (D) and pore size distribution curves (E) of various products. The scale is 100 nm. The copyright of the figure was obtained from Elsevier under license No: 5277971347796, Mar 28, 2022 [74].

nanofiber, a concentration gradient is formed between the cavity inside the tube and outside of the tube when it is dispersed in a high-concentration of BSA solution. This leads to the formation of an osmotic pressure difference, which drives the BSA molecules to spontaneously bind to the imprinting sites on the surface and further migrate to the internal cavity through the pores on the tube wall. Subsequently, SIPTCFs possess a high adsorption capacity and a fast adsorption speed for BSA. The nature of self-driven is fundamentally derived from the osmotic pressure difference.

Fig. 3.24 gives the SEM and TEM electron micrographs of SIPTCFs and the nitrogen adsorption–desorption curve and the pore size distribution curve of the prepared samples. After coating with an imprinted polymer layer on the surface of tubular nanofibers, SIPTCFs still had a hollow tubular nanofiber structure, but the tube wall became rough (Fig. 3.24A). According to the slight differences in the mass thickness contrast and morphology between the imprinted polymers and the nanofibers, the thickness of

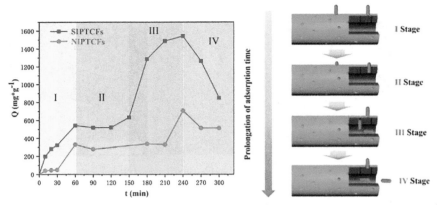

Figure 3.25 The adsorption kinetic curves and self-driven adsorption mechanism of SIPTCFs and NIPTCFs on BSA. The copyright of the figure was obtained from Elsevier under license No: 5277971347796, Mar 28, 2022 [74].

the imprinted layer directly measured by TEM photographs was 7 nm (Fig. 3.24B). In addition, the presence of the N element in the energy spectrum of SIPTCFs indicated that the polydopamine imprinted layer was successfully coated on the surface of tubular nanofibers (Fig. 3.24C). The nitrogen adsorption–desorption curves of the materials belonged to type IV, and the hysteresis loop type was H4, as shown in Fig. 3.24D. This indicated the presence of slit-like pores in the materials. As the modification and coating process was conducted, the specific surface area and pore volume of nanofibers decreased. The reason was described as that the functional groups and imprinted polymers blocked part of the pores. The order of specific surface area and pore volume from high to low was TCFs (251.54 m^2 g^{-1} and $0.4022 \text{ cm}^3 \text{ g}^{-1}$), CMTCFs ($205.03 \text{ m}^2\text{g}^{-1}$ and $0.3708 \text{ cm}^3 \text{ g}^{-1}$), SIPTCFs ($195.75 \text{ m}^2 \text{ g}^{-1}$ and $0.2882 \text{ cm}^3 \text{ g}^{-1}$), NIPTCFs ($174.80 \text{ m}^2 \text{ g}^{-1}$ and $0.2587 \text{ cm}^3 \text{ g}^{-1}$). Owing to the presence of imprinting sites, SIPTCFs were higher than NIPTCFs. The pore size distribution curves of nanofibers were all bimodal (Fig. 3.24E). The pores in the range of 2–20 nm were small holes in the tube wall of the nanofibers. This part of the pores can provide solution and protein molecules to enter the tube cavity. The hole in the interval of 20–50 nm corresponded to the diameter of the cavity in the tube. This large-scale cavity created an osmotic pressure difference between the solution inside and outside the tube, which in turn gave the imprinted fibers a self-driven specific adsorption performance.

When the initial protein concentration was 0.1 mg mL^{-1}, the adsorption capacity of SIPTCFs on BSA was measured at different time intervals. The relationship curve of adsorption capacity with time is presented in Fig. 3.25.

The adsorption process was divided into four stages. The first stage was before 60 min, and the adsorption capacity increased gradually with time. The BSA molecules in the solution at this stage are firstly bound to the imprinting sites on the fiber surface under the action of charge effect and memory effect. This process was the adsorption of BSA on the surface of SIPTCFs. At 60 min, the adsorption capacity of SIPTCF on BSA was 541.99 mg g^{-1}. In the second stage, the adsorption capacity remained stable from 60 min to 150 min. At this stage, the adsorption sites on the fiber surface reached saturation, and the surface no longer bound BSA molecules. Notably, BSA molecules adsorbed on the surface would migrate into the cavity through the pores of the tube wall due to the osmotic pressure difference between the inside and outside of the tube. The equilibrium of adsorption indicated that the adsorbed BSA molecules still occupied the imprinting sites. The third stage lasted from 150 min to 240 min, and the adsorption capacity increased again with time. This meant that the BSA molecules on the surface began to detach from the imprinted sites and diffused into the cavity through pores on the tube wall of fiber. At the same time, the BSA molecules in the solution rebind to the imprinted sites on the fiber surface. At 240 min, the adsorption capacity of SIPTCFs on BSA was as high as 1543.07 mg g^{-1}. In the fourth stage, the adsorption capacity gradually decreased with time after 240 min. This was due to the BSA molecules migrating into the cavity escaping from the fiber port. When the outward migration rate was greater than the adsorption rate, the adsorption amount will decrease. The above adsorption results strongly confirmed the self-driven specific adsorption performance of surface imprinted fibers. The adsorption kinetics curve of NIPTCFs was similar to that of SIPTCFs. However, the adsorption capacity of NIPTCFs was significantly lower than that of SIPTCFs because imprinting sites and no functional groups were matching with BSA on the surface of NIPTCFs. At the same time, as there was no imprinted cavity connected with the fiber surface, the diffusion time of BSA into the fiber through the tube wall lagged behind that of SIPTCFs. The time for the third stage of adsorption capacity to rise was 210 min, which was 60 min later than SIPTCFs. This also proved that SIPTCFs were surface imprinted materials.

3.5 Self-driven surface imprinted magnetic tubular carbon nanofibers

To solve the problems of low adsorption capacity, difficult mass transfer, and slow adsorption and separation speed of protein imprinted materials, tubular

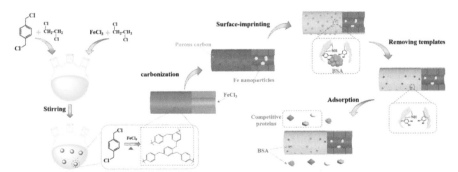

Figure 3.26 Schematic diagram of the preparation process of SIPMTFs. Reprinted (adapted) with permission from [75]. Copyright 2022 American Chemical Society.

nanofiber carriers with self-driven properties were combined with magnetic nanoparticles to obtain a new carrier material [75]. The preparation method of magnetic tubular carbon nanofibers (MTFs) with porous on the tube wall is presented in Fig. 3.26. Fe nanoparticles were loaded inside the tubular fibers. The preparation of BSA surface imprinted materials (SIPMTFs) was realized by using dopamine as functional monomers and MTFs as carriers. Its distinctive feature was that the polydopamine imprinted layer was coated on the surface of MTFs loaded with some Fe nanoparticles inside the tube. Consequently, the imprinted materials can be quickly separated from the solution under the action of an external magnetic field. During the adsorption process, an osmotic pressure difference formed between the internal cavity of MTFs and the external solution to self-drive the binding of the target proteins to the imprinted sites. The separation process was simplified while increasing the adsorption capacity and accelerating the adsorption process.

The electron micrographs of TPFs, MTFs, SIPMTFs, and NIPMTFs and the energy dispersive spectroscopy (EDS) analysis results of SIPMTFs are shown in Fig. 3.27. TPFs presented a solid fibrous structure (Fig. 3.27A). After calcination, the high mass thick contrast filler $FeCl_3$ inside the fiber was converted into Fe_3O_4 nanoparticles. At the same time, the products showed an obvious hollow structure, as shown in the TEM picture of MTFs (Fig. 3.27B). Compared with the smooth surface of TPFs and MTFs (Fig. 3.27D and E), the surface of SIPMTFs became significantly rougher, and the fiber diameter increased slightly, as shown in Figs. 3.27C and F. This was attributed to the coating of the PDA imprinting layers on the surface of MTFs. The dark-field photos of SIPMTFs (Fig. 3.27G) presented different degrees of brightness and darkness on the tube wall, which also

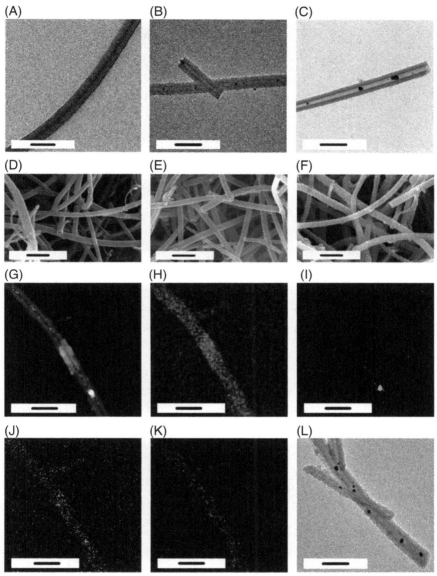

Figure 3.27 TEM (A–C) and SEM (D–F) images of TPFs (A, D), MTFs (B, E), SIPMTFs (C, F); EDS analysis of SIPMTFs: Full spectrum(G), C element (H), Fe (I), O element (J), N element (K); TEM images of NIPMTFs(L). Scale bars of micrographs represent 200 nm. Reprinted (adapted) with permission from [75]. Copyright 2022 American Chemical Society.

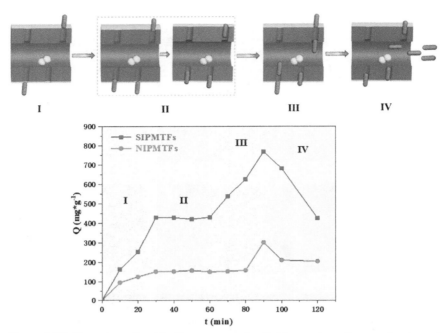

Figure 3.28 The adsorption kinetic curves and self-driven adsorption mechanism of SIPMTFs and NIPMTFs on BSA. Reprinted (adapted) with permission from [75]. Copyright 2022 American Chemical Society.

proved the successful coating of the polydopamine imprinted layers. EDS analysis results displayed that SIPMTFs were rich in C, Fe, O, and N elements (Fig. 3.27H–K). The presence of N and O indicated that the polydopamine imprinted layer was coated on the surface of MTFs. Fe element belonged to the magnetic Fe_3O_4 nanoparticles. Fig. 3.27L is a TEM photograph of NIPMTFs. It can be seen that the morphology of NIPMTFs was not significantly different from SIPMTFs, showing the morphology of MTFs outsourced PDA.

The changes in the adsorption capacity of BSA by SIPMTFs and NIPMTFs over time are given in Fig. 3.28. The adsorption kinetics curve of SIPMTFs for BSA showed the same trend as that of SIPTCFs, which was consistent with the adsorption characteristics of self-driven imprinted materials. However, due to the presence of Fe_3O_4 nanoparticles in the fiber cavity, the density of SIPMTFs was slightly increased relative to that of SIPTCFs. Therefore, the equilibrium adsorption capacity of SIPMTFs to BSA was reduced. The equilibrium adsorption capacity was 430.39 mg g^{-1}.

Specific adsorption performance is an important index to evaluate the recognition ability of imprinted materials. Four proteins with different

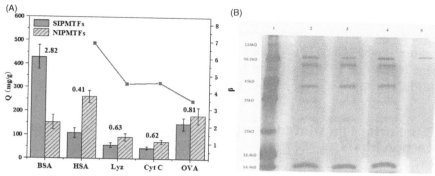

Figure 3.29 Selective adsorption of BSA, HSA, Lyz, Cyt C, and OVA on the SIPMTFs and NIPMTFs (A); SDS-PAGE analysis of adsorption of BSA by SIPMTFs from protein mixture (B). Lane 1: marker; lane 2: protein mixed solution; lane 3: remaining protein mixed solution after adsorption by SIPMTFs; lane 4: the eluent from SIPMTFs; The dosage of protein solution was 10 µL. Experimental conditions:1.0mg SIPMTFs or NIPMTFs were incubated with 20.0 mL mixed proteins at 25°C. Reprinted (adapted) with permission from [75]. Copyright 2022 American Chemical Society.

isoelectric points and molecular weights were selected as competing proteins to investigate the specific adsorption properties of SIPMTFs to BSA. The molecular weight and isoelectric point of BSA are 66.4 kDa and 4.9, respectively. The competitive proteins were HSA (66 kDa, pI 4.64), OVA (43 kDa, pI 4.7) Lyz (14.4 kDa, pI 10.8), and Cyt C (12.4 kDa, pI 10.2), respectively. The surface of SIPMTFs possessed imprinting sites matching the size, shape, and functional group arrangement with BSA molecules. Therefore, SIPMTFs had the highest adsorption capacity for BSA. The order of adsorption capacity for the other four proteins was OVA>HSA>Lyz>Cyt C (Fig. 3.29A). The IF values were marked at the top of the bar chart. The IF of SIPMTFs for BSA was the highest. HSA and OVA exhibited an isoelectric point similar to that of BSA, but the molecular weight of OVA was smaller than that of HSA. Meanwhile, there were differences in the arrangement of functional groups between both BSA. Although Lyz and Cyt C had smaller molecular weights, their isoelectric points were quite different from BSA. The selectivity factors (β) of SIPMTFs for HSA, Lyz, Cyt C, and OVA were 6.88, 4.48, 4.55, and 3.49, respectively, calculated by IF, indicating the specific recognition of SIPMTFs for BSA. The specific recognition ability of SIPMTFs for BSA was further evaluated by using the mixed proteins of BSA, HSA, Lyz, and OVA as adsorption objects. The results of gel electrophoresis SDS–PAGE analysis are shown in Fig. 3.29B. Comparing channel 2 and channel 3, it can be seen that the intensity of the BSA band was weakened after adsorption with SIPMTFs, while the intensity of Lyz, HSA, and OVA bands hardly changed. Compared to channel 2 with channel 4, the strength of the BSA band changed little after adsorption by NIPMTFs. Channel 5 gives the solution components after the first elution of SIPTCFs. It can be seen that only BSA existed. Therefore, SIPMTFs can selectively capture BSA in mixed biological components.

3.6 Surface imprinted manganese dioxide-loaded tubular carbon fibers

In recent years, many attempts have been made by researchers to overcome the difficulties in the process of protein imprinting. The development of surface imprinting and epitope imprinting techniques enabled the imprinting sites of the protein imprinted polymers to be located on the surface of the carriers. It is conducive to the elution and recombination of template protein and has the characteristics of low mass transfer resistance and fast adsorption rate. However, there was no strong interaction between the surface of general carrier materials and proteins. Accordingly, the ability to immobilize or adsorb the proteins was weak, which makes the number of surface imprinting sites relatively limited. As a result, the adsorption capacity was affected. According to the literature, MnO_2 nanosheets can form strong interaction with the amino groups of protein, thus enhancing the affinity between both. Inspired by this, we designed and prepared a novel lamellar MnO_2-coated tubular nanofiber (FTCFs@MnO_2). And with dopamine as the functional monomer, BSA was imprinted on the surface of MnO_2 nanosheets to get FTCFs@MnO_2@MIPs [76]. The preparation process is shown in Fig. 3.30. The introduction of the lamellar MnO_2 layer significantly increased the external surface area of the carrier. There was a strong interaction between MnO_2 and BSA molecules, which was conducive to immobilizing more template molecules and increasing the amount of imprinting. The imprinted polymer layer was coated on the surface of non-absolutely dense MnO_2, and the inside was TCFs with a special hierarchical pore structure. Part of the imprinted layers overlapped between different MnO_2 sheets will produce self-driven specific adsorption performance due to the difference in internal and external osmotic pressure, to accelerate mass transfer and shorten the adsorption time.

The morphology characterization of FTCFs@MnO_2 and FTCFs@MnO_2@MIPs is given in Fig. 3.31. In the presence of FTCFs as templates, flaky MnO_2 grew densely on its surface (Fig. 3.31A and B). The average diameter of FTCFs@MnO_2 was about 230 nm. The thickness of the MnO_2 coating layer ranged from 50 to 65 nm. After coating the

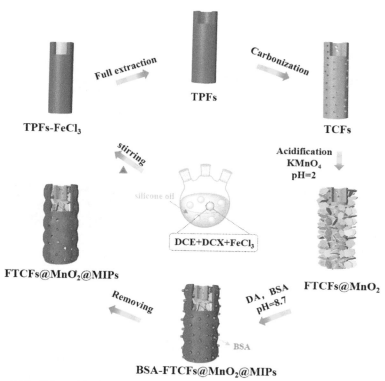

Figure 3.30 Schematic diagram of the preparation process of FTCFs@MnO₂@MIPs. The copyright of the figure was obtained from Elsevier under license No: 5277981270914, Mar 28, 2022 [76].

PDA imprinted layer, the diameter of the fiber increased. Compared with FTCFs@MnO₂, the lamellar sharp edges of MnO₂ on the surface of FTCFs@MnO₂@MIPs disappeared and were replaced by irregular smooth protrusions, as observed in Fig. 3.31C and D. In the dark field photos of FTCFs@MnO₂@MIPs (Fig. 3.31E), the tube wall presented a different degree of brightness and darkness. The high brightness was MnO₂, and the low brightness on the outermost side was polymers. As shown in the EDS analysis results of Fig. 3.31F–H, a large number of Mn and O elements were attributed to layered inorganic component MnO₂. The only source of N element was PDA, whose distribution diameter was significantly higher than Mn and O elements. This verified that PDA was coated on the outer surface of MnO₂.

XPS was used to clarify the interaction between BSA and MnO₂. The occurrence of Mn 2p, C 1s, N1s, and O1s binding energies in the full spectrum of XPS (Fig. 3.32A and B) illustrated the existence of Mn, C,

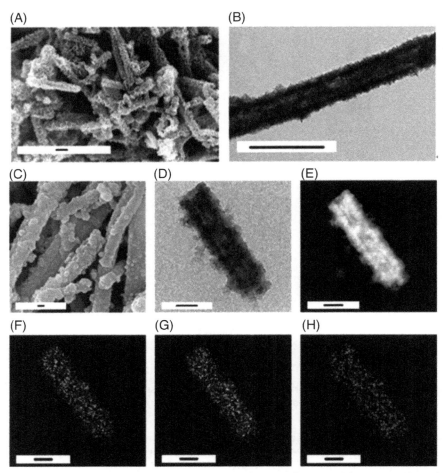

Figure 3.31 SEM (A, C) and TEM (B, D) photos of FTCFs@MnO$_2$ (A, B) and FTCFs@MnO$_2$@MIPs (C, D); EDS analysis results of FTCFs@MnO$_2$@MIPs: full-spectrum (E), Mn element (F), O element (G), N element (H). The scales in Figure A and B are 500 nm, and the rest are 100 nm. The copyright of the figure was obtained from Elsevier under license No: 5277981270914, Mar 28, 2022 [76].

N, and O elements. The binding energy of S 2p also appeared in the full spectrum of BSA-FTCFs@MnO$_2$@MIPs. After elution, the binding energy of S 2p disappeared, indicating that the BSA molecules were completely eluted. There were four types of carbon bonds in the high-resolution XPS spectra of C 1S (Fig. 3.32C and D), among which the existence of C–N and C–O suggested the existence of PDA. Comparing the O 1s high-resolution XPS spectra of BSA-FTCFs@MnO$_2$@MIPs and FTCFs@MnO$_2$@MIPs

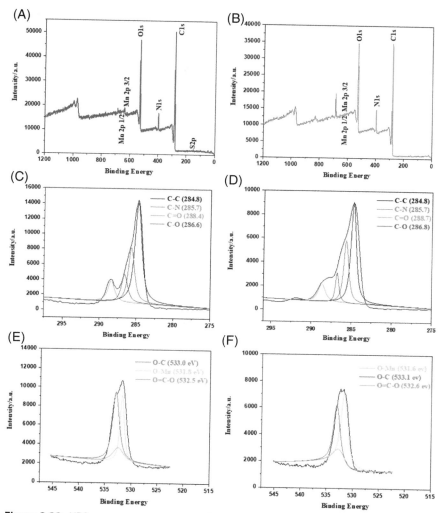

Figure 3.32 XPS survey spectrum of BSA-FTCFs@MnO$_2$@MIPs (A) and FTCFs @MnO$_2$@MIPs (B); deconvoluted XPS C 1s spectra of BSA-FTCFs@MnO$_2$@MIPs (C) and FTCFs@MnO$_2$@MIPs (D); deconvoluted XPS O 1s spectra of BSA-FTCFs@MnO$_2$@MIPs (E) and FTCFs@MnO$_2$@MIPs (F). The copyright of the figure obtained from Elsevier under the license No: 5277981270914, Mar 28, 2022 [76].

(Fig. 3.32E and F), it was not difficult to find that the binding energies of O–Mn, O–C and O=C–O were all biased. The interaction between –COOH of BSA and –OH of MnO$_2$ surface was indicated.

The adsorption kinetic curves of FTCFs@MnO$_2$@MIPs and FTCFs@MnO$_2$@NIPs for BSA are depicted in Fig. 3.33. It was not

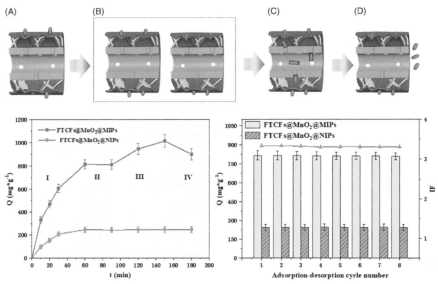

Figure 3.33 The adsorption mechanism, adsorption kinetics, and reusability of FTCFs@MnO$_2$@MIPs for BSA. The copyright of the figure was obtained from Elsevier under license No: 5277981270914, Mar 28, 2022 [76].

difficult to find that the adsorption process was also divided into four stages. Nevertheless, due to the affinity between MnO$_2$ and BSA, the saturated adsorption capacity of FTCFs@MnO$_2$@MIPs for BSA was significantly higher than that of SIPTCFs and SIPMTFs. The adsorption capacity of FTCFs@MnO$_2$@MIPs on BSA was as high as 816.44 mg g^{-1} in 60 min, and the imprinting factor was 3.31. Furthermore, different from the adsorption mechanism of SIPTCFs and SIPMTFs, the migration of BSA to the internal cavity at the second stage 60–90 min was attributed to the osmotic pressure difference between the external solution and the internal cavity formed by different MnO$_2$ sheets and polydopamine imprinted layers. The developed FTCFs@MnO$_2$@MIPs showed excellent reusable performance. As seen in Fig. 3.33, after eight adsorption-desorption cycles, the adsorption capacity of FTCFs@MnO$_2$@MIPs for BSA only lost 1.03%. The reason was described as that the polymer imprinted layer was tightly bound to the MnO$_2$ coating layer during the repeated elution process. The skeleton maintained high stability, and few imprinted sites were damaged. The binding of BSA to FTCFs@MnO$_2$@NIPs relied on non-specific adsorption. So, the elution process had little effect on it, and the adsorption capacity had no obvious change.

3.7 Summary and prospect

Protein imprinted polymer materials exhibit unique characteristics in the purification of biological macromolecules. High adsorption capacity, stability, and excellent selectivity make it possess the advantages of column chromatography separation and bio-specific separation. With the continuous development of material preparation techniques, more and more protein imprinted materials have been developed, many of which show outstanding properties. In the future, the main development directions in the field of protein imprinting may focus on the following aspects:

(1) The design and preparation of new carrier materials, especially micro-nano-scale materials with suitable pores, high surface area, and favorable immobilization of template molecules. The enrichment of carriers can produce self-driven effects.

(2) The synthesis of new functional monomers is used to prepare a more intelligent imprinting layer during the process of imprinting, elution, and adsorption. The development of polymerization methods of imprinted polymer is equally important.

(3) The research on the function mechanism of imprinted polymer needs to be carried out more in-depth. How to characterize and clarify the arrangement of functional groups at the imprinting sites? Furthermore, the specific interaction sites of these functional groups reacting with template molecules should be clarified.

(4) At present, the separation targets of biological macromolecular imprinting are still relatively limited. It is necessary to expand the study of the separated objects. At the same time, it is still difficult to separate a target from a system with many structures similar to the target.

(5) Protein imprinting has made some progress in the field of sensing. The next step that needs to be carried out in the research on the integration of diagnosis and treatment.

(6) The ultimate goal of basic research is valuable practical application. The real application of protein imprinted materials in the industry is advancing slowly, which is the focus of follow-up work.

References

[1] S.A. Bhakta, E. Evans, T.E. Benavidez, C.D. Garcia, Protein adsorption onto nanomaterials for the development of biosensors and analytical devices: A review, Anal. Chim. Acta. 872 (2015) 7–25.

[2] P. Silva-Bermudez, S.E. Rodil, An overview of protein adsorption on metal oxide coatings for biomedical implants, Surface and Coatings Technol. 233 (2013) 147–158.

[3] J. Ashley, M.A. Shahbazi, K. Kant, V.A. Chidambara, A. Wolff, D.D. Bang, et al., Molecularly imprinted polymers for sample preparation and biosensing in food analysis: Progress and perspectives, Biosens. Bioelectron. 91 (2017) 606–615.

[4] L. Chen, X. Wang, W. Lu, X. Wu, J. Li, Molecular imprinting: perspectives and applications, Chem. Soc. Rev. 45 (8) (2016) 2137–2211.

[5] T. Sajini, M.G. Gigimol, B. Mathew, A brief overview of molecularly imprinted polymers supported on titanium dioxide matrices, Materials Today Chemistry 11 (2019) 283–295.

[6] L.D., W.H. Cameron Alexander, Imprinted polymers: artificial molecular recognition materials with applications in synthesis and catalysis, Tetrahedron 59 (2003) 2025–2057.

[7] M.M. Minqiang He, J. Wan, J. He, Y. Yan, A new molecularly imprinted polymer prepared by surface imprinting technique for selective adsorption towards kaempferol, Polym. Bull. 68 (2012) 1039–1052.

[8] A.M. Chrzanowska, M. Diaz-Alvarez, P.P. Wieczorek, A. Poliwoda, A. Martin-Esteban, The application of the supported liquid membrane and molecularly imprinted polymers as solid acceptor phase for selective extraction of biochanin A from urine, J. Chromatogr. A 1599 (2019) 9–16.

[9] M.K.S. Ansari, Synthesis and application of molecularly imprinted polymer for highly selective solid phase extraction trace amount of sotalol from human urine samples: Optimization by central composite design, Chem. Res. 26 (2017) 2477–2490.

[10] S.O.J. Sanchez-Gonzalez, A.M. Bermejo, P. Bermejo-Barrera, F.S. Romolo, S.S.-R.A. Moreda-Pineiro, Development of a micro-solid-phase extraction molecularly imprinted polymer technique for synthetic cannabinoids assessment in urine followed by liquid chromatographytandem mass spectrometry, J. Chromatogr. A 1550 (2018) 8–20.

[11] M.J.G. Chen, P. Du, C. Zhang, X. Cui, Y. Zhang, Y. She, et al., A sensitive chemiluminescence enzyme immunoassay based on molecularly imprinted polymers solid-phase extraction of parathion, Anal. Biochem. 530 (2017) 87–93.

[12] A.N.S. Ghasemi, Molecularly imprinted ultrafiltration polysulfone membrane with specific nano-cavities for selective separation and enrichment of paclitaxel from plant extract, React. Funct. Polym. 126 (2018) 9–19.

[13] J.B.G. Székely, W. Heggie, B. Sellergren, F.C. Ferreira, A hybrid approach to reach stringent low genotoxic impurity contents in active pharmaceutical ingredients: Combining molecularly imprinted polymers and organic solvent nanofiltration for removal of 1,3-diisopropylurea, Sep. Purif. Technol. 86 (2012) 79–87.

[14] X.W.H.L. Qin, W. Zhang, W.Y. Li, Y.K. Zhang, Surface-modified polystyrene beads as photografting imprinted polymer matrix for chromatographic separation of proteins, J. Chromatogr. A 1216 (2009) 807–814.

[15] I.Y.K. Kitahara, T. Hanada, H. Kokuba, S. Arai, Synthesis of monodispersed molecularly imprinted polymer particles for highperformance liquid chromatographic separation of cholesterol using templating polymerization in porous silica gel bound with cholesterol molecules on its surface, J. Chromatogr. A 1217 (2010) 7249–7254.

[16] W.-l. Yang, S.-m. Huang, Q.-z. Wu, J.-f. He, Properties evaluation and separation application of naringin-imprinted polymers prepared by a covalent imprinting method based on boronate ester, J. Polym. Res. 21 (4) (2014) 383–390.

[17] L. Li, L. Chen, H. Zhang, Y. Yang, X. Liu, Y. Chen, Temperature and magnetism bi-responsive molecularly imprinted polymers: Preparation, adsorption mechanism and properties as drug delivery system for sustained release of 5-fluorouracil, Mater Sci Eng C Mater Biol Appl 61 (2016) 158–168.

[18] S.K.-B.H. Hashemi-Moghaddam, M. Jamili, S. Zavareh, Evaluation of magnetic nanoparticles coated by 5-fluorouracil imprinted polymer for controlled drug delivery in mouse breast cancer model, Int. J. Pharm. 497 (2016) 228–238.

[19] S.Z.H. Hashemi-Moghaddam, S. Karimpour, H. Madanchi, Evaluation of molecularly imprinted polymer based on HER2 epitope for targeted drug delivery in ovarian cancer mouse model, React. Funct. Polym. 121 (2017) 82–90.

[20] C.B.V.M. Ekomo, R. Bikanga, A.M. Florea, G. Istamboulie, C. Calas-, T.N. Blanchard, et al., Detection of Bisphenol A in aqueous medium by screen printed carbon electrodes incorporating electrochemical molecularly imprinted polymers, Biosens. Bioelectron 112 (2018) 156–161.

[21] S. Huang, M. Guo, J. Tan, Y. Geng, J. Wu, Y. Tang, et al., Novel fluorescence sensor based on all-inorganic perovskite quantum dots coated with molecularly imprinted polymers for highly selective and sensitive detection of omethoate, ACS Appl. Mater. Interfaces 10 (45) (2018) 39056–39063.

[22] Y.Y.K. Yan, J. Zhang, self-powered sensor based on molecularly imprinted polymer-coupled graphitic carbon nitride photoanode for selective detection of bisphenol A, Sens. Actuators, B 259 (2018) 394–401.

[23] Y.Z.L. Fan, X. Li, C. Luo, F. Lu, H. Qiu, Removal of alizarin red from water environment using magnetic chitosan with Alizarin Red as imprinted molecules, Colloids Surf. B Biointerfaces 91 (2012) 250–257.

[24] B.S.K.H.Y. Lee, Grafting of molecularly imprinted polymers on iniferter-modified carbon nanotube, Biosens. Bioelectron. 25 (2009) 587–591.

[25] Y.H.Y. Liu, J. Liu, W. Wang, G. Liu, R. Zhao, Superparamagnetic surface molecularly imprinted nanoparticles for water-soluble pefloxacin mesylate prepared via surface initiated atom transfer radical polymerization and its application in egg sample analysis, J. Chromatogr. A 1246 (2012) 15–21.

[26] R.D.F. Breton, D. Jegourel, D. Deville-Bonne, L.A. Agrofoglio, Selective adenosine-5′-monophosphate uptake by water-compatible molecularly imprinted polymer, Anal. Chim. Acta. 616 (2008) 222–229.

[27] S.D.Q. Yu, G. Yu, Selective removal of perfluorooctane sulfonate from aqueous solution using chitosan-based molecularly imprinted polymer adsorbents, Water Res. 42 (2008) 3089–3097.

[28] R.J.S. Madhumanchi, R. Suedee, Efficient adsorptive extraction materials by surface protein-imprinted polymer over silica gel for selective recognition/separation of human serum albumin from urine, J. Appl. Polym. Sci. 136 (2019) 46894–46907.

[29] Z.H.G.R. Bardajee, Probing the interaction of a new synthesized CdTe quantum dots with human serum albumin and bovine serum albumin by spectroscopic methods, Mater. Sci. Eng. C Mater. Biol. Appl. 62 (2016) 806–815.

[30] X. Xie, Q. Hu, R. Ke, X. Zhen, Y. Bu, S. Wang, Facile preparation of photonic and magnetic dual responsive protein imprinted nanomaterial for specific recognition of bovine hemoglobin, Chem. Eng. J. 371 (2019) 130–137.

[31] R.-T.M. Xiao-Yu Sun, J. Chen, Y.-.P. Shi, Synthesis of magnetic molecularly imprinted nanoparticles with multiple recognition sites for the simultaneous and selective capture of two glycoproteins, J. Mater. Chem. B. 6 (2018) 688–696.

[32] P. He, H. Zhu, Y. Ma, N. Liu, X. Niu, M. Wei, et al., Rational design and fabrication of surface molecularly imprinted polymers based on multi-boronic acid sites for selective capture glycoproteins, Chem. Eng. J. 367 (2019) 55–63.

[33] Z. Yang, K. Yang, Y. Cui, T. Shah, M. Ahmad, Q. Zhang, et al., Synthesis of surface imprinted polymers based on wrinkled flower-like magnetic graphene microspheres with favorable recognition ability for BSA, J. Materials Sci. Technol. 74 (2021) 203–215.

[34] X. Li, B. Zhang, W. Li, X. Lei, X. Fan, L. Tian, et al., Preparation and characterization of bovine serum albumin surface-imprinted thermosensitive magnetic polymer microsphere and its application for protein recognition, Biosens. Bioelectron. 51 (2014) 261–267.

[35] X. Li, J. Zhou, L. Tian, Y. Wang, B. Zhang, H. Zhang, et al., Preparation of anti-nonspecific adsorption polydopamine-based surface protein-imprinted magnetic microspheres with the assistance of 2-methacryloyloxyethyl phosphorylcholine and its application for protein recognition, Sensors and Actuators B: Chem. 241 (2017) 413–421.

[36] Q. Li, K. Yang, S. Li, L. Liu, L. Zhang, Z. Liang, Y. Zhang, Preparation of surface imprinted core-shell particles via a metal chelating strategy: Specific recognition of porcine serum albumin, Microchimica. Acta. 183 (1) (2015) 345–352.

[37] Y. Wang, J. Zhou, H. Gu, X. Jia, K. Su, B. Zhang, et al., Preparation of anti-nonspecific adsorption chitosan-based bovine serum albumin imprinted polymers with Outstanding adsorption capacity and selective recognition ability based on magnetic microspheres, Macromolecular Materials and Eng. 304 (4) (2019) 800731–1800740.

[38] Y. Wang, J. Zhou, C. Wu, L. Tian, B. Zhang, Q. Zhang, Fabrication of micron-sized BSA-imprinted polymers with outstanding adsorption capacity based on poly(glycidyl methacrylate)/polystyrene (PGMA/PS) anisotropic microspheres, J. Mater. Chem. B 6 (37) (2018) 5860–5866.

[39] Z. Zhang, H. Wang, H. Wang, C. Wu, M. Li, L. Li, Fabrication and evaluation of molecularly imprinted magnetic nanoparticles for selective recognition and magnetic separation of lysozyme in human urine, Analyst 143 (23) (2018) 5849–5856.

[40] B.Q. Yujie Su, C. Chang, X. Li, M. Zhang, B. Zhou, Y. Yang, Separation of bovine hemoglobin using novel magnetic molecular imprinted nanoparticles, RSC Adv. 8 (2018) 6192–6199.

[41] F. Chen, M. Mao, J. Wang, J. Liu, F. Li, A dual-step immobilization/imprinting approach to prepare magnetic molecular imprinted polymers for selective removal of human serum albumin, Talanta 209 (2020) 120509.

[42] J. Zhou, Y. Wang, Y. Ma, B. Zhang, Q. Zhang, Surface molecularly zimprinted 20hermos-sensitive polymers based on light-weight hollow magnetic microspheres for specific recognition of BSA, Applied Surface Sci. 486 (2019) 265–273.

[43] J.W. Zuoting Yang, . Shah, L.a. Pei, M. Ahmad, Q. Zhang, B. Zhang, Development of surface imprinted heterogeneous nitrogen-doped magnetic carbon nanotubes as promising materials for protein separation and purification, Talanta 224 (2021) 121760.

[44] W.Z. Fangfang Chen, J. Zhang, J. Kong, Magnetic two-dimensional molecularly im-printed materials for recognition and separation of proteins, Phys. Chem. Chem. Phys. 18 (2015) 718–725.

[45] X. Zhang, S. Yang, L. Sun, A. Luo, Surface-imprinted polymer coating L-cysteine-capped ZnS quantum dots for target protein specific recognition, J. Mater. Sci. 51 (2016) 6075–6085.

[46] A.M.L. Pilotoa, D.S.M. Ribeiro, S.S.M. Rodrigues, Label-free quantum dot conjugates for human protein IL-2 based on molecularly imprinted polymers, Sensors and Actua-tors B: Chem. 304 (2020) 127343.

[47] X.-W.H. Wei Zhang, Y. Chen, W.-Y. Li, Y.-K. Zhang, Composite of CdTe quantum dots and molecularly imprinted polymer as a sensing material for cytochrome c, Biosens. Bioelectron. 26 (2011) 2553–2558.

[48] P. Lv, D. Xie, Z. Zhang, Magnetic carbon dots based molecularly imprinted polymers for fluorescent detection of bovine hemoglobin, Talanta 188 (2018) 145–151.

[49] Y. Zhao, Y. Chen, M. Fang, Y. Tian, G. Bai, K. Zhuo, Silanized carbon dot-based thermo-sensitive molecularly imprinted fluorescent sensor for bovine hemoglobin detection, Anal Bioanal Chem 412 (23) (2020) 5811–5817.

[50] T. Guo, Q. Deng, G. Fang, D. Gu, Y. Yang, S. Wang, Upconversion fluorescence metal-organic frameworks thermo-sensitive imprinted polymer for enrichment and sensing protein, Biosens. Bioelectron. 79 (2016) 341–346.

[51] H.W. Lifang Chang, X. He, L. Chen, Y. Zhang, A highly sensitive fluorescent turn-on biosensor for glycoproteins based on boronic acid functional polymer capped Mn-doped ZnS quantum dots, Analytica. Chimica. Acta. 995 (2017) 91–98.

[52] Q.D. Ting Guo, G. Fang, C. Liu, X. Huang, S. Wang, Molecularly imprinted upconversion nanoparticles for highly selective and sensitive sensing of Cytochrome c, Biosens. Bioelectron. 74 (2015) 498–503.

[53] Z.-H.Z. Hong-Jun Chen, Li-J Luo, S.-.Z. Yao, Surface-imprinted chitosan-coated magnetic nanoparticles modified multi-walled carbon nanotubes biosensor for detection of bovine serum albumin, Sensors and Actuators B: Chem. 163 (2012) 76–83.

[54] Q.Z. Yubo Wei, J. Huang, Q. Hu, X. Guo, L. Wang, An electro-responsive imprinted biosensor with switchable affinity toward proteins, Chem. Commun 54 (2018) 9163–9166.

[55] B.L. Wei Zhao, S. Xu, X. Huang, J. Luo, Ye Zhu, X. Liu, Electrochemical protein recognition based on macromolecular self-assembly of molecularly imprinted polymer: A new strategy to mimic antibody for label-free biosensing, J. Mater. Chem. B 7 (2019) 2311–2319.

[56] A.H. Kouhei Orihara, T. Arita, H. Muguruma, Y. Yoshimi, Heparin molecularly imprinted polymer thin film on gold electrode byplasma-induced graft polymerization for label-free biosensor, J. Pharmaceutical and Biomed Analysis 151 (2018) 324–330.

[57] M.M.P. Palladino, S. Scarano, Cardiac Troponin T capture and detection in real-time via epitope-imprinted polymer and optical biosensing, Biosens. Bioelectron. 106 (2018) 93–98.

[58] A.S. Bahareh Babamiri, R. Hallaj, A molecularly imprinted electrochemiluminescence sensor for ultrasensitive HIV-1 gene detection using EuS nanocrystals as luminophore, Biosens. Bioelectron. 117 (2018) 332–339.

[59] J.L. Qian Yang, X. Wang, H. Xiong, L. Chen, Ternary emission of a blue–, green–, and red-based molecular imprinting fluorescence sensor for the multiplexed and visual detection of bovine hemoglobin, Anal. Chem 91 (2019) 6561−6568.

[60] M.G.F.S. Ana, P.M. Tavares, Novel electro-polymerized protein-imprinted materials using Eriochrome black T: Application to BSA sensing, Electrochim. Acta 262 (2018) 214e225.

[61] C.Z. Yuxuan Lai, Y. Deng, G. Yang, S. Li, C. Tang, Nongyue He, A novel α-fetoprotein-MIP immunosensor based on AuNPs/PTh modified glass carbon electrode, Chin. Chem. Lett. 30 (2019) 160–162.

[62] K.Z. Meng Qi, Q. Bao, P. Pan, Y. Zhao, Z. Yang, H. Wang, et al., Adsorption and electrochemical detection of bovine serum albumin imprinted calcium alginate hydrogel membrane, Polymers 11 (2019) 622–633.

[63] B.Z. Xiangjie Li, L. Tian, W. Li, T. Xin, H. Zhang, Q. Zhang, Effect of carboxyl density at the core–shell interface ofsurface-imprinted magnetic trilayer microspheres onrecognition properties of proteins, Sens. Actuators B 196 (2014) 265–271.

[64] Y. Wang, J. Zhou, B. Zhang, L. Tian, Z. Ali, Q. Zhang, Fabrication and characterization of glutathione-imprinted polymers on fibrous SiO 2 microspheres with high specific surface, Chem. Eng. J. 327 (2017) 932–940.

[65] X.J. Xinlong Fan, H. Zhang, B. Zhang, C. Li, Q. Zhang, Synthesis of raspberry-like poly(styrene−glycidyl methacrylate) particles via a one-step soap-free emulsion polymerization process accompanied by phase separation, Langmuir 29 (2013) 11730–11741.

[66] J. Zhou, Y. Wang, J. Bu, B. Zhang, Q. Zhang, Ni(2+)-BSA directional coordination-assisted magnetic molecularly imprinted microspheres with enhanced specific rebinding to target proteins, ACS Appl. Mater. Interfaces 11 (29) (2019) 25682–25690.

[67] M.W. Jingjing Zhou, B. Zhang, Q. Zhang, Metal coordination assisted thermo-sensitive magnetic imprinted microspheres for selective adsorption and efficient elution of proteins, Colloids and Surfaces A: Physicochem Eng Aspects 612 (2021) 125981.

[68] X. Li, J. Zhou, L. Tian, W. Li, Z. Ali, N. Ali, et al., Effect of crosslinking degree and thickness of thermosensitive imprinted layers on recognition and elution efficiency of protein imprinted magnetic microspheres, Sensors and Actuators B: Chem. 225 (2016) 436–445.

[69] X. Dong, Y. Ma, C. Hou, B. Zhang, H. Zhang, Q. Zhang, Preparation of pH and temperature dual-sensitive molecularly imprinted polymers based on chitosan and N-isopropylacrylamide for recognition of bovine serum albumin, Polymer Int. 68 (5) (2019) 955–963.

[70] Y.S. Xue Yang, Y. Xiang, F. Qiu, G. Fu, Controlled synthesis of PEGylated surface proteinimprinted nanoparticles, Analyst 144 (2019) 5439–5448.

[71] X. Li, B. Zhang, L. Tian, W. Li, H. Zhang, Q. Zhang, Improvement of recognition specificity of surface protein-imprinted magnetic microspheres by reducing nonspecific adsorption of competitors using 2-methacryloyloxyethyl phosphorylcholine, Sensors and Actuators B: Chem. 208 (2015) 559–568.

[72] Y.Z. Zuoting Yang, J. Ren, Q. Zhang, B. Zhang, Cobalt-iron double ion-bovine serum albumin chelation-assisted thermo-sensitive surface-imprinted nanocage with high specificity, ACS Appl. Mater. Interfaces 13 (2021) 34829–34842.

[73] Z.S. Jingjing Zhou, M. Wang, Y. Wang, J. Wang, B. Zhang, Q. Zhang, Thiolactone-based conjugation assisted magnetic imprinted microspheres for specific capturing target proteins, Chem. Eng. J. 300 (2020) 125767.

[74] Z. Yang, J. Xu, J. Wang, Q. Zhang, B. Zhang, Design and preparation of self-driven BSA surface imprinted tubular carbon nanofibers and their specific adsorption performance, Chem. Eng. J. 373 (2019) 923–934.

[75] Z. Yang, J. Chen, J. Wang, Q. Zhang, B. Zhang, Self-Driven BSA surface imprinted magnetic tubular carbon nanofibers: Fabrication and adsorption performance, ACS Sustainable Chem Eng 8 (8) (2020) 3241–3252.

[76] Z. Yang, J. Chen, K. Yang, Q. Zhang, B. Zhang, Preparation of BSA surface imprinted manganese dioxide-loaded tubular carbon fibers with excellent specific rebinding to target protein, J. Colloid Interface Sci. 570 (2020) 182–196.

CHAPTER 4

Functional carbon nano-material as efficient metal ion adsorbent from wastewater

Mudasir Ahmad[a,b] and Baoliang Zhang[a,b]
[a]School of Chemistry and Chemical Engineering, Northwestern Polytechnical University, Xi'an, China
[b]Xi'an Key Laboratory of Functional Organic Porous Materials, Northwestern Polytechnical University, Xi'an, China

4.1 Introduction

The increasing industrialization and the growing energy demand have been a worldwide concern [1]. Increasing energy needs large mining and drilling operations all around the world. These operations cause environmental pollution due to release of toxic pollutants in water. For several decades radioactive elements have been used as an energy source in nuclear power plants. Among all of them, uranium has been considered a sustainable energy resource. Due to the environment and the resources limits. First, uranium is highly toxic and nondegradable even at low concentrations. Secondly, it is estimated with current consumption rate, the uranium availability is not enough for more than 70 years. Therefore, it is necessary to make availability of uranium from a conservative source such as seawater/wastewater.

For several decades various adsorbents including biopolymers [112–114], clay materials [2,3], metal hydroxides [4,5], metal-organic frameworks, and carbon nanomaterials [6,7] are used for the removal of metal ions from water. However, substantial improvements have been achieved in recent years, but the removal of some metal ions at a low concentration from the seawater is still challenging because of limited adsorption performance. Thus, it is necessary to develop new adsorbent materials for the adsorption of various metal ions from water under harsh conditions with higher rates, better selectivity, and maximum adsorption capacities.

Recently, the invention of nanomaterials has attracted the interest of researchers due to their large surface area, high mechanical strength, and

Fabrication and Functionalization of Advanced Tubular Nanofibers and their Applications.
DOI: https://doi.org/10.1016/B978-0-323-99039-4.00001-2

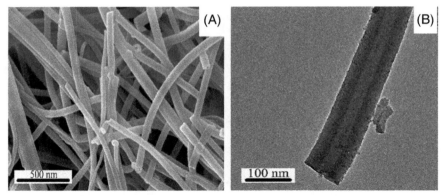

Figure 4.1 SEM micrograph (A) and TEM micrograph (B) prepared from α, α'-dichloro-p-xylene. Reprinted (adapted) with permission from. Copyright {2022} American Chemical Society [9].

ease of modifying to a specific functional group. Carbon nanomaterials are hydrophobic and combine due to high wander walls interaction in the water. The functionalization not only enhances the dispersibility of nanomaterials in an aqueous medium [8] but also increases the metal ion adsorption efficiency in aqueous media. In this chapter, we have described the functionalization and the methods of preparation of different materials for efficient adsorption of metal ions from wastewater.

4.2 Functional group modifications and synthetic methods

4.2.1 Carbon nanotubes

Carbon nanotubes (CNT) are hollow tubes composed of carbon. The cylindrical carbon features, a high aspect ratio length to a diameter up to 10^3 with a diameter from 1 nm to tens of nanometers and length up to nanometers. The SEM and TEM images in Fig. 4.1, show the morphology and structure of carbon CNT obtained from α, α'-dichloro-p-xylene by self-polymerization method [9]. This unique one-dimensional structure and associated properties endow the special nature of CNT. From the last decade, CNT has attracted considerable interest due to its small structure, high surface area, high ability of functionalization, and chemical stability at different pH. Due to these properties, CNT are used in the medical field, nanoelectronics, catalysis, and adsorption [10]. However, CNT are hydrophobic and easily aggregate in an aqueous solution due to the Wander wall's force of attraction, which affects the adsorption capacity of

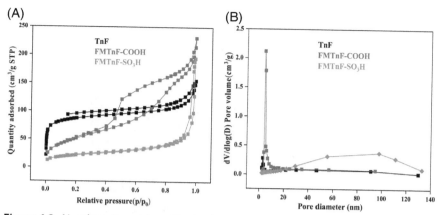

Figure 4.2 N_2 adsorption-desorption isotherm (A) and pore size distribution curves (B) of the samples. Reprinted (adapted) with permission from. Copyright {2022} American Chemical Society [9].

the adsorbent [11]. Therefore, chemical modification is a significant way to enhance various properties such as functionality, dispersibility, stability [12]. Chemical modification of CNT with various groups such as aminopropyl groups [13], amidoxime [14], polyoxime [15], oleyl-phosphate [16], tributyl phosphate [17], and oleic acid bilayer [18]. Phosphate [19] and organo-phosphorous modified CNT have been used for uranium adsorption from an aqueous solution with a maximum adsorption capacity of 666.7 (mg g^{-1}) at pH = 7 within 24 h [20]. CNT have been functionalized by strong acids (HNO_3/H_2SO_4), to enhance uranium adsorption on CNT. The colloidal stability depends on the cationic stability of CNT. The tubular structure of CNT is confirmed from SEM image and TEM image analysis (Fig. 4.1). The surface area of CNT is a significant factor for the adsorption of metal ions. The functionalization of CNT with HNO_3 increases surface area than H_2SO_4 (Fig. 4.2) [21]. Multiwalled carbon nanotubes (MCNT) with large surface area and dense pore structure attract interest to modify the surface of MCNT for the adsorption of organic and inorganic pollutants from the aqueous media [22,23]. However, some of the limitations have restricted its use as adsorbents such as hydrophobic surfaces, agglomeration, and less selectivity for toxic pollutants from the aqueous solution. However, the inclusion of functionality was important to increase hydrophilicity and removal efficiency. Wu et al. has modified the surface of MCNT with amidoxime for selective adsorption of uranium from the wastewater. Firstly, MCNT were irradiated with 200 kGY and subsequent treated with

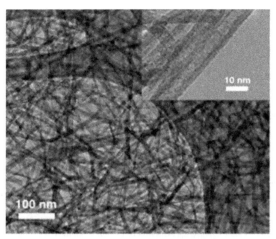

Figure 4.3 TEM image of TNT obtained from the hydrothermal method. The copyright of Figure was obtained from Elsevier under license No: 5234210908509, Jan 22, 2022, [31].

acrylonitrile to obtain amidoxime modified materials. The Adsorption kinetics followed pseudo second order kinetics and the adsorption data followed Langmuir adsorption isotherm with maximum adsorption capacity 67.9 (mg g^{-1}) at 298 K and the equilibrium of adsorption was achieved within 60 min at a pH of 3.0–5.0 [24]. Grafting polymerization method was practiced to carry the amidoxime functionalization of an adsorbent. Two techniques were reported for grafting polymerization of amidoxime as follows; (1) chemical grafting [25,26] and (2) plasma technique [27]. Chemical grafting requires an initiator, catalyst, and high temperature, which is not easy and might introduce impurity on the surface [28]. The plasma technique introduces the electron beam to activate surface for the formation of free radicals for grafting with high purity.

4.2.2 Titanate nanotubes

In recent, titanate nanotubes (TNT) have exhibited various properties and attracted the attention of scientists. The common and cost-effective hydrothermal method was used to synthesize TNT structures [29–30] with large surface area and pore volume [31]. The TEM images (Fig. 4.3) and SEM images (Fig. 4.4) of TNT showed a tubular structure. Also, there are numerous hydroxyl and Na^{+} groups on the surface of the TNT, which take part in ion exchange with the metal ions in aqueous media

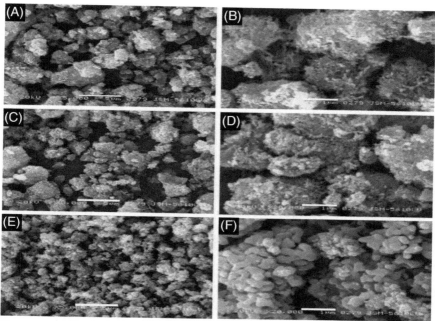

Figure 4.4 SEM images of the titanate nanotubes calcined at 400 °C (A and B), 600°C (C and D), and 700 °C (E and F). The copyright of Figure was obtained from Elsevier under license No: 5234210908509, Jan 22, 2022, [31].

[32]. For example, Ni et al. and Hu et al. showed Cd(II) and uranium adsorption on TNT from wastewater [33,34]. The limitation of TNT as an adsorbent, as it requires extra filtration for regeneration of adsorbent which increases the adsorption cost that restricts its use for the adsorption of metal ions. Later, TNT are modified with magnetic cobalt ferrite ($COFe_2O_4$) reported by Zhu et al. $COFe_2O_4$ is widely used as catalyst [35], magnetic memory [36] and absorbing materials [37]. Hence $COFe_2O_4$ has been used to separate adsorbent from the solution using an external magnetic field. To increase the adsorption capacity, $COFe_2O_4$/TNT is grafted with ethylenediamine [38], tetraethylenpentamine [39], polyethyleneimine [40], and triethylenetetramine. Among all of these, only amine groups show coordination between metal ions and NH_2 group [41]. Zhu et al. reports tetraethylenepentamine modified $COFe_2O_4$/TNT for adsorption of metal ions from wastewater [42]. Owing to the fact, the researchers are in search of materials that are easily available, environmentally friendly, low–cost material selects the materials used for the adsorption of metal ions from water.

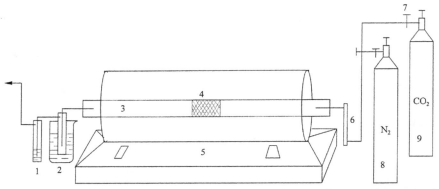

Figure 4.5 Experimental set-up of physicochemical activation for the production of activated carbon (1), trap (2), cooling bath (3), quartz tube (4), carbon sample (5), furnace (6), flow meter (7), valve (8), N-cylinder (9), and CO-cylinder (44). The copyright of Figure was obtained from Elsevier under license No 5234200129325, Jan 22, 2022 [44].

4.2.3 Waste paper carbon

Waste paper-derived carbon (WPC) is low-cost and easily available material, modified with dopamine for the adsorption of uranium from the wastewater [43]. At initial, WPC is activated with KOH, carbonized and subsequent polymerization with dopamine as adsorbent for metal adsorption due to high surface area, environmentally friendly and easily available at a low cost [44] (Fig. 4.5). The increased surface area (364.6 m^2 g^{-1}) of WPC enhances uranium extraction not only from wastewater but also enhances in seawater. To improve the porosity, WPC is treated with KOH to activate the carbon surface. Dopamine is secreted by muscle cells and itself undergoes an oxidation reaction to form dopaquinone. It might undergo self-polymerization by the dismutation reaction, which might produce efficient attachment on the surface of the substrate, as well as to polydopamine [45]. It has a strong affinity, adhesion, well functionality, and biocompatibility. The presence of catechol and imine/amine groups can improve the adsorption of metal ions. Researchers modify several adsorbents by dopamine, such as graphene oxide [46] and SBA-15 [47]. However, these adsorbents do not show sufficient adsorption at the low concentration of uranium. WPC modified dopamine show efficient adsorption at a low concentration range (3.3 μg L^{-1}) present in seawater [43] with a distribution coefficient of 1.97×4^{10}, which is much better than other adsorbents that demonstrates the highest selectivity for the uranium with the highest adsorption capacity 384.6 (mg g^{-1}) at pH 7.

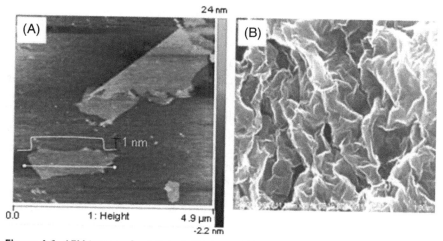

Figure 4.6 AFM image of exfoliated GO sheets (A) and SEM image of aggregated RGO (B). The copyright of Figure was obtained from Elsevier under license No 5245131218544, Feb 09, 2022 [49].

Lamb et al., reports a series of hydrophobic and hydrophilic materials such as carbon black and oxides of titanium and silica are analyzed for adsorption of uranium from wastewater. Hydrous titanium oxide is considered a better selective adsorbent and polar groups are responsible for uranium adsorption. The surface of colloidal silica is negatively charged thorough out the whole pH range is studied and revealed a strong attraction with the positive uranium in the solution. The use of carbon black as a cost–effective adsorbent for uranium [48].

4.2.4 Graphene oxide-nano sheets

Graphene is a single atom carbon structure with sp^2 carbon. Graphene oxide nano–sheet is prepared by Hummer's method used for adsorption of the uranium from the wastewater. Since 2004, the first report of the electric property of graphene has attracted the attention of researchers due to its two-dimensional structure. Graphene oxide nanosheets show excellent physical and chemical properties such as high mechanical properties, lightweight, and large surface area are reported (Fig. 4.6). Chemical treatment of graphene improves the properties in various fields for advanced applications. In recent the use of graphene for the environmental cleanup of heavy metal detoxification from industrial wastewater such as Pb(II), Cr(VI), Co(II), Hg(II), and AS(II) and several organic pollutants and phenolic products.

Graphene oxide (GO) is a reduced form of graphene mostly containing the groups of epoxides and hydroxyl groups. The presence of carboxyl and carbonyl groups on the edge results in hydrophilic nature and is easily dispersed in the aqueous solution. The GO preparation by the Hummers method adds more oxygen groups than CNT, which needs special treatment for the introduction of hydrophilic groups. Few layered GO materials are selected as better adsorbents than other nanomaterials for the adsorption of Pb(II) Cd(II) and Co(II) from the aqueous media. Li et al. have developed a single-layer GO nano-sheets for the adsorption of uranium from industrial wastewater. This kind of adsorption is studied by batch methods such as pH, ionic strength, and concentration of uranium. The highest adsorption capacity is achieved (299 mg g^{-1}) at pH 4.0. The highest adsorption of GO is due to the presence of oxygen groups [49].

4.2.5 Activated sludge modified graphene-oxide

The most common adsorbent used for adsorption is activated carbon. However, the main key issue is the separation of powder from the huge water, which restricts the practical applicability of activated carbon [50,51]. Lately activated sludge is used for the adsorption of dyes, organic, inorganic pollutants from the wastewater [52,53]. Activated sludge contains a large number of microorganisms such as bacteria, fungi, algae, yeasts. These microorganisms have a cell wall that is made of lipids, proteins, chitin, etc., carrying numerous functional groups that can attract pollutants [54]. Moreover, the other component of activated sludge is extracellular polymer substance (EPS), which is mainly produced by bacteria secretion. EPS mainly possess carbohydrates, proteins, nucleic acids which carry numerous functional groups, which tend to chelate a large amount of pollutants [55]. However, the chemical structure of activated sludge is typical and the reuse of the spent adsorbent is too difficult. Hence large effort is devoted to modifying the activated sludge for the adsorption of the pollutants and its regeneration after successful treatment. Chemical modification of activated sludge sometimes needs extra treatment to remove the byproducts hence it is imperative to adopt a green process for the modification of activating sludge. Immobilization technology is the most advanced to overcome these drawbacks and improve the adsorption capacity of the adsorbent. The immobilization of activated sludge offers numerous advantages such as easy separation, numerous functionality, easy recyclability, and reuse of the adsorbent [56,57]. Activated sludge-graphene oxide (AS-GO) is synthesized for adsorption of uranium (180.6 mg g^{-1}) and the adsorption data is best fitted with the Langmuir model described the monolayer adsorption onto AS–GO. The important aspect of this work is that a little amount of adsorbent dosage is used 0.02 mg L^{-1}, which was the lowest reported for graphene-based adsorbents. The adsorption mechanism showed that electrostatic interaction and inner-sphere complexation occurred between uranium and hetero groups like oxygen and nitrogen [58].

In past years layered double hydroxy (LDH) was used for the adsorption of heavy metals. It possesses transition metals and their general formula [M$_1^{2+}$-x M^{3+}x(OH)$_2$] (A^{n-}]$_{x/m}$.mH$_2$O, where M^{2+} and M^{3+} are the di-valent and trivalent metal cations and A^{n-} represents the interlayer anion. LDH has high interlayer space and a large number of active sites are allowed as a candidate for ion exchange [59] in adsorption [60] and magnetic properties [61]. However, LDH shows a low surface area due to pill up together, hence adsorption capacity is very low. Taking high interest due to high ion exchange capacity, three-dimensional hydroxy/graphene hybrid material is used for the adsorption of uranium from wastewater. A simple and low-cost in situ growth procedure is used for the fabrication of three-dimensional LDH/graphene composite with numerous advantages such as high surface area and numerous active sites. The highest adsorption capacity (277.8 mg g^{-1}) is because of increasing active sites [62]. GO activated carbon felt (GO-ACF) is reported by Chen et al. for the adsorption of uranium from wastewater. This kind of material is prepared by electrodeposition and subsequent thermal annealing. The adsorption capacity (170 mg g^{-1}) of the activated felt, while as in the case of GO-ACF the adsorption capacity (298 mg g^{-1}) at pH 5.5 revealed the carboxylic groups present on the surface of the GO-ACF have played an important role in the adsorption of uranium from the wastewater [63].

4.2.6 Metal-oxides

Tin oxide nanoparticles as an adsorbent for the adsorption of uranium and thorium from the wastewater is reported by Nilchi et al. [64]. These nanoparticles are developed by several methods such as micro emulsion [65], sol-gel method [66], gel combustion technique [67], spray pyrolysis [68], hydrothermal analysis [69], strike precipitation [70], and two-step solid-state synthesis [71]. Among all of these methods, tin oxide nanoparticles are synthesized from hydroxide precipitation of NH$_4$OH with SnCl$_4$ to obtain a large amount of powder at a low cost. The high surface powder showed an irregular shape and huge agglomeration. The failure of the physical characteristics of nanoparticles due to non–uniform saturation. If the saturation of the solution is physical then the physical characteristics are controlled. Homogenous precipitation is the best way to control the precipitation in the reaction. Urea decomposition controls kinetics of the release of the super saturation species [72,73]. The significant characteristics of the homogenous precipitation is to bring the uniformity in pH of the solution which controls the urea decomposition. Thus when a metal salt is heated at 85°C in acid solution in presence of urea. The urea decompose ammonia and carbonate ions slowly. The uniform rise in the pH will increase the nucleation growth with suitable size of metal oxy-basic carbonates. This kind of method is useful for the formation of metal mono dispersed oxide ceramics materials [74].

4.3 Preparation and functionalization of adsorbents from biowastes

Of course, activated carbon has a large surface area which enhances the adsorption [75]. But the chemical activation makes its huge scale application expensive [76]. To reduce the cost, agricultural waste is used for the preparation of activated carbon [77]. Bio wastes are the main sources like fruit peels, edible waste, and shells of nuts. Reportedly bio waste-based adsorbents played a major role in adsorption [78]. For example, corn cob is an agro-waste from maize crop waste, almost 16% of the maize plant contains corn cobs [79]. There are several reports from the corn cob for the preparation of activated carbon for the adsorption of dyes by the chemical activation process. However, the activated carbon obtained from the corn cob is not sufficient for the adsorption of heavy metal ions [80]. To fill the gap, another adsorbent with high surface areas is nanoporous silica, which enhances the adsorption [81]. Its thermal and hydrothermal advantages make it a good candidate [82]. The high adsorption capacity of mesoporous silica is reported in early [83]. The mesoporous silica follows the Stobes process, which has been easily modified to yield nanotubes, and nanorods [84]. In nature, silica is prepared by an easy process under normal conditions. Corn cobs mainly consist of 40%–45% silica by mass [85]. Dutta et al. have synthesized silica carbon nanocomposites from corn cob-based adsorbents. The obtained nanocomposites are used for the adsorption of dyes methyl blue) and uranium and Cr from wastewater with adsorption capacity of 255.12 and 90.1 mg g^{-1} respectively [86].

4.3.1 Extracellular polymer matrix

Extracellular polymer matrix (EPM) based adsorbent for uranium adsorption from wastewater obtained from the activated sludge (aerobic bacteria). EPM particularly contains tryptophan and humus, which carries functional groups such as $-COOH, -CONH_2, -H_2PO_4, -OH, -NH_2$, etc. In recent, bio–adsorbents have gained attention for the adsorption of uranium. Several species such as *Eichhornia crassipes, Elodea*, and *green algae* are studied by different batch methods. However, all these materials showed less adsorption efficiency of uranium at low concentrations [87,88]. The zeta potential of the EPM is negative in the pH range of 3–8. The electronegativity increases with increasing the pH. The reason for the poor adsorption for the EPM at low pH means uranyl exists mainly in form of UO_2^{2+}. However, the EPM has a negatively charged surface, a huge number of H_3O^+. The competition between H_3O^+ and UO_2^{2+}, the quick occupation of H_3O^+ to combine with –OH and –COOH groups resulted in decreases in the adsorption of uranium by EPM [89]. Thereafter, the increase of the pH increases the electronegativity of EPM increases uranium adsorption due to a decrease in the amount of H_3O^+. With the hydrolysis of UO_2^{2+} with the increasing pH, there is a morphological alteration that existed in the form of UO_2^{2+} (OH)$_7$, and UO_2^{2+} (OH)$_3$. These forms of uranium do not combine with the functional groups strongly affect the adsorption of uranium onto the EPM [90]. Currently, it has been a trend to use the cells to adsorb the metal ions from the wastewater. He et al. have used ClO_2 modified cell structure for uranium adsorption. ClO_2 has efficient functionality for uranium adsorption. ClO_2 attracts attention as a ligand for adsorption because it is safe and used as an environmental disinfectant without any toxicity. This kind of functionalized material may be applicable to purify the wastewater at the high concentration or low concentration of uranium and could be regenerated after several cycles without the loss of adsorption capacity of less than 90%. For analyzing the adsorption capacity all treated cells and without ClO_2 treated cells show (436 mg g^{-1} and 423 mg g^{-1}) respectively confirmed higher adsorption by ClO_2 treated cells [91].

4.3.2 Biochar

Biochar is a porous material obtained from the pyrolysis of biomass under oxygen-limited conditions. It can act as a strong adsorbent because of its porous structure and functional surface. Of all the thermal technologies, pyrolysis is best for the thermal conversion of biomaterial above 300°C in liquid and solids products. During the pyrolysis process the biomaterial such as cellulose, chitin, gelatin, lignin, fat, starch, etc. In general, the thermally degraded material existed in three forms (1) condensed vapors (bio-oils), (2) solid (biochar), and (3) noncondensed (gases) [92]. In practice more biochar is obtained from the biomaterial which carries the highest ratio of lignin than cellulose and confirms the porosity is increased as the amount of lignin was increased [93]. The use of biochar for the adsorption of uranium from wastewater and also managed the sustainable agro-industrial wastes and wastewater containing uranium [94]. The selection of the temperature for the pyrolysis of biomass depends on the relation between the fixed carbon and the gravimetric mass to maximize the adsorption capacity for uranium from wastewater. With increasing the temperature, the gravimetric yield decreases with progressing the thermal decomposition of the biomass, simultaneously carbon content increases, while as volatile matters decrease in amount. The adsorbent materials depend on the oxygen and hydrogen content for the adsorption process, with the increase the temperature decreases the oxygen and hydrogen content leading to a decrease in the adsorption of uranium from wastewater. The hydrophobicity of the biochar surface can be concluded by the O/C ratio present in the biochar as the O content is a polar moiety. The O/C ratio decreased on the surface of the biochar indicates the surface contains the highest amount of organic content and a smaller amount of hydrophilicity [95]. The FTIR analysis revealed the biochar obtained at low temperature has enough adsorption polar groups for uranium adsorption. Hence the hetero groups such as nitrogen, oxygen, sulfur present on the surface of biochar increase the adsorption by chemisorption process [94].

4.3.3 Phosphate bacteria biochar

The phosphate-containing bacteria is used for the immobilization of uranium from wastewater. To analyze the phosphate-containing bacteria to trap the uranium from uranium mine waste. For bacteria composite preparation, these bacteria are isolated and mingled with the biochar to synthesize bacteria-biochar composite for the adsorption of the uranium. Also, the re-oxidation undergoes subsequently to change uranium(IV) to uranium(VI) by using the HNO_3 to maximize the adsorption capacity [96]. The uranium precipitates were stable in presence of Ca and Mg, as it forms stable complexes with uranium. However, the simultaneous re-oxidation by NO_3 and Na^+ oxidized the uranium and resulted in its fast dissolution and quick release of uranium (Fig. 4.7) [96].

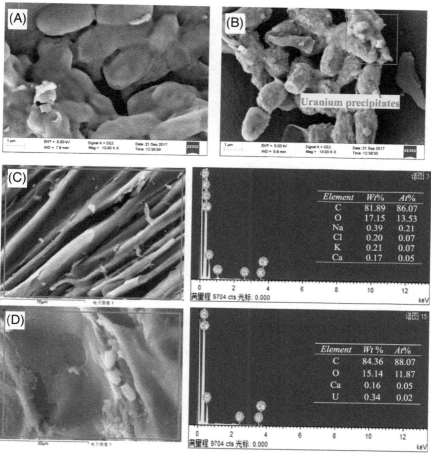

Figure 4.7 The SEM-EDS of PAOs-BC (A,C) and (B,D) before and after uranium respectively. The copyright of Figure was obtained from Elsevier under license No 5247521089621, Feb 14, 2022 [96].

4.3.4 Magnetic biochar composites

The magnetic biochar was synthesized for the regeneration of adsorbent and efficient adsorption of uranium [97,98]. Biochar is prepared from the rice husk and siderite by pyrolysis under the nitrogen atmosphere at 550°C at 30 min. This type of biochar show adsorption of uranium from wastewater due to large surface area and porous structures [99]. The magnetic biochar is synthesized by using Fe^{2+} or Fe^{3+} or by a simplified aqueous method that contains iron chloride solution under alkaline conditions on the surface of the biochar. China is rich in siderite resources and only 10 % is utilized. It

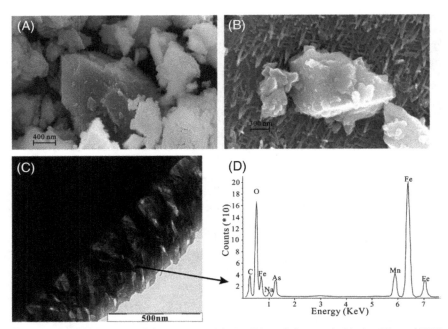

Figure 4.8 SEM images of the pristine siderite (A) and the used siderite (B), and TEM image of the used siderite (C) and EDS spectra for the coating (D). (In the TEM image dark spot is siderite and the light spot is Fe-oxide coating). The copyright of Figure was obtained from Elsevier under license No 5247540180341, Feb 14, 2022 [100].

is transformed into magnetic siderite by magnetic phenomena [100]. The magnetic siderite show adsorption capacity (52.65 mg g^{-1}) obtained from the Langmuir isotherm at low pH 4 of the solution at temperature 318 K. This kind of adsorbent showed the highest adsorption for uranium in comparison to other biochar adsorbents such as biochar (2.2mg g^{-1}) at pH 3.9, sulfonated graphene oxides (45.05 mg g^{-1}), Fe_3O_4/HA (10.5 mg g^{-1}), silicate diatomite(31.54 mg g^{-1}) (Fig. 4.8). The comparison of results confirmed magnetic biochar acted as a potential adsorbent for the adsorption of the uranium from the wastewater [99].

4.4 Preparation and functionalization of inorganic/organic adsorbents

4.4.1 Phosphoryl functionalized silica

Several studies confirm uranium adsorption is generally carried by inorganic or organic adsorbents. The inorganic origin adsorbents exhibited the

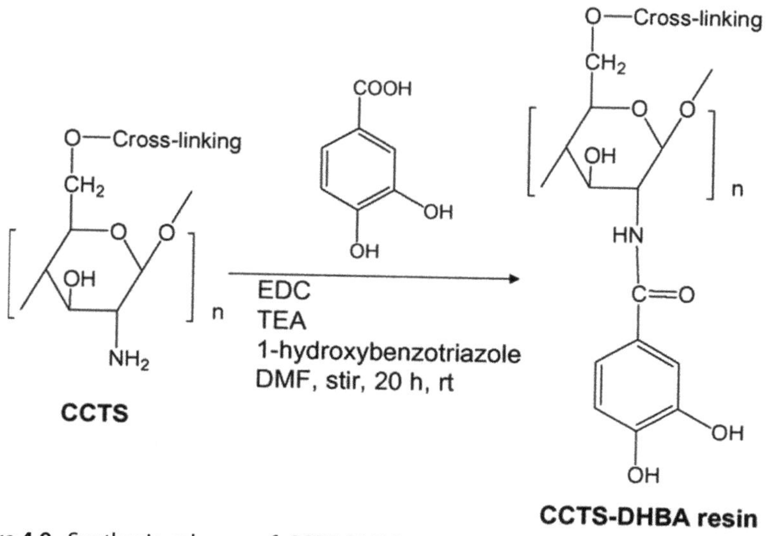

Figure 4.9 Synthesis scheme of CCTS-DHBA resin. The copyright of this Figure was obtained from Elsevier under license No 5247541470995, Feb 14, 2022 [102].

highest kinetics, fast adsorption, easy regeneration, and elution. However, the complex adsorption and poor direction properties have limited its use for the adsorption purpose [101]. While in case of organic adsorbents showed rich direction adsorption capacity, which attributes to the presence of hetero elements such as N, P, O, S and other elements present in the functional groups of an adsorbent (Fig. 4.9) [47,102]. But, organic adsorbent shows adsorption capacity very low. The hot topic in research is currently the preparation of inorganic and organic composites for the adsorption of uranium. The strong chelating ability of phosphorous to uranium attracts the use of phosphorous as functional groups [103]. Of all these tributyl phosphate is chosen as the best chelating agent in several studies [103,104]. Here in this study highly dense phosphate reagent is intruded in silica that shows a large surface area and pore size, regular structure, and highest thermal stability [105]. Besides, the silica material has the highest number of hydroxyl groups which can combine with the other organic groups. Silica is modified by q amino triethoxy silane and tributyl phosphate functionalities for the adsorption of uranium.

4.5 Our contribution

The emerging class of carbon tubular nanofibers CNF are promising materials used to remove the organic [106,107] and inorganic [108,109] pollutants

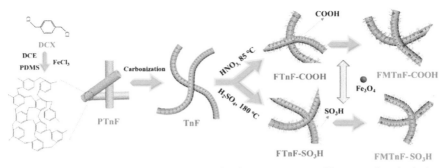

Figure 4.10 Systematic reaction pathway for the preparation of functionalized tubular nanotubes. Reprinted (adapted) with permission from Copyright 2022 American Chemical Society [9].

from the water. For example, Sun et al. report that radioactive elements such as uranium and europium(VI) are adsorbed onto CNF with maximum adsorption capacities (125.0 and 91.0 mg g^{-1}) at pH 4.5 respectively [110]. CNF shows adsorption results for other metal ions such as Pb(II) and Cr(III) are (423.7 mg g^{-1} and 221.3 (mg g^{-1}) [111]. Low cost and high adsorption efficient adsorbents are highly interesting for the removal of uranium from aqueous solutions. In our contribution, we have prepared bifunctional magnetic tubular nano-fibers (FTnF-SO$_3$H and HTnF-COOH) from α, α'-dichloro-p-xylene(DCX) via self-polymerization by two oil solvent method. The polymerized material is followed by carbonization and subsequent functionalization of sulphonic and carboxylic groups and the loading of Fe$_3$O$_4$ for the preparation of magnetic functionalized adsorbent for uranium adsorption. The complete preparation process (Fig. 4.10) [9]. The functionalized adsorbent showed a wide diameter of approximately 160 nm and magnetization saturation of 21.61 and 19.44 emu g^{-1} respectively.

References

[1] Y. Sun, X. Wang, S. Lu, X. Wang, C. Xu, J. Li, et al., Plasma-facilitated synthesis of amidoxime/carbon nanofiber hybrids for effective enrichment of [238]U(VI) and [241]Am(III), Environ. Sci. Technol. (2017), doi:10.1021/acs.est.7b02745.

[2] Y. Sun, J. Li, X. Wang, The retention of uranium and europium onto sepiolite investigated by macroscopic, spectroscopic and modeling techniques, Geochim. Cosmochim. Acta. (2014), doi:10.1016/j.gca.2014.06.001.

[3] A. Van Veelen, J.R. Bargar, G.T.W. Law, G.E. Brown, R.A. Wogelius, Uranium immobilization and nanofilm formation on magnesium-rich minerals, Environ. Sci. Technol. (2016), doi:10.1021/acs.est.5b06041.

[4] C. Ding, W. Cheng, Y. Sun, X. Wang, Effects of Bacillus subtilis on the reduction of U(VI) by nano-Fe0, Geochim. Cosmochim. Acta. (2015), doi:10.1016/j.gca.2015.05.036.

[5] M.C. Duff, J.U. Coughlin, D.B. Hunter, Uranium co-precipitation with iron oxide minerals, Geochim. Cosmochim. Acta. (2002), doi:10.1016/S0016-7037(02)00953-5.

[6] A. Mellah, S. Chegrouche, M. Barkat, The removal of uranium(VI) from aqueous solutions onto activated carbon: Kinetic and thermodynamic investigations, J. Colloid Interface Sci. (2006), doi:10.1016/j.jcis.2005.09.045.

[7] S.J. Coleman, P.R. Coronado, R.S. Maxwell, J.G. Reynolds, Granulated activated carbon modified with hydrophobic silica aerogel-potential composite materials for the removal of uranium from aqueous solutions, Environ. Sci. Technol. (2003), doi:10.1021/es020929e.

[8] Q. Chen, C. Saltiel, S. Manickavasagam, L.S. Schadler, R.W. Siegel, H. Yang, Aggregation behavior of single-walled carbon nanotubes in dilute aqueous suspension, J. Colloid Interface Sci. (2004), doi:10.1016/j.jcis.2004.07.028.

[9] M. Ahmad, F. Wu, Y. Cui, Q. Zhang, B. Zhang, Preparation of novel bifunctional magnetic tubular nanofibers and their application in efficient and irreversible uranium trap from aqueous solution, ACS Sustain. Chem. Eng. (2020), doi:10.1021/acssuschemeng.0c00332.

[10] M. Ahmad, J. Wang, Z. Yang, Q. Zhang, B. Zhang, Ultrasonic-assisted preparation of amidoxime functionalized silica framework via oil-water emulsion method for selective uranium adsorption, Chem. Eng. J. (2020), doi:10.1016/j.cej.2020.124441.

[11] M. Sundararajan, S.K. Ghosh, Designing novel materials through functionalization of carbon nanotubes for application in nuclear waste management: Speciation of uranyl, J. Phys. Chem. A. (2011), doi:10.1021/jp203723t.

[12] A. Barhoum, K. Pal, H. Rahier, H. Uludag, I.S. Kim, M. Bechelany, Nanofibers as new-generation materials: From spinning and nano-spinning fabrication techniques to emerging applications, Appl. Mater. Today. (2019), doi:10.1016/j.apmt.2019.06.015.

[13] D. Li, S. Egodawatte, D.I. Kaplan, S.C. Larsen, S.M. Serkiz, J.C. Seaman, Functionalized magnetic mesoporous silica nanoparticles for U removal from low and high pH groundwater, J. Hazard. Mater. (2016), doi:10.1016/j.jhazmat.2016.05.093.

[14] Y. Zhao, J. Li, L. Zhao, S. Zhang, Y. Huang, X. Wu, X. Wang, Synthesis of amidoxime-functionalized $Fe_3O_4@SiO_2$ core-shell magnetic microspheres for highly efficient sorption of U(VI), Chem. Eng. J. (2014), doi:10.1016/j.cej.2013.09.034.

[15] M. Xu, X. Han, D. Hua, Polyoxime-functionalized magnetic nanoparticles for uranium adsorption with high selectivity over vanadium, J. Mater. Chem. A. (2017), doi:10.1039/c7ta02684f.

[16] S.S. Lee, W. Li, C. Kim, M. Cho, J.G. Catalano, B.J. Lafferty, et al., Engineered manganese oxide nanocrystals for enhanced uranyl sorption and separation, Environ. Sci. Nano. (2015), doi:10.1039/c5en00010f.

[17] S. Mishra, J. Dwivedi, A. Kumar, N. Sankararamakrishnan, The synthesis and characterization of tributyl phosphate grafted carbon nanotubes by the floating catalytic chemical vapor deposition method and their sorption behavior towards uranium, New J. Chem. (2016), doi:10.1039/c5nj02639c.

[18] W. Li, L.D. Troyer, S.S. Lee, J. Wu, C. Kim, B.J. Lafferty, et al., Engineering nanoscale iron oxides for uranyl sorption and separation: Optimization of particle core size and bilayer surface coatings, ACS Appl. Mater. Interfaces. (2017), doi:10.1021/acsami.7b01042.

[19] S. Bachmaf, B. Planer-Friedrich, B.J. Merkel, Effect of sulfate, carbonate, and phosphate on the uranium(VI) sorption behavior onto bentonite, Radiochim. Acta. (2008), doi:10.1524/ract.2008.1496.

[20] S.S. Lee, W. Li, C. Kim, M. Cho, B.J. Lafferty, J.D. Fortner, Surface functionalized manganese ferrite nanocrystals for enhanced uranium sorption and separation in water, J. Mater. Chem. A. (2015), doi:10.1039/c5ta04406e.

[21] X. Wang, C. Chen, W. Hu, A. Ding, D. Xu, X. Zhou, Sorption of [243]Am(III) to multiwall carbon nanotubes, Environ. Sci. Technol. (2005), doi:10.1021/es048287d.

[22] G.P. Rao, C. Lu, F. Su, Sorption of divalent metal ions from aqueous solution by carbon nanotubes: A review, Sep. Purif. Technol. (2007), doi:10.1016/j.seppur.2006.12.006.

[23] E.A. Santos, A.C.Q. Ladeira, Recovery of uranium from mine waste by leaching with carbonate-based reagents, Environ. Sci. Technol. (2011), doi:10.1021/es2002056.

[24] J. Wu, K. Tian, J. Wang, Adsorption of uranium (VI) by amidoxime modified multiwalled carbon nanotubes, Prog. Nucl. Energy. (2018), doi:10.1016/j.pnucene.2018.02.020.

[25] A. Gopalan, M.F. Philips, J.-H. Jeong, K.-P. Lee, Synthesis of Novel Poly(amidoxime) grafted multiwall carbon nanotube gel and uranium adsorption, J. Nanosci. Nanotechnol. (2014), doi:10.1166/jnn.2014.8507.

[26] W.-L. Tang, L. Xie, H.-J. Wang, S. Wang, Removal of Pb by adsorption of amidoxime group modified carbon nanotubes, in: 2015. 10.2991/icmsa-15.2015.128.

[27] Y. Wang, Z. Gu, J. Yang, J. Liao, Y. Yang, N. Liu, J. Tang, Amidoxime-grafted multiwalled carbon nanotubes by plasma techniques for efficient removal of uranium(VI), Appl. Surf. Sci. (2014), doi:10.1016/j.apsusc.2014.08.182.

[28] R. Rengarajan, M. Vicic, S. Lee, Solid phase graft copolymerization. I. Effect of initiator and catalyst, J. Appl. Polym. Sci. (1990), doi:10.1002/app.1990.070390815.

[29] Q. Chen, L.M. Peng, Structure and applications of titanate and related nanostructures, Int. J. Nanotechnol. (2007), doi:10.1504/ijnt.2007.012314.

[30] N. Liu, X. Chen, J. Zhang, J.W. Schwank, A review on TiO$_2$-based nanotubes synthesized via hydrothermal method: Formation mechanism, structure modification, and photocatalytic applications, Catal. Today. (2014), doi:10.1016/j.cattod.2013.10.090.

[31] J. Yu, H. Yu, B. Cheng, C. Trapalis, Effects of calcination temperature on the microstructures and photocatalytic activity of titanate nanotubes, J. Mol. Catal. A Chem. (2006), doi:10.1016/j.molcata.2006.01.003.

[32] R. Yoshida, Y. Suzuki, S. Yoshikawa, Effects of synthetic conditions and heat-treatment on the structure of partially ion-exchanged titanate nanotubes, Mater. Chem. Phys. (2005), doi:10.1016/j.matchemphys.2004.12.010.

[33] T. Wang, W. Liu, N. Xu, J. Ni, Adsorption and desorption of Cd(II) onto titanate nanotubes and efficient regeneration of tubular structures, J. Hazard. Mater. (2013), doi:10.1016/j.jhazmat.2013.02.016.

[34] G. Sheng, J. Hu, A. Alsaedi, W. Shammakh, S. Monaquel, F. Ye, et al., Interaction of uranium(VI) with titanate nanotubes by macroscopic and spectroscopic investigation, J. Mol. Liq. (2015), doi:10.1016/j.molliq.2015.10.018.

[35] Y. Yao, Z. Yang, D. Zhang, W. Peng, H. Sun, S. Wang, Magnetic CoFe$_2$O$_4$-graphene hybrids: Facile synthesis, characterization, and catalytic properties, Ind. Eng. Chem. Res. (2012), doi:10.1021/ie300271p.

[36] X. Gao, L. Liu, B. Birajdar, M. Ziese, W. Lee, M. Alexe, et al., High-density periodically ordered magnetic cobalt ferrite nanodot arrays by template-assisted pulsed laser deposition, Adv. Funct. Mater. (2009), doi:10.1002/adfm.200900422.

[37] J. Wei, X. Zhang, Q. Liu, Z. Li, L. Liu, J. Wang, Magnetic separation of uranium by CoFe$_2$O$_4$ hollow spheres, Chem. Eng. J. (2014), doi:10.1016/j.cej.2013.12.035.

[38] X. jiang Hu, J. song Wang, Y. guo Liu, X. Li, G. ming Zeng, Z. lei Bao, et al., Adsorption of chromium (VI) by ethylenediamine-modified cross-linked magnetic chitosan resin: Isotherms, kinetics and thermodynamics, J. Hazard. Mater. (2011), doi:10.1016/j.jhazmat.2010.09.034.

[39] W. Cai, L. Tan, J. Yu, M. Jaroniec, X. Liu, B. Cheng, et al., Synthesis of amino-functionalized mesoporous alumina with enhanced affinity towards Cr(VI) and CO2, Chem. Eng. J. (2014), doi:10.1016/j.cej.2013.11.011.

[40] G. Bayramoğlu, M. Yakup Arica, Adsorption of Cr(VI) onto PEI immobilized acrylate-based magnetic beads: Isotherms, kinetics and thermodynamics study, Chem. Eng. J. (2008), doi:10.1016/j.cej.2007.07.068.

[41] S.P. Kuang, Z.Z. Wang, J. Liu, Z.C. Wu, Preparation of triethylene-tetramine grafted magnetic chitosan for adsorption of Pb(II) ion from aqueous solutions, J. Hazard. Mater. (2013), doi:10.1016/j.jhazmat.2013.05.019.

[42] J. Zhu, Q. Liu, Z. Li, J. Liu, H. Zhang, R. Li, et al., Efficient extraction of uranium from aqueous solution using an amino-functionalized magnetic titanate nanotubes, J. Hazard. Mater. (2018), doi:10.1016/j.jhazmat.2018.03.042.

[43] J. Zhu, Q. Liu, J. Liu, R. Chen, H. Zhang, M. Zhang, et al., Investigation of uranium (VI) adsorption by poly(dopamine) functionalized waste paper derived carbon, J. Taiwan Inst. Chem. Eng. (2018), doi:10.1016/j.jtice.2018.05.024.

[44] Z. Hu, M.P. Srinivasan, Y. Ni, Novel activation process for preparing highly micro-porous and mesoporous activated carbons, Carbon N. Y. (2001), doi:10.1016/S0008-6223(00)00198-6.

[45] J. Ou, J. Wang, D. Zhang, P. Zhang, S. Liu, P. Yan, et al., Fabrication and biocompatibility investigation of TiO$_2$ films on the polymer substrates obtained via a novel and versatile route, Colloids Surfaces B Biointerfaces (2010), doi:10.1016/j.colsurfb.2009.10.024.

[46] Z. Zhao, J. Li, T. Wen, C. Shen, X. Wang, A. Xu, Surface functionalization graphene oxide by polydopamine for high affinity of radionuclides, Colloids Surfaces A Physic-ochem. Eng. Asp. (2015), doi:10.1016/j.colsurfa.2015.05.020.

[47] J.K. Gao, L.A. Hou, G.H. Zhang, P. Gu, Facile functionalized of SBA-15 via a biomimetic coating and its application in efficient removal of uranium ions from aqueous solution, J. Hazard. Mater. (2015), doi:10.1016/j.jhazmat.2014.12.061.

[48] A.C.M. Lamb, F. Grieser, T.W. Healy, The adsorption of uranium (VI) onto colloidal TiO$_2$, SiO$_2$ and carbon black, Colloids Surfaces A Physicochem. Eng. Asp. (2016), doi:10.1016/j.colsurfa.2016.04.003.

[49] Z. Li, F. Chen, L. Yuan, Y. Liu, Y. Zhao, Z. Chai, W. Shi, Uranium(VI) adsorption on graphene oxide nanosheets from aqueous solutions, Chem. Eng. J. (2012), doi:10.1016/j.cej.2012.09.030.

[50] Z. Han, B. Sani, W. Mrozik, M. Obst, B. Beckingham, H.K. Karapanagioti, et al., Magnetite impregnation effects on the sorbent properties of activated carbons and biochars, Water Res (2015), doi:10.1016/j.watres.2014.12.016.

[51] W. Qiu, D. Yang, J. Xu, B. Hong, H. Jin, D. Jin, et al., Efficient removal of Cr(VI) by magnetically separable CoFe$_2$O$_4$/activated carbon composite, J. Alloys Compd. (2016), doi:10.1016/j.jallcom.2016.03.304.

[52] O. Gulnaz, A. Kaya, F. Matyar, B. Arikan, Sorption of basic dyes from aqueous solution by activated sludge, J. Hazard. Mater. (2004), doi:10.1016/j.jhazmat.2004.02.012.

[53] Z. Al-Qodah, Biosorption of heavy metal ions from aqueous solutions by activated sludge, Desalination (2006), doi:10.1016/j.desal.2005.12.012.

[54] B.W. Pang, C.H. Jiang, M. Yeung, Y. Ouyang, J. Xi, Removal of dissolved sulfides in aqueous solution by activated sludge: mechanism and characteristics, J. Hazard. Mater. (2017), doi:10.1016/j.jhazmat.2016.11.048.

[55] L. Wei, Y. Li, D.R. Noguera, N. Zhao, Y. Song, J. Ding, et al., Adsorption of Cu^{2+} and Zn^{2+} by extracellular polymeric substances (EPS) in different sludges: Effect of EPS fractional polarity on binding mechanism, J. Hazard. Mater. (2017), doi:10.1016/j.jhazmat.2016.05.016.

[56] R. Paliwal, S. Uniyal, J.P.N. Rai, Evaluating the potential of immobilized bacterial con-sortium for black liquor biodegradation, Environ. Sci. Pollut. Res. (2015), doi:10.1007/s11356-014-3872-x.

[57] P. Sun, C. Hui, S. Wang, R.A. Khan, Q. Zhang, Y.H. Zhao, Enhancement of al-gicidal properties of immobilized Bacillus methylotrophicus ZJU by coating with magnetic Fe$_3$O$_4$ nanoparticles and wheat bran, J. Hazard. Mater. (2016), doi:10.1016/j.jhazmat.2015.08.048.

[58] C. Zhao, J. Liu, G. Yuan, J. Liu, H. Zhang, J. Yang, et al., A novel activated sludge-graphene oxide composites for the removal of uranium(VI) from aqueous solutions, J. Mol. Liq. (2018), doi:10.1016/j.molliq.2018.09.069.

[59] Z. Liu, R. Ma, M. Osada, N. Iyi, Y. Ebina, K. Takada, et al., Synthesis, anion exchange, and delamination of Co-Al layered double hydroxide: Assembly of the exfoliated nanosheet/polyanion composite films and magneto-optical studies, J. Am. Chem. Soc. (2006), doi:10.1021/ja0584471.

[60] Y. Guo, Z. Zhu, Y. Qiu, J. Zhao, Enhanced adsorption of acid brown 14 dye on calcined Mg/Fe layered double hydroxide with memory effect, Chem. Eng. J. (2013), doi:10.1016/j.cej.2012.12.084.

[61] J. Han, Y. Dou, M. Wei, D.G. Evans, X. Duan, Erasable nanoporous antireflection coatings based on the reconstruction effect of layered double hydroxides, Angew. Chemie - Int. Ed. (2010), doi:10.1002/anie.200907005.

[62] L. Tan, Y. Wang, Q. Liu, J. Wang, X. Jing, L. Liu, et al., Enhanced adsorption of uranium (VI) using a three-dimensional layered double hydroxide/graphene hybrid material, Chem. Eng. J. (2015), doi:10.1016/j.cej.2014.08.015.

[63] S. Chen, J. Hong, H. Yang, J. Yang, Adsorption of uranium (VI) from aqueous solution using a novel graphene oxide-activated carbon felt composite, J. Environ. Radioact. (2013), doi:10.1016/j.jenvrad.2013.09.002.

[64] A. Nilchi, T. Shariati Dehaghan, S.Rasouli Garmarodi, Kinetics, isotherm and thermodynamics for uranium and thorium ions adsorption from aqueous solutions by crystalline tin oxide nanoparticles, Desalination (2013), doi:10.1016/j.desal.2012.06.022.

[65] K.C. Song, J.H. Kim, Preparation of nanosize tin oxide particles from water-in-oil microemulsions, J. Colloid Interface Sci. (1999), doi:10.1006/jcis.1998.6022.

[66] G. Zhang, M. Liu, Preparation of nanostructured tin oxide using a sol-gel process based on tin tetrachloride and ethylene glycol, J. Mater. Sci. (1999), doi:10.1023/A:1004685907751.

[67] M. Bhagwat, P. Shah, V. Ramaswamy, Synthesis of nanocrystalline SnO_2 powder by amorphous citrate route, Mater. Lett. (2003), doi:10.1016/S0167-577X(02)01040-6.

[68] S. wen WEI, B. PENG, L. yuan CHAI, Y. chao LIU, Z. ying LI, Preparation of doping titania antibacterial powder by ultrasonic spray pyrolysis, Trans. Nonferrous Met. Soc. China (English Ed. (2008), doi:10.1016/S1003-6326(08)60196-X.

[69] N.S. Baik, G. Sakai, N. Miura, N. Yamazoe, Preparation of stabilized nano-sized tin oxide particles by hydrothermal treatment, J. Am. Ceram. Soc. (2000), doi:10.1111/j.1151-2916.2000.tb01670.x.

[70] K.C. Song, Y. Kang, Preparation of high surface area tin oxide powders by a homogeneous precipitation method, Mater. Lett. (2000), doi:10.1016/S0167-577X(99)00199-8.

[71] F. Li, L. Chen, Z. Chen, J. Xu, J. Zhu, X. Xin, Two-step solid-state synthesis of tin oxide and its gas-sensing property, Mater. Chem. Phys. (2002), doi:10.1016/S0254-0584(01)00357-1.

[72] C. Mahr, Precipitation from Homogeneous Solution, von L. Gordon, M.L. Salutsky und H.H. Willard. John Wiley & Sons, Inc., New York; Chapman & Hall, Ltd., London 1959. 1. Aufl., VIII, 187 S., geb. $ 7.50, Angew. Chemie. (2007), doi:10.1002/ange.19610731426.

[73] H.H. Willard, N.K. Tang, A study of the precipitation of aluminum basic sulfate by urea, J. Am. Chem. Soc. (1937), doi:10.1021/ja01286a010.

[74] B. Djuričić, S. Pickering, D. McGarry, P. Glaude, P. Tambuyser, K. Schuster, The properties of zirconia powders produced by homogeneous precipitation, Ceram. Int. (1995), doi:10.1016/0272-8842(95)90910-B.

[75] J.M. Dias, M.C.M. Alvim-Ferraz, M.F. Almeida, J. Rivera-Utrilla, M. Sánchez-Polo, Waste materials for activated carbon preparation and its use in aqueous-phase treatment: A review, J. Environ. Manage. (2007), doi:10.1016/j.jenvman.2007.07.031.

[76] M. Song, B. Jin, R. Xiao, L. Yang, Y. Wu, Z. Zhong, et al., The comparison of two activation techniques to prepare activated carbon from corn cob, Biomass and Bioenergy (2013), doi:10.1016/j.biombioe.2012.11.007.

[77] M.A. Yahya, Z. Al-Qodah, C.W.Z. Ngah, Agricultural bio-waste materials as potential sustainable precursors used for activated carbon production: A review, Renew. Sustain. Energy Rev. (2015), doi:10.1016/j.rser.2015.02.051.

[78] S. Wong, N. Ngadi, I.M. Inuwa, O. Hassan, Recent advances in applications of activated carbon from biowaste for wastewater treatment: A short review, J. Clean. Prod. (2018), doi:10.1016/j.jclepro.2017.12.059.

[79] S. Preethi, A. Sivasamy, S. Sivanesan, V. Ramamurthi, G. Swaminathan, Removal of safranin basic dye from aqueous solutions by adsorption onto corncob activated carbon, Ind. Eng. Chem. Res. (2006), doi:10.1021/ie0604122.

[80] X.X. Hou, Q.F. Deng, T.Z. Ren, Z.Y. Yuan, Adsorption of Cu^{2+} and methyl orange from aqueous solutions by activated carbons of corncob-derived char wastes, Environ. Sci. Pollut. Res. (2013), doi:10.1007/s11356-013-1792-9.

[81] W. Yantasee, C.L. Warner, T. Sangvanich, R.S. Addleman, T.G. Carter, R.J. Wiacek, et al., Removal of heavy metals from aqueous systems with thiol functionalized superparamagnetic nanoparticles, Environ. Sci. Technol. (2007), doi:10.1021/es0705238.

[82] M. Zarezadeh-Mehrizi, A. Badiei, Highly efficient removal of basic blue 41 with nanoporous silica, Water Resour. Ind. (2014), doi:10.1016/j.wri.2014.04.002.

[83] W. Zhu, J. Wang, D. Wu, X. Li, Y. Luo, C. Han, et al., Investigating the heavy metal adsorption of mesoporous silica materials prepared by microwave synthesis, Nanoscale Res. Lett. (2017), doi:10.1186/s11671-017-2070-4.

[84] X. Wang, Y. Zhang, W. Luo, A.A. Elzatahry, X. Cheng, A. Alghamdi, et al., Synthesis of ordered mesoporous silica with tunable morphologies and pore sizes via a non-polar solvent-assisted stöber method, Chem. Mater. (2016), doi:10.1021/acs.chemmater.6b00499.

[85] E. Bäuerlein, Biomineralization of unicellular organisms: An unusual membrane biochemistry for the production of inorganic nano- and microstructures, Angew, Chemie-Int. Ed. (2003), doi:10.1002/anie.200390176.

[86] D.P. Dutta, S. Nath, Low cost synthesis of SiO_2/C nanocomposite from corn cobs and its adsorption of uranium (VI), chromium (VI) and cationic dyes from wastewater, J. Mol. Liq. (2018), doi:10.1016/j.molliq.2018.08.028.

[87] Z. ji Yi, J. Yao, H. lun Chen, F. Wang, Z. min Yuan, X. Liu, Uranium biosorption from aqueous solution onto Eichhornia crassipes, J. Environ. Radioact. (2016), doi:10.1016/j.jenvrad.2016.01.012.

[88] Z. ji Yi, J. Yao, M. jia Zhu, H. lun Chen, F. Wang, Z. min Yuan, X. Liu, Batch study of uranium biosorption by Elodea canadensis biomass, J. Radioanal. Nucl. Chem. (2016), doi:10.1007/s10967-016-4839-9.

[89] T.T. More, J.S.S. Yadav, S. Yan, R.D. Tyagi, R.Y. Surampalli, Extracellular polymeric substances of bacteria and their potential environmental applications, J. Environ. Manage. (2014), doi:10.1016/j.jenvman.2014.05.010.

[90] L. Domínguez, M. Rodríguez, D. Prats, Effect of different extraction methods on bound EPS from MBR sludges Part II: Influence of extraction methods over molecular weight distribution, Desalination (2010), doi:10.1016/j.desal.2010.06.001.

[91] S. He, B. Ruan, Y. Zheng, X. Zhou, X. Xu, Immobilization of chlorine dioxide modified cells for uranium absorption, J. Environ. Radioact. (2014), doi:10.1016/j.jenvrad.2014.06.016.

[92] D. Mohan, C.U. Pittman, P.H. Steele, Pyrolysis of wood/biomass for bio-oil: A critical review, Energy and Fuels (2006), doi:10.1021/ef0502397.

[93] A.V. Bridgwater, G.V.C. Peacocke, Fast pyrolysis processes for biomass, Renew. Sustain. Energy Rev. (2000), doi:10.1016/S1364-0321(99)00007-6.

[94] S.N. Guilhen, N. Ortiz, D.A. Fungaro, O. Masek, Pyrolytic temperature evaluation of macauba biochar for uranium adsorption from aqueous solutions, Biochar Prod. Charact. Appl. (2017), doi:10.1210/jc.2008-0104.

[95] X. Chen, G. Chen, L. Chen, Y. Chen, J. Lehmann, M.B. McBride, et al., Adsorption of copper and zinc by biochars produced from pyrolysis of hardwood and corn straw in aqueous solution, Bioresour. Technol. (2011), doi:10.1016/j.biortech.2011.06.078.

[96] L. Ding, W. fa Tan, S. bo Xie, K. Mumford, J. wen Lv, H. qiang Wang, et al., Uranium adsorption and subsequent re-oxidation under aerobic conditions by Leifsonia sp-Coated biochar as green trapping agent, Environ. Pollut. (2018), doi:10.1016/j.envpol.2018.07.050.

[97] Y. Han, X. Cao, X. Ouyang, S.P. Sohi, J. Chen, Adsorption kinetics of magnetic biochar derived from peanut hull on removal of Cr(VI) from aqueous solution: Effects of production conditions and particle size, Chemosphere (2016), doi:10.1016/j.chemosphere.2015.11.050.

[98] K.R. Thines, E.C. Abdullah, N.M. Mubarak, M. Ruthiraan, Synthesis of magnetic biochar from agricultural waste biomass to enhancing route for waste water and polymer application: A review, Renew. Sustain. Energy Rev. (2017), doi:10.1016/j.rser.2016.09.057.

[99] Y. Zhou, B. Gao, A.R. Zimmerman, H. Chen, M. Zhang, X. Cao, Biochar-supported zerovalent iron for removal of various contaminants from aqueous solutions, Bioresour. Technol. (2014), doi:10.1016/j.biortech.2013.11.021.

[100] H. Guo, D. Stüben, Z. Berner, Adsorption of arsenic(III) and arsenic(V) from groundwater using natural siderite as the adsorbent, J. Colloid Interface Sci. (2007), doi:10.1016/j.jcis.2007.06.010.

[101] E. Liger, L. Charlet, P. Van Cappellen, Surface catalysis of uranium(VI) reduction by iron(II), Geochim. Cosmochim. Acta. (1999), doi:10.1016/S0016-7037(99)00265-3.

[102] A. Sabarudin, M. Oshima, T. Takayanagi, L. Hakim, K. Oshita, Y.H. Gao, S. Motomizu, Functionalization of chitosan with 3,4-dihydroxybenzoic acid for the adsorption/collection of uranium in water samples and its determination by inductively coupled plasma-mass spectrometry, Anal. Chim. Acta. (2007), doi:10.1016/j.aca.2006.08.024.

[103] O. Abderrahim, M.A. Didi, A. Villemin, A new sorbent for uranium extraction: Polyethyleniminephenylphosphonamidic acid, J. Radioanal. Nucl. Chem. (2009), doi:10.1007/s10967-007-7270-z.

[104] T. Scott, D. Horner, R. Yates, J. Mailen, S. Thiel, Interphase transfer kinetics of uranium using the drop method, lewis cell, and kenics mixer, Ind. Eng. Chem. Fundam. (2005), doi:10.1021/i160073a018.

[105] X. Wang, G. Zhu, C. Gao, Adsorption of uranium (VI) on silica mesoporous material SBA-15 with short channels, Huagong Xuebao/CIESC J (2013), doi:10.3969/j.issn.0438-1157.2013.07.024.

[106] N. Xiao, Y. Zhou, Z. Ling, J. Qiu, Synthesis of a carbon nanofiber/carbon foam composite from coal liquefaction residue for the separation of oil and water, Carbon N. Y. (2013), doi:10.1016/j.carbon.2013.03.051.

[107] K.T. Peter, J.D. Vargo, T.P. Rupasinghe, A. De Jesus, A.V. Tivanski, E.A. Sander, et al., Synthesis, optimization, and performance demonstration of electrospun carbon nanofiber-carbon nanotube composite sorbents for point-of-use water treatment, ACS Appl. Mater. Interfaces. (2016), doi:10.1021/acsami.6b01253.

[108] C. Ding, W. Cheng, X. Wang, Z.Y. Wu, Y. Sun, C. Chen, et al., Competitive sorption of Pb(II), Cu(II) and Ni(II) on carbonaceous nanofibers: A spectroscopic and modeling approach, J. Hazard. Mater (2016), doi:10.1016/j.jhazmat.2016.04.002.

[109] R. Zhang, C. Chen, J. Li, X. Wang, Investigation of interaction between U(VI) and carbonaceous nanofibers by batch experiments and modeling study, J. Colloid Interface Sci. (2015), doi:10.1016/j.jcis.2015.08.073.

[110] Y. Sun, Z.Y. Wu, X. Wang, C. Ding, W. Cheng, S.H. Yu, et al., Macroscopic and microscopic investigation of U(VI) and Eu(III) adsorption on carbonaceous nanofibers, Environ. Sci. Technol (2016), doi:10.1021/acs.est.6b00058.

[111] H.W. Liang, X. Cao, W.J. Zhang, H.T. Lin, F. Zhou, L.F. Chen, et al., Robust and highly efficient free-standing carbonaceous nanofiber membranes for water purification, Adv. Funct. Mater. (2011), doi:10.1002/adfm.201100983.

[112] M. Ahmad, B. Zhang, J. Wang, J. Xu, K. Manzoor, S. Ahmad, S. Ikram, New method for hydrogel synthesis from diphenylcarbazide chitosan for selective copper removal, Int. J. Biol. Macromol. (2019), doi:10.1016/j.ijbiomac.2019.06.084.

[113] M. Ahmad, K. Manzoor, S. Ikram, Versatile Nature of Hetero-Chitosan Based Derivatives as Biodegradable Adsorbent for Heavy Metal Ions; A Review, Int. J. Bio. Macromol. (2017), doi:10.1016/j.ijbiomac.2017.07.008.

[114] M. Ahmad, S. bAhmed, B.L. Swami, S. Ikram, Characterization of Antibacterial Thiosemicarbazide Chitosan as Efficient Cu(II) Adsorbent, Carbohydrate Polymers (2015), doi:10.1016/j.carbpol.2015.06.034.

CHAPTER 5

Preparation of carbon tubular nanofibers and their application for efficient enrichment of uranium from aqueous solution

Mudasir Ahmad[a,b] and Baoliang Zhang[a,b]
[a]School of Chemistry and Chemical Engineering, Northwestern Polytechnical University, Xi'an, China
[b]Xi'an Key Laboratory of Functional Organic Porous Materials, Northwestern Polytechnical University, Xi'an, China

5.1 Introduction

Carbon tubular nanofibers (CNF) are filament-like nanostructured materials with diameter size ranges from 3 to 100 nm and organized by staked graphene layers with certain orientations along with the direction of the fiber axis [1]. This kind of material is classified into three categories according to the angle between the graphene layers and axis are as follows; fishbone (angle < 90°), parallel (angle = 0°), and platelets (angle = 90°) (Fig. 5.1) [2–4]. The carbon presence of CNF in the turbostratic nature with an average gap between the layers is around 0.34 nm, which is approximately close to the graphite 0.33 nm, which is why sometimes in the literature CNF have been called graphite CNF [5]. It is very important to distinguish between carbon nanotubes and hallow carbon nanotubes constituted the smaller size diameter and not be distinguished by surface analysis. In theory, carbon nanotubes are formed by wrapping single graphite in a cylindrical shape (single-walled carbon nanotubes) or multiple sheets are combined to form a nest (multiwalled carbon nanotubes). Whereas CNF is a common concept that consists of several graphite layers but not continuous and may present as a hallow core structure [6]. In terms of properties, carbon nanotubes showed good electric and mechanical properties however compared with the CNF the main disadvantage is difficult scalability to produce huge groups and high cost. Carbon materials showed high electrical conductivity due to this CNF has been used for different applications from electronics to composites.

Fabrication and Functionalization of Advanced Tubular Nanofibers and their Applications.
DOI: https://doi.org/10.1016/B978-0-323-99039-4.00012-7
115

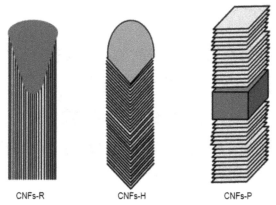

CNFs-R   CNFs-H   CNFs-P

Figure 5.1 Various shapes of carbon nanofibers. The copyright of Figure was obtained from Elsevier under license No: 5275150594774 dated Mar 24, 2022 [2].

CNF is electrical conductor material due to the presence of delocalized electrons that can free rotate through the whole structure. In graphite, the conductivity is anisotropic with a value range of 2.5×10^6 S m^{-1} and 1×10^4 S m^{-1} in parallel and perpendicular to the graphite layer [7]. However, the porous carbon materials showed lower electrical conductivity due to the presence of resistive groups and the same trend was followed by CNF [8]. The mechanical and electrical property of CNF depends on the degree of graphitization and the simple way to increase the graphitization is simply done by heat treatment [9]. A highly developed graphite structure characterizes the structures of CNF and treatment of temperature can induce the graphite alignments to increase the graphitization process of CNF [10]. Further CNF can be chemically treated to increase the functionalization and surface area. Functionalization in acidic media (nitric acid/sulphuric acid) increases the hydrophobicity due to presence of hetero groups on the surface that might induce a well dispersion of CNF in aqueous media. Despite increasing the hydrophilicity due to oxidation of CNF, the surface oxygen of CNF might decrease the electric conductivity, whereas the chemical coordination with metal ions increases. Due to exceptional properties, CNF have been used for several applications such as catalyst, catalytic support, sensors, and adsorbents for the removal of various metal ions [11]. CNF is considered a promising material that can be prepared with different morphology, shape, diameter and size. The mechanical and electrical properties are simulated with thermal treatment or adding the dopants in the matrix. The surface functionality can be enhanced by thermal and chemical treatment [12]. The surface functionality can be enhanced by the grafting or oxidation process,

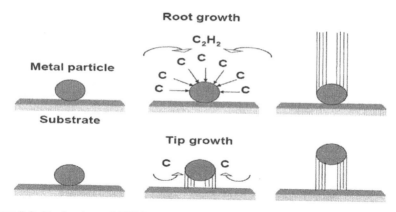

Figure 5.2 Mechanism of CNF formation. The copyright of Figure was obtained from Elsevier under license No: 5275160443086 dated Mar 24, 2022 [16].

which might enhance the chemical coordination between the metal ions and the active surface sites [13]. CNF have been studied broadly in the literature and several review articles were published that cover the major findings of CNF from the last two decades. Hence, the purpose of this chapter is to collect some recent studies that mainly focused on new synthetic pathways of CNF and their potential in new applications.

5.2 Synthesis of CNF

5.2.1 Synthesis of CNF using catalyst

Synthesis of CNF from carbon-containing gases in presence of a metal catalyst and the carbon sources may not be restricted and mostly the hydrocarbons from C1 to C6 have been reported. The starting materials for CNF were produced by the chemical vapor dispersion (CVD) technique which is mostly used the gases of methane, ethylene, acetylene, and propylene [14]. Generally, the metal catalysts are iron, nickel and cobalt are used to catalyze the growth of CNF [12]. In addition, some other metals are molybdenum, chromium, and vanadium were used in several studies [15]. A special advantage for the preparation of CNF in mass production is economically efficient than other carbon nanomaterials. The mechanism of CNF starts by decomposition of hydrocarbon and carbon is carried by metal particles and the hexagonal Sp^2 carbon arrangement occurs. The complete mechanism of CNF formation is shown in Fig. 5.2 [16,4]. Particularly CNF are short and very difficult to assemble, align, the process for a particular

Table 5.1 Catalytic decomposition at various parameters for preparation of CNF.

Catalyst	Carbon precursor	Structure	Temperature	Reference
Ni	$CO, CO_2, H_2=$	Fish bone	400–600	[22]
Ni	n-hexane	-	500	[23]
Ni foam	$C_2H_4 \, H_2^{-1} \, N_2^{-1}$	-	600	[24]
Ni/SiO2	C_2H_4	-	450-750	[25]
Ni-Cr	Chlorobenzene/H_2	-	600	[12]
Ni_3C	n-Hexane	Twisted	350	[26]
Ni/Al_2O_3	$C_2H_2 \, H_2^{-1} \, N_2^{-1}$	Fishbone	550	[27]
Co/graphite	$C2H4 \, H2^{-1}$ (4:1)	-	500	[12]
Co-Mo(9:1)/	CO/H_2 (4:1)	Hallow Fishbow	460	[28]
Fe	$CO \, H2^{-1}$	Tubular	480	[29]
Fe	Isopropyl alcohol,	-	680	[30]
Fe	Benzylchloride	Tubular	500	[13]

application, requires expensive equipment to synthesize and needs a significant amount of catalysts [17]. The difference between CNF and CNT synthesized by CVD is only in morphology. A CNF showed a staggering cone shape with a finite cone angle while a CNT showed a cone angle equal to zero. The control process is done to obtain the special shape, size and morphology, which can be obtained by understanding the kinetics of nucleation and growth of CNF/CNT. Kinetics based on thermogravimetric analysis suggested different carbon diffusion mechanisms from the reaction sites to the deposition place and the diffusion of carbon does not alter the properties of Ni, hence the deposition of carbon occurs on the surface [18]. Different types of carbon-based materials have been synthesized using the CVD technique and the morphology of the CNF depends on the shape of a metal particle (catalyst). The diameter mostly depends on the size of the metal catalyst particle [19,12]. The catalyst structures can be modified to obtained similar size carbon nanomaterial by monolayer deposition of metal catalysts on supporting materials (silica, alumina) and followed by nucleation to get small catalyst particles [20]. The pyrolytic temperature significantly affects the morphology of CNF. A straight CNF were obtained at 500°C and with an increased temperature greater than 500°C the twisted CNF were obtained [21]. The synthesis of CNF by catalytic decomposition is listed in Table 5.1.

5.2.2 Synthesis of CNF by electro-spinning

Electro-spinning technique applied strong electrostatic for the synthesis of CNF from polymer solution and beneficial to prepare polymeric nanofibers,

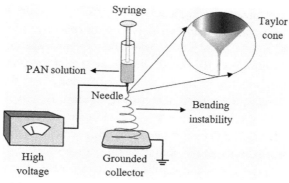

Figure 5.3 Graphic representation of electrospinning setup for nanofiber fabrication. The copyright of Figure was obtained from Elsevier under license No: 5275120600292, Mar 23, 2022 [34].

allowing to altering the morphology, functionality, and mechanical strength. This technique has been used to prepare the CNF from several tens of nanometers to a few micrometers in various like as nonwoven mats, yarns, etc. it is a simple and cost–efficient technique to produce CNF from polymer solutions [31]. The parameters that affect the morphology of CNF include the concentration of polymer solution, voltage, working distance, and solution feeding rate. The synthesis possesses the spinning of polymer precursor which is followed by a thermal process. The increasing thermal treatment decreases the fiber size [32] and commonly used solvents are tetrahydrofuran (THF), C_2S, water, cyclohexane, toluene, dimethylformamide (DMF) [33].

5.2.3 Mechanism and control

In the mechanism, a high charge of either AC/DC is applied to polymer solution which created a large potential difference between the feed syringe and the target place. The complete diagrammatic representation is shown in Fig. 5.3 [34]. The charge repulsion of the drop competes with the surface charges, once the critical reached when the surface charges dominate a jet drawn from the syringe (spinneret). The accelerating jet decreases in diameter with increasing the external potentials and the continuous charge is drowned from the surface on the jet and the point a reached where the jet starts to bend and the jet starts whipping. After solvent evaporation, the jet solidifies in the form of nanofibers which are grounded on the target surface. The obtained electrospun nanofibers have to undergo a carbonization process at high temperature (1000°C) to produce CNF and during carbonization weight loss and shrinkage occurs which usually results in the reduction of diameter size. The morphology of CNF depends on three basic things (1) solution properties, (2) processing type and (3) environment conditions [31]. Mostly the properties of the polymer solution are the dominant factor to control the morphology and the influencing property is viscosity, dielectrics, and surface tension. At the same time, the viscosity of the solution depends on the additives, molecular weight, and solvent used. Mainly the diameter decreases with decreasing the viscosity and the electric conductivity of the solution. The processing conditions include the hydrostatic pressure in the capillary, electric potential at the tip of the capillary and the distance between the tip of the capillary and the target place [35]. The atmospheric conditions include temperature, air velocity and the humidity of the chamber. Fabrication of CNF using electrospinning technique, it is necessary to prepare polymer nanofibers as a precursor for CNF. The final CNF depends on the polymer solution and three types of polymers that have been widely used for the preparation of CNF are polyacrylonitrile (PAN), cellulose and pitch. Mostly PAN is used to prepare CNF, because of the large amount of carbon is produced from PAN and approximately 90% of CNF were produced from PAN. In addition, polyvinyl alcohols, polyimides, phenolic resin, lignin etc., have been also investigated for the preparation of CNF [36]. Different strategies have been practiced for tuning the porosity and morphology of polymer nanotubes to produce the CNF. Electrospun poly(methylmethacrylate) in presence of THF and $CHCl_3$ with a diameter in between 100 and 1000 nm and surface area 10–1000 m^2 g^{-1}. CNF produced from PAN have been activated by physical or chemical treatments such as steam, CO_2 and alkali hydroxide to obtain porous and microporous CNF [37,38]. The effect of temperature on length and morphology of PAN CNF was studied by Kim et.al. this study revealed at 700°C the shape changes from straight to undulated due to maximum weight loss along with gas escape [39]. The length of CNF can be up to several kilometers, which depends on the solution of polymer and the time of electrospinning. To yield the smaller size of diameter, porous morphology, core-shell structure, post preparation modification in combination with mixed technology to fabricate electrospun blended polymers. Electrospinning is a simple technique for fabricating highly porous and larger surface area nonwoven nanofiber mats. Several precursors have been analyzed for electrospinning and are responsible for the properties of CNF. Table 5.2 summarizes experimental conditions and precursors for the fabrication of CNF.

Table 5.2 Shows various materials with different parameters for the preparation of CNF.

Precursor	Solvent	Temperature	Morphology	References
PAN	DMF	1000	Wed	[40]
PAN	DMF	900	Align in one direction	[41]
PAN	DMF	500 and 800	Mat	[42]
PAN/PVP/TiCl$_4$	DMF	1000	Web	[43]
Pitch	DMF/THF	600	Dumbbell	[44]
PAN/PMMA/pitch	DMF/THF	800	Web	[45]
PVA	Distill H2O	600-1000	–	[46]
Lignin	ethanol	800	web	[47]
Lignin/H$_3$PO$_4$	ethanol	900	web	[48]
Alkali lignin/PVA	water	1200	membrane	[49]
PVP/PANi/graphene	water	1200	fiber	[50]
PVP/PMMA/ZrB	DMF	1200	fiber	[51]

5.2.4 Crosslinking method

Hypercrosslinked polymers (HPs) are the class of materials whose synthetic route follows Friedel craft alkylation reaction mechanism [52,53]. The high crosslinking degree and high polymer density of HPs stop the tight shrinkage of polymer chains and therefore the gaps between the polymers chains become the permanent pores [54,55]. Also, HPs are prepared by self polycondensation reaction [56,57] but are very difficult to prepare porous material with regular morphology because of the randomness between the monomers during the crosslinking reactions [58]. Recently, some advancements have been carried for the preparation of hyper-crosslinked polymers with regular morphology. Jiang et al. have developed polymer from aromatic hydrocarbons as a monomer and formaldehyde dimethyl acetal as a crosslinking agent using the braiding method, however, the mechanism is still unknown [59]. After that, another study was carried out for the preparation of tubular polymer nanotubes (TPs) with regular morphology. These TPs were prepared self polycondensation of monomer in dual oil phase system. Polysubstituted benzyl halides widely used self polycondensed monomer has been used for the preparations of TPs. Tubular carbon nanofibers (TCs) were obtained from TPs by carbonization at high temperatures (500°C) were modified to TCs. Further, these nanofibers were modified magnetic tubular nanofibers with a large surface area after carbonization [60]. The dual–phase system consists of 1,2-dichloroethane (DCE) and polydimethylsiloxane (PDMS) as a solvent in presence of Alpha,alpha-dichloro-p-xylene (DCX)(monomer) and FeCl$_3$ (catalyst). Both DCX and FeCl$_3$ were added in the dual oil phase drop by drop and collided and coalesced with the monomers. The hypercrosslinking reactions occurs in

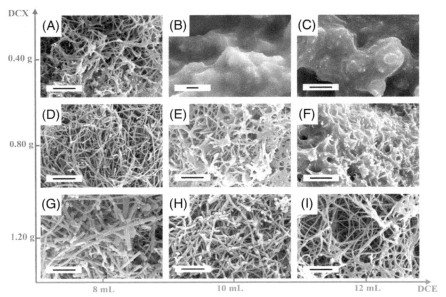

Figure 5.4 SEM images of CNF: A–C, the dosage of DCX was 0.40 g, at DCX concentrations of 50, 40, and 30 mg mL^{-1} (C); D–F, the dosage of DCX was 0.80 g, at DCX concentrations of 100, 80, and 67 mg mL^{-1}; G–I, the dosage of DCX was 1.20 g, at DCX concentrations of 150,120, and 100 mg mL^{-1} (I). The scale bar is 2 μm. Reproduced from Ref. [60] with permission from the Royal Society of Chemistry.

the droplets in dual oil system compared with the traditional polymerization of hypercrosslinked polymers. The monomer concentration was increased with the volatilization of DCE in dual where as in traditional monomer concentration is unchanged due to refluxing of DCE. The concentration of monomer and catalyst in dual phase was determined by amount of DCE, because the droplets of both monomer and catalyst came from the DCE. Therefore the influence of DCX, FeCl$_3$, and amount of DCE was analyzed by analyzing the morphology of various products obtained at different concentration. Fig. 5.4 shows the effect of solvents dosage and concentration of monomer and catalyst. The tradition polymerization of DCX yield the hypercrosslinked mats of granules, while as the concatenation of droplets of monomers and catalyst in dual oil phase yield hypercrosslinked polymers with regular morphology. The same dosage of DCE with higher monomer concentration yields fibrous morphology. However the morphology gets detorated with increasing the amount of DCE. The wall thickness of TPs were analyzed at various concentrations of 50, 100, and 150 mg mL^{-1} using TEM analysis and distribution curves shown in Fig. 5.5 and the corresponding wall thinks at all concentration are approximately 30 μm.

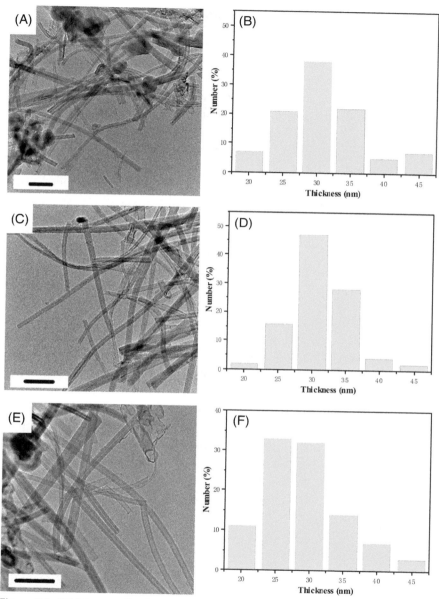

Figure 5.5 TEM images (A, C and E) and the wall thickness distributions (B, D, and F) of CNF at various dosages of DCE and DCX, the dosage of DCE was 8 mL, at DCX concentrations of 50 mg mL^{-1} (A, B), 100 mg mL^{-1} (C, D), 150 mg mL^{-1} (E, F). The scale bar is 200 nm. Reproduced from Ref. [60] with permission from the Royal Society of Chemistry.

5.3 Applications of carbon tubular nanofibers

5.3.1 Catalyst

Carbon is widely used as a support material for heterogamous catalysis, especially in liquid phase catalysis. The use of carbon materials in catalysis is because of low cost when compared with other precious metal catalysts such as platinum and gold [61]. Different properties including the porosity, surface area, electrical conductivity, and corrosion resistivity greatly influence the performances of catalysts such as heat dissipation, activity, transport of electrons and stability [62]. For CNF these features were obtained by stacking graphite planes [63]. The structure of CNF can be tuned using the type of precursor, temperature, catalyst and the type of gas composition supplied during the carbonization process. Several methodologies have been synthesized for CNF-supported platinum catalysts reported in the literature. There are some variations in synthetic strategies, but all of them have been considered with different synthetic methodologies with the same characteristics [62]. Among all of the impregnation and microemulsion methods have been reported for platinum deposition, the impregnation with reduction with sodium borohydride, impregnation with and reduction with formic acid [64]. Four component oil, water microemulsion consists n–heptane, water, surfactant and surfactant and three-component oil-water system and reduction with sodium borohydride [65].

Other methodologies have been synthesized for CNF supported catalysts for specific catalytic reactions. Mesoporous CNF with larger pores has been synthesized by thermal treatment of electrospun fibers that can act as support for platinum [66]. Functionalized CNF have been used as a support material for Ni-Mo catalyst for hydrogenation of heavy oil fractionalization [67]. CNF supported cobalt catalysts have been used for the Fischer-Tropsch process [68]. CNF incorporated manganese dioxide have been prepared by electrospinning technique for vanadium redox battery flow for battery application to obtain high energy [69]. The fabrication of electrospun metal oxide-CNF composites have been reported as an electrochemical catalyst. The electrospinning technique has been used for the preparation of V_2O_3-CNF and pristine-CNF composites for redox flow for battery applications [70].

5.3.2 Functionalized nanocomposites

CNF have been incorporated in a polymer matrix to enhance electric conductivity, mechanical strength, surface area, etc. CNF provides a good interfacial area and reduces the possibility of defects because of the high aspect ratio and nanoscale diameter [71]. Two different ways are adapted to produce the functional nanocomposites are as follows; (1) the incorporation of nonmaterial into the polymer matrix and (2) the deposition of nanomaterial on wall surfaces of CNF by grafting or chemical modifications [72]. The properties of CNF/polymers composites depend on the distribution of CNF in the polymer matrix. Among all of the technique, the widest dispersion technique is melt mixing have been reported in the literature. CNF does not possess a perfect structure in comparison to carbon nanotubes, it has been showing promising properties in polymer nanocomposites, and metal-polymer matrix composites [73]. The incorporation of CNF in a polymer can boost the thermal conductivity of CNF-polymer composites [74]. The carbon nanomaterial highly influences the properties of nanocomposites. In epoxy resins-based nanocomposites consist monofilaments obtained by Ni or Fe, it has been analyzed the electric resistivity was decreased when it was obtained by Fe-based CNF [75].

5.3.3 Uranium capture

To fulfill the increasing energy demands, the use of nuclear power sources has been considered a sustainable power source with negligible carbon emission. Uranium a key element play an important role in energy production in nuclear power plant. For the environment and resources cause, the uranium at low concentration is toxic and nondegradable and at the current usage rate, the uranium in terrestrial ores is enough for a few decades. Therefore it is necessary to capture uranium from a renewable source such as seawater. The presence of uranium in seawater is very low concentration and different competing ions present makes it difficult to capture uranium from seawater. Different adsorbents have been used for uranium extraction from seawater. Recently tubular nanofibers (CNF) have gained interest for uranium capture because of their larger surface area, durability and stability at a wide range of pH. However, CNF undergoes agglomeration due to hydrophobic interactions. The dispersion of CNF in an aqueous solution was boosted by inserting hetero-groups by chemical modification. The various groups are incorporated on the CNF surface by various techniques. The widest method used for oxidation of CNF uses nitric acid to insert the functionality. The addition of functional groups on the surface of CNF enhances the dispersion in an aqueous solution due to the presence of various hydrophilic groups. The insertion of phosphate and organo-phosphorous groups onto CNF enhances uranium adsorption from

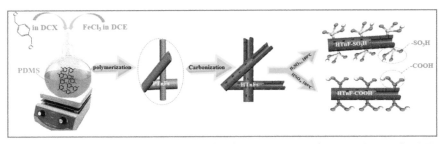

Figure 5.6 Systematic reaction pathway for the preparation of HCNF-SO$_3$H and HCNF-COOH. Reprinted (adapted) with permission from [77]. Copyright {2022} American Chemical Society.

the liquid phase with an adsorption capacity of 666 mg g^{-1} within a time of 24 h. Amidoxime-modified CNF have been used for uranium extraction. Wang et al. have synthesized Amidoxime modified carbon nanotubes using plasma technique for uranium extraction from nuclear effluents with adsorption capacity 145 mg g^{-1} at pH 4.5 [76].

5.4 Our contribution

Recently CNF were modified with nitric acid and sulphuric acid for grafting of carboxyl and sulphonic groups for uranium extraction from aqueous solution and step by step preparation shown in Fig. 5.6 [77–79]. CNF were obtained from polymer tubular nanofibers (PCNFs) from a dual oil phase system by a self-polymerization of a monomer DCX. Carboxyl and sulphonic groups have been grafted by chemical modification for uranium extraction with adsorption capacity of 1828 mg g^{-1} and 1929 mg g^{-1} from simulated water. Another study revealed that these CNF were functionalized with HNO$_3$ and subsequent loading of Fe$_3$O$_4$ was reported. The magnetic CNF were used for uranium extraction with maximum uranium extraction from an aqueous solution. The loading of Fe$_3$O$_4$ on the surface of CNF enhances the uranium extraction and separation of CNF from an aqueous solution. The SEM images showed a linear shape with two opening ends and the average diameter was about 100 nm (Fig. 5.7 A and D). TEM images of FMCNF-SO$_3$H and FMCNF-COOH showed porous tubular structure with a wall thickness of approximately 32 nm (Fig. 5.7 B and D). Also, the deposition of Fe$_3$O$_4$ on FMCNF-SO$_3$H and FMCNF-COOH does not alter morphology or damage the structure of CNF. The diameter of FMCNF-SO$_3$H and FMCNF-COOH was 160 nm and saturation magnetization is 21.61 and 19.44 emu g^{-1}.

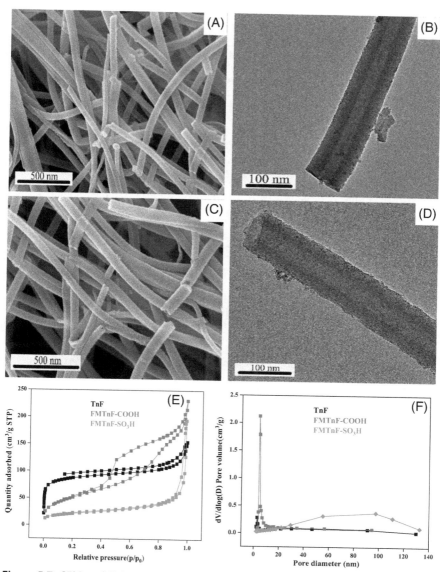

Figure 5.7 SEM and TEM images of FMCNF-SO₃H (A, B) and FMCNF-COOH (C, D); N₂ adsorption-desorption isotherm (E) and pore size distribution curves (F) of the samples. Reprinted (adapted) with permission from [77]. Copyright {2022} American Chemical Society.

References

[1] K.P. De Jong, J.W. Geus, Carbon nanofibers: Catalytic synthesis and applications, Catal. Rev.-Sci. Eng. (2000), doi:10.1081/CR-100101954.

[2] I.U. Din, M.S. Shaharun, A. Naeem, M.A. Alotaibi, A.I. Alharthi, M.A. Bakht, et al., Carbon nanofibers as potential materials for catalysts support, a mini-review on recent advances and future perspective, Ceram. Int. (2020), doi:10.1016/j.ceramint.2020.04.275.

[3] R.T.K. Baker, Catalytic growth of carbon filaments, Carbon N. Y. (1989), doi:10.1016/0008-6223(89)90062-6.

[4] J.C. Ruiz-Cornejo, D. Sebastián, M.J. Lázaro, Synthesis and applications of carbon nanofibers: A review, Rev. Chem. Eng. (2020), doi:10.1515/revce-2018-0021.

[5] Z. Osváth, A. Darabont, P. Nemes-Incze, E. Horváth, Z.E. Horváth, L.P. Biró, Graphene layers from thermal oxidation of exfoliated graphite plates, Carbon N. Y. (2007), doi:10.1016/j.carbon.2007.09.033.

[6] A. Aqel, K.M.M.A. El-Nour, R.A.A. Ammar, A. Al-Warthan, Carbon nanotubes, science and technology part (I) structure, synthesis and characterisation, Arab. J. Chem. (2012), doi:10.1016/j.arabjc.2010.08.022.

[7] O.R. Brown, Carbon-electrochemical and physicochemical properties, Electrochim. Acta. (1989), doi:10.1016/0013-4686(89)87066-5.

[8] D. Sebastián, A.G. Ruiz, I. Suelves, R. Moliner, M.J. Lázaro, On the importance of the structure in the electrical conductivity of fishbone carbon nanofibers, J. Mater. Sci. (2013), doi:10.1007/s10853-012-6893-1.

[9] T. Ozkan, M. Naraghi, I. Chasiotis, Mechanical properties of vapor grown carbon nanofibers, Carbon N. Y. (2010), doi:10.1016/j.carbon.2009.09.011.

[10] H. Zhu, X. Li, L. Ci, C. Xu, D. Wu, Z. Mao, Hydrogen storage in heat-treated carbon nanofibers prepared by the vertical floating catalyst method, Mater. Chem. Phys. (2003), doi:10.1016/S0254-0584(02)00233-X.

[11] N. Bayat, M. Rezaei, F. Meshkani, Methane decomposition over Ni-Fe/Al$_2$O$_3$ catalysts for production of COx-free hydrogen and carbon nanofiber, Int. J. Hydrogen Energy. (2016), doi:10.1016/j.ijhydene.2015.10.053.

[12] N.M. Rodriguez, A review of catalytically grown carbon nanofibers, J. Mater. Res. (1993), doi:10.1557/JMR.1993.3233.

[13] M. Ahmad, K. Yang, L. Li, Y. Fan, T. Shah, Q. Zhang, B. Zhang, Modified tubular carbon nanofibers for adsorption of uranium(VI) from water, ACS Appl. Nano Mater (2020), doi:10.1021/acsanm.0c00837.

[14] M.O. Danilov, A.V. Melezhyk, G.Y. Kolbasov, Carbon nanofibers as hydrogen adsorbing materials for power sources, J. Power Sources. (2008), doi:10.1016/j.jpowsour.2007.10.037.

[15] M. Pudukudy, Z. Yaakob, Methane decomposition over Ni, Co and Fe based monometallic catalysts supported on sol gel derived SiO$_2$ microflakes, Chem. Eng. J. (2015), doi:10.1016/j.cej.2014.10.077.

[16] R. Purohit, K. Purohit, S. Rana, R.S. Rana, V. Patel, Carbon nanotubes and their growth methods, Procedia Mater. Sci. (2014), doi:10.1016/j.mspro.2014.07.088.

[17] L. Zhang, A. Aboagye, A. Kelkar, C. Lai, H. Fong, A review: Carbon nanofibers from electrospun polyacrylonitrile and their applications, J. Mater. Sci. (2014), doi:10.1007/s10853-013-7705-y.

[18] L.S. Lobo, Nucleation and growth of carbon nanotubes and nanofibers: Mechanism and catalytic geometry control, Carbon N. Y. (2017), doi:10.1016/j.carbon.2016.12.005.

[19] D. Chen, K.O. Christensen, E. Ochoa-Fernández, Z. Yu, B. Tøtdal, N. Latorre, et al., Synthesis of carbon nanofibers: Effects of Ni crystal size during methane decomposition, J. Catal. (2005), doi:10.1016/j.jcat.2004.10.017.

[20] R.T.K. Baker, P.S. Harris, R.B. Thomas, R.J. Waite, Formation of filamentous carbon from iron, cobalt and chromium catalyzed decomposition of acetylene, J. Catal. (1973), doi:10.1016/0021-9517(73)90055-9.

[21] F. Yuan, H. Ryu, The synthesis, characterization, and performance of carbon nanotubes and carbon nanofibres with controlled size and morphology as a catalyst support material for a polymer electrolyte membrane fuel cell, Nanotechnology (2004), doi:10.1088/0957-4484/15/10/017.

[22] J. Jiao, P.E. Nolan, S. Seraphin, A.H. Cutler, D.C. Lynch, Morphology of carbon nanoclusters prepared by catalytic disproportionation of carbon monoxide, J. Electrochem. Soc. (1996), doi:10.1149/1.1836561.

[23] M. Li, N. Li, W. Shao, C. Zhou, Synthesis of carbon nanofibers by CVD as a catalyst support material using atomically ordered Ni$_3$C nanoparticles, Nanotechnology (2016), doi:10.1088/0957-4484/27/50/505706.

[24] Y. Hyun, J.Y. Choi, H.K. Park, C.S. Lee, Synthesis and electrochemical performance of ruthenium oxide-coated carbon nanofibers as anode materials for lithium secondary batteries, Appl. Surf. Sci. (2016), doi:10.1016/j.apsusc.2016.01.095.

[25] J.A. Díaz, M. Martínez-Fernández, A. Romero, J.L. Valverde, Synthesis of carbon nanofibers supported cobalt catalysts for Fischer-Tropsch process, Fuel (2013), doi:10.1016/j.fuel.2013.04.003.

[26] M. Li, R. Carter, A.P. Cohn, C.L. Pint, Interconnected foams of helical carbon nanofibers grown with ultrahigh yield for high capacity sodium ion battery anodes, Carbon N. Y. (2016), doi:10.1016/j.carbon.2016.05.051.

[27] G. Bin Zheng, K. Kouda, H. Sano, Y. Uchiyama, Y.F. Shi, H.J. Quan, A model for the structure and growth of carbon nanofibers synthesized by the CVD method using nickel as a catalyst, Carbon N. Y. (2004), doi:10.1016/j.carbon.2003.12.077.

[28] S. Lim, A. Shimizu, S.H. Yoon, Y. Korai, I. Mochida, High yield preparation of tubular carbon nanofibers over supported Co-Mo catalysts, Carbon N. Y. (2004), doi:10.1016/j.carbon.2004.01.027.

[29] A. Tanaka, S.H. Yoon, I. Mochida, Formation of fine Fe-Ni particles for the non-supported catalytic synthesis of uniform carbon nanofibers, Carbon N. Y. (2004), doi:10.1016/j.carbon.2004.01.029.

[30] Z. He, J.L. Maurice, A. Gohier, C.S. Lee, D. Pribat, C.S. Cojocaru, Iron catalysts for the growth of carbon nanofibers: Fe, Fe$_3$C or both? Chem. Mater. (2011), doi:10.1021/cm202315j.

[31] M. Inagaki, Y. Yang, F. Kang, Carbon nanofibers prepared via electrospinning, Adv. Mater. (2012), doi:10.1002/adma.201104940.

[32] R. Ruiz-Rosas, J. Bedia, M. Lallave, I.G. Loscertales, A. Barrero, J. Rodríguez-Mirasol, T. Cordero, The production of submicron diameter carbon fibers by the electrospinning of lignin, Carbon N. Y. (2010), doi:10.1016/j.carbon.2009.10.014.

[33] S. Megelski, J.S. Stephens, D.Bruce Chase, J.F. Rabolt, Micro- and nanostructured surface morphology on electrospun polymer fibers, Macromolecules (2002), doi:10.1021/ma020444a.

[34] L. Zhang, S. Gbewonyo, A. Aboagye, A.D. Kelkar, Development of carbon nanofibers from electrospinning, Nanotub. Superfiber Mater. Sci. Manuf. Commer. (2019), doi:10.1016/B978-0-12-812667-7.00033-1.

[35] J. Doshi, D.H. Reneker, Electrospinning process and applications of electrospun fibers, J. Electrostat. (1995), doi:10.1016/0304-3886(95)00041-8.

[36] Y. Yang, A. Centrone, L. Chen, F. Simeon, T. Alan Hatton, G.C. Rutledge, Highly porous electrospun polyvinylidene fluoride (PVDF)-based carbon fiber, Carbon N. Y. (2011), doi:10.1016/j.carbon.2011.04.015.

[37] B. Xu, F. Wu, R. Chen, G. Cao, S. Chen, Y. Yang, Mesoporous activated carbon fiber as electrode material for high-performance electrochemical double layer capacitors with ionic liquid electrolyte, J. Power Sources. (2010), doi:10.1016/j.jpowsour.2009.09.077.
[38] X. Mao, F. Simeon, G.C. Rutledge, T.A. Hatton, Electrospun carbon nanofiber webs with controlled density of states for sensor applications, Adv. Mater. (2013), doi:10.1002/adma.201203045.
[39] C. Kim, K.S. Yang, M. Kojima, K. Yoshida, Y.J. Kim, Y.A. Kim, et al., Fabrication of electrospinning-derived carbon nanofiber webs for the anode material of lithium-ion secondary batteries, Adv. Funct. Mater. (2006), doi:10.1002/adfm.200500911.
[40] G. Wei, X. Fan, J. Liu, C. Yan, Investigation of the electrospun carbon web as the catalyst layer for vanadium redox flow battery, J. Power Sources. (2014), doi:10.1016/j.jpowsour.2014.07.161.
[41] M. Kim, Y. Kim, K.M. Lee, S.Y. Jeong, E. Lee, S.H. Baeck, et al., Electrochemical improvement due to alignment of carbon nanofibers fabricated by electrospinning as an electrode for supercapacitor, Carbon N. Y. (2016), doi:10.1016/j.carbon.2015.12.068.
[42] L. Sabantina, M.Á. Rodríguez-Cano, M. Klöcker, F.J. García-Mateos, J.J. Ternero-Hidalgo, Al Mamun, et al., Fixing PAN nanofiber mats during stabilization for carbonization and creating novel metal/carbon composites, Polymers (Basel) (2018), doi:10.3390/polym10070735.
[43] G. Zhou, T. Xiong, S. Jiang, S. Jian, Z. Zhou, H. Hou, Flexible titanium carbide-carbon nanofibers with high modulus and high conductivity by electrospinning, Mater. Lett. (2016), doi:10.1016/j.matlet.2015.11.119.
[44] S.H. Park, C. Kim, K.S. Yang, Preparation of carbonized fiber web from electrospinning of isotropic pitch, Synth. Met. (2004), doi:10.1016/j.synthmet.2003.11.006.
[45] J.H. Jeong, B.H. Kim, Synergistic effects of pitch and poly(methyl methacrylate) on the morphological and capacitive properties of MnO2/carbon nanofiber composites, J. Electroanal. Chem. (2018), doi:10.1016/j.jelechem.2017.12.063.
[46] J. Ju, W. Kang, N. Deng, L. Li, Y. Zhao, X. Ma, et al., Preparation and character-ization of PVA-based carbon nanofibers with honeycomb-like porous structure via electro-blown spinning method, Microporous Mesoporous Mater. (2017), doi:10.1016/j.micromeso.2016.10.024.
[47] F.J. García-Mateos, T. Cordero-Lanzac, R. Berenguer, E. Morallón, D. Cazorla-Amorós, J. Rodríguez-Mirasol, et al., Lignin-derived Pt supported carbon (submi-cron)fiber electrocatalysts for alcohol electro-oxidation, Appl. Catal. B Environ. (2017), doi:10.1016/j.apcatb.2017.04.008.
[48] F.J. García-Mateos, R. Berenguer, M.J. Valero-Romero, J. Rodríguez-Mirasol, T. Cordero, Phosphorus functionalization for the rapid preparation of highly nanoporous submicron-diameter carbon fibers by electrospinning of lignin solutions, J. Mater. Chem. A. (2018), doi:10.1039/c7ta08788h.
[49] R.J. Beck, Y. Zhao, H. Fong, T.J. Menkhaus, Electrospun lignin carbon nanofiber membranes with large pores for highly efficient adsorptive water treatment applications, J. Water Process Eng. (2017), doi:10.1016/j.jwpe.2017.02.002.
[50] A.M. Al-Enizi, A.A. Elzatahry, A.M. Abdullah, A. Vinu, H. Iwai, S.S. Al-Deyab, High electrocatalytic performance of nitrogen-doped carbon nanofiber-supported nickel oxide nanocomposite for methanol oxidation in alkaline medium, Appl. Surf. Sci. (2017), doi:10.1016/j.apsusc.2017.01.038.
[51] Y. Dai, G. Zhu, X. Shang, T. Zhu, J. Yang, J. Liu, Electrospun zirconia-embedded carbon nanofibre for high-sensitive determination of methyl parathion, Electrochem. Commun. (2017), doi:10.1016/j.elecom.2017.05.017.
[52] Z.A. Qiao, S.H. Chai, K. Nelson, Z. Bi, J. Chen, S.M. Mahurin, et al., Polymeric molecular sieve membranes via in situ cross-linking of non-porous polymer membrane templates, Nat. Commun. (2014), doi:10.1038/ncomms4705.

[53] M. Seo, S. Kim, J. Oh, S.J. Kim, M.A. Hillmyer, Hierarchically porous polymers from hyper-cross-linked block polymer precursors, J. Am. Chem. Soc. (2015), doi:10.1021/ja511581w.

[54] L. Tan, B. Tan, Hypercrosslinked porous polymer materials: Design, synthesis, and applications, Chem. Soc. Rev. (2017), doi:10.1039/c6cs00851h.

[55] J.S.M. Lee, M.E. Briggs, T. Hasell, A.I. Cooper, Hyperporous carbons from hyper-crosslinked polymers, Adv. Mater. (2016), doi:10.1002/adma.201603051.

[56] C.D. Wood, T. Bien, A. Trewin, N. Hongjun, D. Bradshaw, M.J. Rosseinsky, et al., Hydrogen storage in microporous hypercrosslinked organic polymer networks, Chem. Mater. (2007), doi:10.1021/cm070356a.

[57] C.H. Lau, X. Mulet, K. Konstas, C.M. Doherty, M.A. Sani, F. Separovic, M.R. Hill, C.D. Wood, Hypercrosslinked additives for ageless gas-separation membranes, Angew. Chemie-Int. Ed. (2016), doi:10.1002/anie.201508070.

[58] J. Wang, Y. Huyan, Z. Yang, A. Zhang, Q. Zhang, B. Zhang, Tubular carbon nanofibers: Synthesis, characterization and applications in microwave absorption, Carbon N. Y. (2019), doi:10.1016/j.carbon.2019.06.048.

[59] Z. Zhang, J. Sun, M. Dou, J. Ji, F. Wang, Nitrogen and Phosphorus codoped meso-porous carbon derived from polypyrrole as superior metal-free electrocatalyst toward the oxygen reduction reaction, ACS Appl. Mater. Interfaces. (2017), doi:10.1021/acsami.7b03375.

[60] J. Wang, Z. Yang, M. Ahmad, H. Zhang, Q. Zhang, B. Zhang, A novel synthetic method for tubular nanofibers, Polym. Chem. (2019), doi:10.1039/c9py00612e.

[61] P. Serp, J.L. Figueiredo, Carbon Materials for Catalysis (2008), doi:10.1002/9780470403709.

[62] P. Serp, M. Corrias, P. Kalck, Carbon nanotubes and nanofibers in catalysis, Appl. Catal. A Gen. (2003), doi:10.1016/S0926-860X(03)00549-0.

[63] M. Tsuji, M. Kubokawa, R. Yano, N. Miyamae, T. Tsuji, M.S. Jun, et al., Fast preparation of PtRu catalysts supported on carbon nanofibers by the microwave-polyol method and their application to fuel cells, Langmuir (2007), doi:10.1021/la062223u.

[64] J.R.C. Salgado, E. Antolini, E.R. Gonzalez, Structure and activity of carbon-supported Pt - Co electrocatalysts for oxygen reduction, J. Phys. Chem. B. (2004), doi:10.1021/jp0486649.

[65] S. Eriksson, U. Nylén, S. Rojas, M. Boutonnet, Preparation of catalysts from microemul-sions and their applications in heterogeneous catalysis, Appl. Catal. A Gen. (2004), doi:10.1016/j.apcata.2004.01.014.

[66] Z. Liu, D. Fu, F. Liu, G. Han, C. Liu, Y. Chang, et al., Mesoporous carbon nanofibers with large cage-like pores activated by tin dioxide and their use in supercapacitor and catalyst support, Carbon N. Y. (2014), doi:10.1016/j.carbon.2014.01.011.

[67] J.L. Pinilla, I. Suelves, M.J. Lázaro, R. Moliner, J.M. Palacios, Influence of nickel crystal domain size on the behaviour of Ni and NiCu catalysts for the methane decomposition reaction, Appl. Catal. A Gen. (2009), doi:10.1016/j.apcata.2009.05.009.

[68] Z. Yu, Ø. Borg, D. Chen, B.C. Enger, V. Frøseth, E. Rytter, et al., Carbon nanofiber supported cobalt catalysts for Fischer-Tropsch synthesis with high activity and selectivity, Catal. Letters. (2006), doi:10.1007/s10562-006-0054-6.

[69] A. Di Blasi, C. Busaccaa, O. Di Blasia, N. Briguglio, G. Squadrito, V. Antonuccia, Synthesis of flexible electrodes based on electrospun carbon nanofibers with Mn_3O_4 nanoparticles for vanadium redox flow battery application, Appl. Energy (2017), doi:10.1016/j.apenergy.2016.12.129.

[70] C. Busacca, O. Di Blasi, N. Briguglio, M. Ferraro, V. Antonucci, A. Di Blasi, Electro-chemical performance investigation of electrospun urchin-like V_2O_3-CNF composite nanostructure for vanadium redox flow battery, Electrochim. Acta. (2017), doi:10.1016/j.electacta.2017.01.193.

[71] S. Rana, R. Alagirusamy, M. Joshi, Effect of carbon nanofiber dispersion on the tensile properties of epoxy nanocomposites, J. Compos. Mater. (2011), doi:10.1177/0021998311401076.

[72] A. Bhattacharyya, S. Rana, S. Parveen, R. Fangueiro, R. Alagirusamy, M. Joshi, Mechanical and thermal transmission properties of carbon nanofiber-dispersed carbon/phenolic multiscale composites, J. Appl. Polym. Sci. (2013), doi:10.1002/app.38947.

[73] E. Hammel, X. Tang, M. Trampert, T. Schmitt, K. Mauthner, A. Eder, et al., Carbon nanofibers for composite applications, Carbon N. Y. (2004), doi:10.1016/j.carbon.2003.12.043.

[74] A.L. Moore, A.T. Cummings, J.M. Jensen, L. Shi, J.H. Koo, Thermal conductivity measurements of nylon 11-carbon nanofiber nanocomposites, J. Heat Transfer. (2009), doi:10.1115/1.3139110.

[75] I. Suelves, R. Utrilla, D. Torres, S. De Llobet, J.L. Pinilla, M.J. Lázaro, et al., Preparation of polymer composites using nanostructured carbon produced at large scale by catalytic decomposition of methane, Mater. Chem. Phys. (2013), doi:10.1016/j.matchemphys.2012.10.026.

[76] Y. Wang, Z. Gu, J. Yang, J. Liao, Y. Yang, N. Liu, et al., Amidoxime-grafted multiwalled carbon nanotubes by plasma techniques for efficient removal of uranium(VI), Appl. Surf. Sci. (2014), doi:10.1016/j.apsusc.2014.08.182.

[77] M. Ahmad, F. Wu, Y. Cui, Q. Zhang, B. Zhang, Preparation of novel bifunctional magnetic tubular nanofibers and their application in efficient and irreversible Uranium Trap from Aqueous Solution, ACS Sustain. Chem. Eng. (2020), doi:10.1021/acssuschemeng.0c00332.

[78] M. Ahmad, R. Jianquan, X. Tao, N. Mehraj-ud-din, Z. Qiuyu, Z. Baoliang, A novel preparation and vapor phase modification of 2D-open channel bioadsorbent for uranium separation, J. AICHE (2022), doi:10.1002/aic.17884.

[79] M. Ahmad, R. Jianquan, Z. Yunfei, K. Hao, N. Mehraj-ud-din, Z. Qiuyu, Z. Baoliang, Simple and facile preparation of tunable chitosan tubular nanocomposite microspheres for fast uranium(VI) removal from seawater, J. Chem. Eng. (2022), doi:10.1016/j.cej.2021.130934.

CHAPTER 6

Uranium adsorption property of carboxylated tubular carbon nanofibers enhanced chitosan microspheres

Mudasir Ahmad[a,b] and Baoliang Zhang[a,b]
[a]School of Chemistry and Chemical Engineering, Northwestern Polytechnical University, Xi'an, China
[b]Xian Key Laboratory of Functional Organic Porous Materials, Northwestern Polytechnical University, Xi'an, China

6.1 Introduction

Modified carbon materials, combining one or more nanoparticles with one-dimensional carbon materials such as tubular carbon nanofibers (TCN) or carbon nanotubes have received much interest because of their unique combination of electric, thermal, mechanical, morphology, and numerous applications supporting high–temperature catalysis, energy storage, and environmental sustainability [1–8]. Carbon materials hold some highly desirable adsorbent characteristics that make them attractive for the extraction of radionuclide in the liquid phase [9]. Previous studies indicated that carbon nanotubes are potential candidates in the field of nuclear technology for preconcentration and solidification of lanthanides and actinides from large-scale water solutions [10,11]. With the significant contribution at the industrial scale application, the toxicity of carbon nanotube is considered [12]. Once they release into the environment, the biological life receive toxic effects and can affect humans, plants, and animals via entering the cell [13–15].

By surface modification, the oxygen-containing functional groups are linked on outer and inner tubular channels widely well-known are carboxylic species [16–18]. These functional groups can serve as attaching points for the chemical entities for molecular entities, polymer substances, and metal nanoparticles by in situ fabrication methods to obtain TCN/carbon nanotube-based modified nanocomposites with numerous functionality along with the tailored properties, while still preserving the intrinsic characteristics of carbon materials [19–21]. In these cases, surface functionalization

Fabrication and Functionalization of Advanced Tubular Nanofibers and their Applications.
DOI: https://doi.org/10.1016/B978-0-323-99039-4.00011-5
133

is of great importance for the subsequent chemical functionalization for a particular requirement. Presently there are various methods for introducing carboxylic (COOH) onto the surfaces of carbon materials. Each method has certain flaws and mostly shows some advantages [22]. The following methods are discussed in the coming text.

With regards to the fabrication of hybrid nanocomposites, numerous examples were found in the literature where carboxyl functional TCN have coupled chemical entities and biopolymers for the fabrication of nanohybrid one-directional and nanospheres respectively [23]. Wu et al. have developed a one-step strategy for carboxylic derivatization of vapor-grown carbon nanofibers (VGCNFs) with oxalyl chloride and their subsequent modification of an amino silsesquioxane. The structure of VGCNFs consists of staked cones and exposes the maximum edges towards the surface wall of the tubular channel walls that can be readily changed with the chemical functional groups to a larger extent. The covalent modification of TCN may lead to nonuniform distribution of functional groups that tend to attach the defective ends of the TCN. A major amount of hydrogen atoms exposed on the edges of benzene rings get replaced on the wall of TCN under Friedel craft acylation. The advantage of Friedel craft acylation in presence of aluminum chloride (catalyst) to obtain carboxylate TCN without the damage of TCN structure receive to strong acid treatments and the problems to the functionalization of TCN. The carboxylic acids are directly connected with surfaces of TCN in their place of separation by a nonconjugated moiety. The carboxyl TCN is then modified with octa-aminophenyl T8 (OASQ) via amide bond linkage which is reflected to be stable. This chemistry leads to the formation of a 3D hybrid nanostructure [22]. In our studies, we have developed a one-pot in situ fabrication method for the preparation of chitosan functionalized tubular carbon nanocomposite microsphere (CsFTnCM) as adsorbent for uranium extraction. The newly designed adsorbent was used for uranium extraction from seawater. In this application, the focus is not only to remove uranium completely from seawater but to recover as much as possible with limited amounts of adsorbent. The carboxyl-modified carbon materials enhance the functionality and form the base for nanospheres with a larger surface area that increases adsorption capacity, kinetics for uranium extraction. The uranium stripping from seawater exposed adsorbent would be highly expected with the loss of the adsorbent mass. In previous studies, we have prepared carboxyl-modified tubular nanofibers (FTn) and used them as an enhancer for uranium adsorption, protein, and microwave absorption. The inclusion of

FTn in chitosan is proven to increase improve the mechanical strength and stability of the chitosan-based microspheres in an aqueous solution [23].

6.2 Preparation and functionalization of tubular carbon nanofibers

6.2.1 Liquid-phase oxidation

Vapor-grown nanofibers (VGCNF) are prepared by catalytic dehydrogenation of hydrocarbons in a flow system under temperature conditions. Small size catalytic transition metal particles are obtained by spraying a metal solution $Fe(CO)_5$ as a mist inflow stream consisting of hydrocarbon, H_2S, and hydrogen [24,25]. The iron particles function as a nucleation center for each TCN and the diameter of the TCN is directly proportional to the diameter of the iron particles (10–15 nm) [26,27]. The iron particle diameter must be <15 nm, the considerable decrease in the activity is observed with the increase in the diameter greater than 15 nm [28]. Thus it is significant to keep the catalyst size smaller and avoid agglomeration in the flowing gas. At temperatures above 900, the iron particles begin extruding long slender hollow filaments of graphitic carbon and at higher temperatures 1050–1300°C the graphite structure follows high order. The CNT grows very fastly 1 mm min^{-1} and continues to grow in length up to several minutes until the iron particle either deactivates or escapes from the reaction zone [29]. The study carried out by Oberlin et al. confirmed the initial filaments consist of concentric conical cylinders of graphitic basal planes and almost in parallel orientation to the fiber axis [30]. The carbon filaments are thickened by continuous chemical vapor deposition which forms the outer turbostratic carbon layer (Fig. 6.1) [31]. Long-term exposure to chemical vapor deposition produced extra-thick carbon nanofibers with a tree-like morphology. In the liquid phase oxidation, Pyrograf III fibers (PR–19–PS) are used with a thickness of 200 nm. The small size VGCNF with a diameter of 0.06–0.3 nm and length 30–100 um requires small time in a flow system which reduces the preparation cost. Since the size of PR–19–PS is small but the analysis of individual properties is very difficult [32]. However, in composites studies, it is demonstrated that they have strong and high modulus reinforcing material [33,34]. In composites processing, the length of the fibers is reduced and the fiber distribution length in the composites is not known now.

Unlike the PAN and pitch-based fibers, VGCNF after chemical treatment the outer CVD turbostratic carbon layer suffers fiber formation with

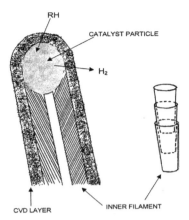

Figure 6.1 Preparation of VGCNF (A) and the concentric cones of graphitic planes (B) The copyright of Figure was obtained from Elsevier under license No: 5274150247233 dated Mar 22, 2022 [31].

the low functional group density. This might be the insufficient bonding with the composite materials made in the polymer. To increase the surface functionality of modification methods including the in situ modification in air and nitric acid at 425–500°C and 100 respectively [35]. The treatment yielded 2.5%, 4.6%, and 7.1% oxygen onto the surface as functional groups. The structural nature of the surface oxygen and the ratio of the sp^3 tetrahedral to the sp^2 aromatic ratio, the order, and size of the graphite mats depends on the surface treatment and temperature conditions [36]. The morphology and structure of VGCNF are very different from that of PAN, pitch, or rayon-based nanofibers. The largest portion of VGCNF consists of a graphite-like structure, however, the one-fifth portion of turbostratic carbon existed in the structure which is different in comparison to the well-ordered graphite carbon of the inner filaments. The turbostratic carbon is less ordered but stable at high temperatures. The surface functionality of VGCNF is less as compared to the TCN obtained from PAN, Pitch nanofibers. In nitric acid solution, VGCNF showed less weight loss as compared to the TCN obtained from PAN and is also stable at high temperatures. To carry the oxidation of VGCNF in nitric acid (69%–71%) in a round neck flask fitted with a condenser. The oxidation reaction temperature was set to 115°C at different time intervals of 10, 20, 30, 40, 50, 60, and 90 min. The long-term oxidation at time intervals 4, 10, and 24 h were also carried out at a set temperature of 115°C. The oxidized fibers were washed continuously with water until the pH of the solution reached neutral. The product is collected and dried under a vacuum oven for 24 h [31].

6.2.2 Oxygen-plasma treatment

The significant property of TCN due to the presence of covalent sp2 carbon bonds and tubular structure to larger lengths make them potential candidates in many fields including electronics, energy, and environmental sustainability. The efficient stability of TCN in an oxidative environment renders them not only catalyst support material but also used as a high-performance catalyst in the oxidation of hydrocarbons [37]. Processing in the oxidative environment or reshaping the carbon framework, or tailoring the physical, chemical, structural characteristics by introducing the oxygen groups in the carbon framework [38–41]. The oxygen-containing TCN can convert the different forms to enhance the adhesion, hydrophilicity, and selectivity for a particular function to meet the application demand [42–44]. In general, the oxidation process is much more complex and the introduction of oxygen groups and the gasification of various products mainly depends on the reaction conditions and the structure of the materials. Another issue is the defective sites of the carbon framework, receive the oxygen groups or other functional groups in the structure. For the optimal density of oxygen functional groups with negligible material loss via gasification, the oxygen loading is controlled very carefully and considering the structural characteristics, for example, graphite is much more stable than TCN, and the curved portion of the TCN receive more oxygen groups as compared to wall surfaces [45]. Although the oxidation of product is widely used to measure the size or the functionality of the material and only fewer studies have been reported to show the advanced oxygen states. The major challenge is the defective sites act as reaction centers or receives the other functional groups. Even high quality TCN consists the defects which can alter the electric, chemical, and transport properties [46–48]. Barinov et al. have used exposure to atomic oxygen at different temperatures to carry the oxidation reaction and the morphological changes of TCN [49]. Gas phase oxidation has advantage over acid, peroxides, or permanganate, as it doesn't generate the liquid as a waste. Using oxygen plasma or U-V photo oxidation instead of molecular oxygen is the efficient way for the oxidation reactions. Since the oxidation reaction happen at low temperature which decrease the level of gasification. The oxygen groups formed on the surface of product and their effect on the electron and morphological properties are analyzed using photo electronic spectroscopy. This study also showed the C1s image which illustrate the initial morphology of CNT/TCN, which is preserved upto the oxygen exposure of 2×10^{18} atoms cm^{-2} at temperature (20°C) [49].

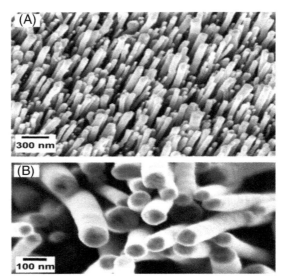

Figure 6.2 SEM images of VACNF at low magnification (A) and high magnification (B). Reprinted (adapted) with permission from [53]. Copyright {2022} American Chemical Society.

This indicated no defect is observed due to desorption of CO and CO_2 and efficient oxygen is loaded on the surface of CNT/CNG evidenced by XPS analysis [49]. To see the changes at 20°C much higher exposure is necessary. According to Mass spectrometry of HOPG, at higher temperature the gasification occurs very fastly, the co and CO_2 get evolved four times larger at 200°C than 20°C.

6.2.3 Photochemical grafting

In the previous studies, it is confirmed the hydrogen-terminated surfaces of diamond [50,51] or glassy carbon [52] will react with the alkene functionality when illuminated with a light at 254 nm. Baker et al. have performed this study with vertically aligned carbon nanofibers (VACNF) (Fig. 6.2) without carrying the hydrogen termination step [53].

All the experiments used the VACNF immediately after the preparation (Fig. 6.3). The experiments that use VAGNF after several days of preparation yield poor results. Previous reports show that the sidewall of VACNF has composed of several exposed graphite edges. In this study, the author supposed that the edges of the VACNF consist of terminated hydrogen. VACNF are placed in a nitrogen-filled quartz vessel and 5 uL of tert-butyloxycarbamate (tBOC) protected 10-aminodec-1-ene is dropped on the

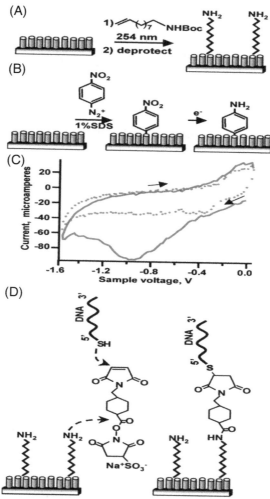

Figure 6.3 VACNF functionalization and amino-terminated surface (A), electrochemical functionalization to obtain amino-terminated surface (B), Cyclic voltammograms (C), Schematic of the method used to covalently link amino-modified VACNFs to thiomodified DNA(D). Reprinted (adapted) with permission from [53]. Copyright {2022} American Chemical Society.

surface forming a liquid film on the surface. Each substrate was subject to a quartz window, and nitrogen is supplied for several minutes and illuminated with a light of 254 nm from a mercury lamp under a continuous nitrogen supply. After a complete photochemical reaction, the nonspecifically attached alkene is removed from the surface of VACNF by dipping in

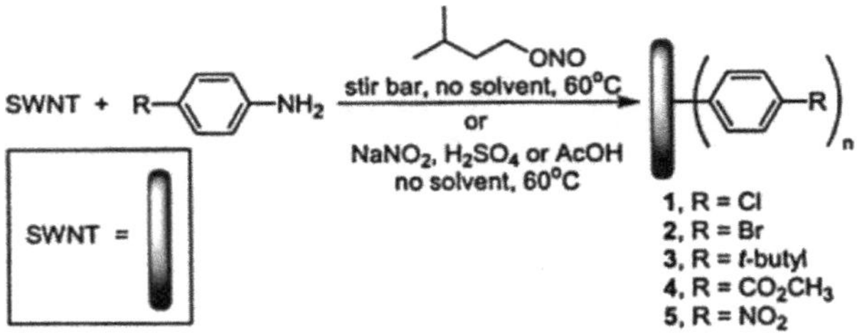

Figure 6.4 Solvent-free functionalization of SWCNT. Reprinted (adapted) with permission from [54], Copyright {2022} American Chemical Society.

chloroform and methanol alternately for 1 h. To avoid the complication of the tBOC-protected 10-aminodec-1-ene due to water or oxygen being stored in a dry place under a continuous nitrogen environment. However, the influence of contamination is still unknown. The photochemical functionalization reaction with the tBOC-protected 10-aminodec-1-ene is characterized using the XPS analysis before and after the functionalization. The N1s and O1s XPS spectra of fibers before functional shows the oxygen and nitrogen peaks in them, the XPS spectra after functionalization reaction reveals the changes in the peaks of N 1s and O 1s that are the characteristics of tBOC protecting group. The tBOC protecting group is detached by adding in a solution of trifluoroacetic acid:CH_2Cl_2 (1:1) for 1 h and renise with chloroform. Finally, the amine groups are neutralized by 10% ammonium hydroxide and keeping the surface at the termination with amine groups [53].

6.2.4 Electrochemical coupling

CNT and SCNT have attracted interest due to their physical, chemical, and as well as potential candidate to play important role in materials chemistry. For their potential to be understood, the intrinsic Wander wall's forces must be overcome to enhance the surface functionality, thus generating more compatible composites and soluble. The reduced diazonium salts and thermally produced diazonium compounds readily functionalized on the surface of CNT. However, the major limitation of all these methods needs a huge volume of solvents and sometimes sonication is necessary to dissolve a small amount of product. Christopher et al. report solvent-free functionalization, thereby leading the way for large-scale functionalization of CNT. The first solvent-free functionalization is reported of SWCNT under inert atmosphere, 6 g of SWCNT and 4-substituted aniline are added in a flask fitted with a condenser on a magnetic stirrer (Fig. 6.4). After that,

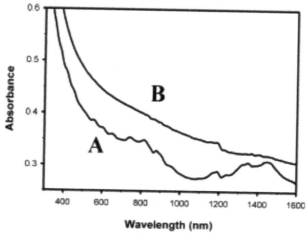

Figure 6.5 Absorption spectra in DMF of SWNT (A) and aryl chloride synthesized solvent-free technique (B). Reprinted (adapted) with permission from [54], Copyright {2022} American Chemical Society.

0.3 mL of isoamyl nitrite is added to the reaction mixture via syringe, heat the solution at 60°C, and stir vigorously of the paste were commenced. Then next, the paste is diluted with DMF and filtered with PTFE filter paper (0.45 um). The solid product is collected and sonicated in DMF and washed with DMF until the filtrate comes out colorless. The solid product is sonicated and filtrated to yield functionalized nanotubes. The DMF from the product is removed with ether and the compound (1–5) is dried in an oven (65°C) and characterized. U–Visible/NIR absorption spectroscopy, Raman spectroscopy, and thermogravimetric (TGA) analysis results are similar to the previously reported analyses obtained using a solvent (Fig. 6.5 and Fig. 6.6) [54].

6.2.5 Addition of carboxylate (coo−) containing alkyl radicals

Chemical modification of TCN by covalent and noncovalent bonding of organic/inorganic molecules have been studied vastly to improve its dispersion in a liquid phase and polymer matrix [55]. To such an extent, TCN can individualize, position, and be oriented to yield applications in various fields such as electronics. For such as reason, TCN have been modified with poly(2-vinyl pyridine) (P2VP) by covalent bonding technique [56]. In acid conditions P2VP gets protonated, which makes it readily soluble, to deposit it selectively on the opposite charge and to use the polyelectrolyte shell as a template for the deposition of the metal cluster. Recently some studies

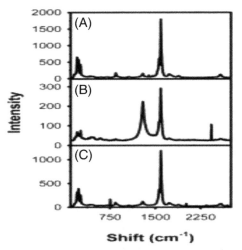

Figure 6.6 Raman spectra of SWNT (A) and aryl chloride synthesized solvent-free technique (B). Reprinted (adapted) with permission from [54], Copyright {2022} American Chemical Society.

have reported the loading of magnetic particles onto the TCN by stacking of carboxylic pyridine followed by magnetic particle loading [57]. Korneva et al. show the new synthetic strategy for the preparation of magnetic TCN by filling magnetic iron oxide [57]. Such materials are easy to handle in a magnetic field. The magnetic carbon nanotube with a larger diameter have been used and the magnetic particle is not available on to the outer surface of the TCN, which makes them unavailable for the other reaction and modification. Stoffelbach et al. reports a simple and cheap method for the decoration of TCN and their orientation under a magnetic environment. In the first step, the grafting of magnetic nanoparticles onto the TCN by radical grafting method. In this method 4,49–azobis(4–cyanovaleric acid) (V501) is used as an initiator, which releases carboxylate(COO^-) comprising alkyl radical under thermolysis. Complexing carboxylic groups TCN for the magnetic particles which ultimately linked with TCN by the radical addition process. In recent some other works reports the addition of carboxylic groups by heating of TCN under concentrated acid to yield COOH- TCN, which results a severe decrease in tube length. Moreover the COOH groups are attached without the spacer. In recent nitrous oxide (initiator) mediated polymerization for the grafting g of polyacrylic acid and amphiphilic diblock on TCN. In comparison to MWCNT heated in cheap V501 under heat at 80°C throughout the night, $C(CH_3)(CN)CH_2CH_2COO^-$, Na$^+$ radicals is directly attached to the CNT surface without any effect such as the shortening in length of the tube [58].

6.2.6 Diels–alder (da) cycloaddition reaction

Carbon nanomaterial shows efficient mechanical, electrical, and thermal properties due to the presence of sp^2 hybridized carbon in the skeleton. However, these materials show aggregation which leads to a decrease in the properties in composites [59]. Different strategies have been followed to increase the dispersion of TCN in a polymer matrix. In thermoset and thermoplastic polymers the dispersion of TCN is carried out using sonication, high-speed mixing, and calendaring [60–64]. Other approaches deal with noncovalent interaction for surface functionalization of TCN. Wrapping of TCN surfaces or the addition of polymer chains enhances the dispersion in different solvents [65]. The attachment of polymer chains to the carbon nanofibers has led to enhancing the reinforcement of polymer nanocomposites [66,67]. The covalent bond creation destroys the sp^2 hybridized carbon skeleton which results from the decrease of optoelectrical property. The control of functionalization on TCN which results in less effect on their properties is significant for the preparation of high-performance nanocomposites. In literature, two methods have been adopted for the covalent bonding of polymer chains to any substrate known as "grafting to" or "grafting from" [68]. In many studies, several steps need to carry the functionalization of TCN with polymer chains need many harmful chemicals including thionyl choloride [69–72]. Single-step polymerization, like in situ polymerization in presence of TCN or "graft to" the functional polymers is the easiest way for the preparation of nanocomposites. However, the functionalization of TCN and their dispersion in a polymer matrix is difficult [73]. Therefore, it is necessary to develop a method for controlled functionalization of TCN and postsynthetic complexation by using the extrusion, or injection method are the ideal methods for the preparation of such materials. One of the facile approaches is adapted for the function of the graphene base structure of TCN is the Diels-Alder addition (DA) reaction, which cannot produce enough byproducts and can undergo mild conditions [74]. The possibility of DA is analyzed from the last two decades. It is well studied the DA reaction is carried on CNT either through micro-wave assisted, or fluorinated TCN, or by metal-catalyzed under high-pressure conditions [75]. However, these reaction methods are carried out by using reactants without further functional groups. The benzocyclobutane based compounds have been used for the DA reaction method for TCN but this method is difficult and increases the chance of side reaction and enough byproducts may occur. Recently the DA reaction is found with MWCNT but the reaction is delicate and vulnerable to temperature and dienophiles like oxygen. Munirasu et al. report the DA for the carbonous materials including CNT, MWCNT, and Herringbone carbon nanofibers using dienophiles and diene functional groups [59]. The reactivity of carbonous materials as diene or dienophile with the suitable components as compared with the small molecules DA reaction for the first time. In this study, the author shows simple and facile furfuryl-based ATRP as an initiator "graft from" polymerization of styrene of MWCNT. In this study CNT (50 mg), furfuryl alcohol (0.5 mL), and then 5 mL anisole are charged in Schlenk tube. The reaction tube is charged with argon for 30 min and closed tightly using the septum. The reaction temp is heated at 150°C for 24 h. after completion the product is washed with THF until furfuryl alcohol is removed [59].

6.3 Chitosan enhanced COOH-TCN for uranium adsorption

Bioadsorbent plays a significant role to trap the metal ions from aqueous solution either covalent bonding or by wander wall's force or by ion exchange process [76,77]. Chitosan (Cs), a natural biopolymer with a linear structure obtained from chitin the second most biopolymer on earth used as an adsorbent for metal, dyes, and other pollutants from water [78]. Chitosan has been used as an adsorbent, however, some of the active sites are restricted for bonding with the metal ions by inter-and/ intramolecular interactions existing in natural Cs [79]. The major disadvantage of chitosan-based adsorbents is less stability and low surface area, which results from slow adsorption kinetics and low adsorption capacity [80]. In several reports, the stability of Cs is increased by crosslinking method and the crosslinker used in some studies are glutaraldehyde, polyacrylonitrile, and tripolyphosphate [81–83]. Due to the toxic effect of crosslinking, such reagents are restricted due to posing serious health problems in the environment [84]. The increased surface area not only increases the adsorption capacity but also increases the adsorption kinetics. Various methods are used for the preparation of chitosan-based microsphere (CsM) including spray drying, cryotropic-gelation, precipitation, and emulsion techniques [85,86].

Recently few studies have reported, the adsorption capacity of chitosan-based adsorbents have been enhanced by the preparation of chitosan-based nanocomposites. The chitosan-based nanocomposites are obtained by mixing various fillers such as Fe_3O_4, TiO_2, ZrO_2, SiO_2 [87,88] to enhance

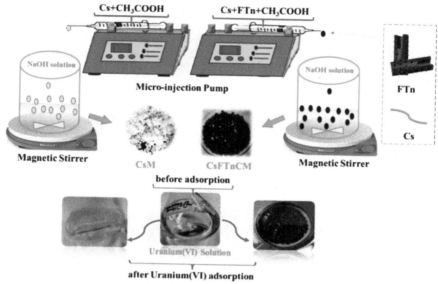

Figure 6.7 Preparation of adsorbent and adsorption of uranium. The copyright of Figure was obtained from Elsevier under license No: 5274270401043, Mar 22, 2022, [23].

the stability in wide pH range, mechanical property to avoid the adsorbent loss and adsorption capacity [89]. It is reported that COOH functionalized TCN (COOH- TCN) as an efficient uranium adsorbent [90], however, the separation of uranium is quite difficult and expensive. Some works show in the literature magnetic TCN have been used for uranium extraction, but the loading of magnetic nanoparticles onto COOH- TCN decreases the adsorption sites [91].

Ahmad et al. develops a simple and facile preparation of functionalized TCN -chitosan nanocomposites (CsFTnCM) by the alkali cracking process. Functional TCN (0.6 g), and chitosan (0.5 g) are dissolved in 50 mL of the dilute acid solution for 1 h. Then after transfer the chitosan solution in dropwise with a flow rate of 0.5 mL min^{-1} in alkaline solution. The obtained nanospheres are kept in an alkaline medium for 24 h and washed several times with distilled water until the pH of a solution containing microsphere equals neutral. The obtained product is dried by the freeze–drying technique. Microspheres are kept in the freezer for 24 h and direct evaporation is done by keeping them in an oven (Fig. 6.7). The structural morphology and the geometry are changed by changing the ratio of functional-TCN (Fig. 6.8). With increasing the functional-TCN ratio 0–0.5 g a smooth surface CsFTnCM was obtained and further increasing results cracks on

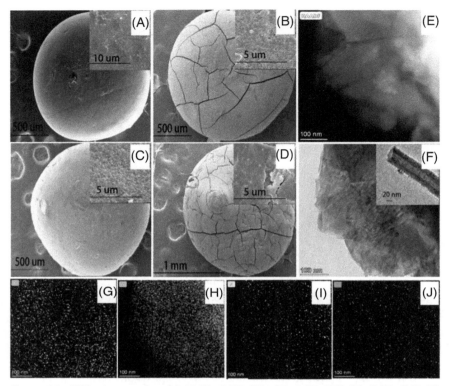

Figure 6.8 SEM micrographs of CsM (A), CsFTnCM1 (B), CsFTnCM2 (C), CsFTnCM3 (D), TEM micrographs of CsM (E), CsFTnCM2 (F) and optical micrographs of CsFTnCM2 (C, O, N and U G–J). The copyright of Figure was obtained from Elsevier under license No: 5274270401043, Mar 22, 2022 [23].

the surface. For uranium adsorption from seawater, adsorption kinetics play a significant role as the uranium is present at low concentration (3.3 ppb), various competing ions, and loading of various organic contents are present. The uranium adsorption from seawater requires advanced adsorbents and the experimental analysis confirms CsFTnCM as an efficient adsorbent for uranium adsorption. A time-dependent study is carried out to determine the kinetics of uranium using one sight ligand saturation model (Fig. 6.9A–C) and uranium adsorption results show adsorption capacity of uranium in 10 days seawater exposure is 0.665 mg g^{-1} with half-saturation of 0.1 (day). The selectivity of uranium in seawater was also observed and the percentage of various metal ions is analyzed with a 10-day adsorption experiment. It is measured uranium is adsorbed at the fourth rank in presence of other competing ions (Fig. 6.9B–C) [23].

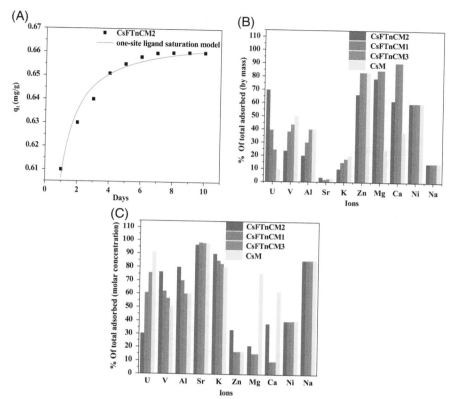

Figure 6.9 Time-dependent uranium adsorption capacity (A), a relative abundance of various ions by mass % (B), and molar % concentration (C). The copyright of Figure was obtained from Elsevier under license No: 5274270401043, Mar 22, 2022 [23].

References

[1] X. Wang, E.C. Landis, R. Franking, R.J. Hamers, Surface chemistry for stable and smart molecular and biomolecular interfaces via photochemical grafting of alkenes, Acc. Chem. Res. (2010), doi:10.1021/ar100011f.

[2] M.Q. Zhao, X.F. Liu, Q. Zhang, G.L. Tian, J.Q. Huang, W. Zhu, F. Wei, Graphene/single-walled carbon nanotube hybrids: One-step catalytic growth and applications for high-rate Li-S batteries, ACS Nano (2012), doi:10.1021/nn304037d.

[3] Y. Yu, L. Gu, C. Zhu, P.A. Van Aken, J. Maier, Tin nanoparticles encapsulated in porous multichannel carbon microtubes: Preparation by single-nozzle electrospinning and application as anode material for high-performance Li-based batteries, J. Am. Chem. Soc. (2009), doi:10.1021/ja906261c.

[4] J. Wu, K. Tian, J. Wang, Adsorption of uranium by amidoxime modified multiwalled carbon nanotubes, Prog. Nucl. Energy. (2018), doi:10.1016/j.pnucene.2018.02.020.

[5] M. Ahmad, R. Jianquan, X. Tao, N. Mehraj-ud-din, Z. Qiuyu, Z. Baoliang, A Novel Preparation and Vapour Phase Modification of 2D-open Channel Bio-adsorbent for Uranium Separation, AICHE Journal (2022), doi:10.1002/aic.17884.

[6] M. Ahmad, C. Junjie, Y. Ke, S. Tariq, N. Mehraj-ud-din, Z. Qiuyu, Z. Baoliang, Preparation of amidoxime modified porous organic polymer flowers for selective uranium recovery from seawater, Chem. Eng. J. (2021), doi:10.1016/j.cej.2021.129370.

[7] M. Ahmad, W. Jiqi, Y. Zuoting, Z. Qiuyu, Z. Baoliang, Ultrasonic-assisted Preparation of Amidoxime Functionalized Silica Framework Via Oil-Water Emulsion Method for Selective Uranium Adsorption, Chem. Eng. J. (2020), doi:10.1016/j.cej.2020.124441.

[8] M. Ahmad, W. Jiqi, X. Jia, Z. Qiuyu, Z. Baoliang, Magnetic tubular carbon nanofibers as efficient Cu(II) ion adsorbent from wastewater, J. Clean. Prod. (2020), doi:10.1016/j.jclepro.2019.119825.

[9] C.M. Hussain, C. Saridara, S. Mitra, Microtrapping characteristics of single and multi-walled carbon nanotubes, J. Chromatogr. A. (2008), doi:10.1016/j.chroma.2008.01.073.

[10] C. Chen, X. Li, D. Zhao, X. Tan, X. Wang, Adsorption kinetic, thermodynamic and desorption studies of Th(IV) on oxidized multi-wall carbon nanotubes, Colloids Surfaces A Physicochem. Eng. Asp. (2007), doi:10.1016/j.colsurfa.2007.03.007.

[11] X. Wang, C. Chen, W. Hu, A. Ding, D. Xu, X. Zhou, Sorption of 243Am(III) to multiwall carbon nanotubes, Environ. Sci. Technol. (2005), doi:10.1021/es048287d.

[12] C. Chen, J. Hu, D. Xu, X. Tan, Y. Meng, X. Wang, Surface complexation modeling of Sr(II) and Eu(III) adsorption onto oxidized multiwall carbon nanotubes, J. Colloid Interface Sci. (2008), doi:10.1016/j.jcis.2008.04.046.

[13] B. Nowack, T.D. Bucheli, Occurrence, behavior and effects of nanoparticles in the environment, Environ. Pollut. (2007), doi:10.1016/j.envpol.2007.06.006.

[14] M. Davoren, E. Herzog, A. Casey, B. Cottineau, G. Chambers, H.J. Byrne, F.M. Lyng, In vitro toxicity evaluation of single-walled carbon nanotubes on human A549 lung cells, Toxicol. Vitr. (2007), doi:10.1016/j.tiv.2006.10.007.

[15] A. Simon-Deckers, B. Gouget, M. Mayne-L'Hermite, N. Herlin-Boime, C. Reynaud, M. Carrière, In vitro investigation of oxide nanoparticle and carbon nanotube toxicity and intracellular accumulation in A549 human pneumocytes, Toxicology (2008), doi:10.1016/j.tox.2008.09.007.

[16] K.L. Klein, A.V. Melechko, T.E. McKnight, S.T. Retterer, P.D. Rack, J.D. Fowlkes, D.C. Joy, M.L. Simpson, Surface characterization and functionalization of carbon nanofibers, J. Appl. Phys. (2008), doi:10.1063/1.2840049.

[17] K. Balasubramanian, M. Burghard, Chemically functionalized carbon nanotubes, Small (2005), doi:10.1002/smll.200400118.

[18] A.J. Plomp, D.S. Su, K.P.D. Jong, J.H. Bitter, On the nature of oxygen-containing surface groups on carbon nanofibers and their role for platinum depositionsan XPS and titration study, J. Phys. Chem. C. (2009), doi:10.1021/jp900637q.

[19] X. Guo, J.P. Small, J.E. Klare, Y. Wang, M.S. Purewal, I.W. Tam, et al., Covalently bridging-gaps in single-walled carbon nanotubes with conducting molecules, Science (80) (2006), doi:10.1126/science.1120986.

[20] S.K. Yadav, S.S. Mahapatra, J.W. Cho, Synthesis and application of conducting polyaniline-Fe_3O_4 nanohybrid by click chemistry reaction, Text. Sci. Eng. (2013), doi:10.12772/tse.2013.50.345.

[21] Y. Jiao, A. Tibbits, A. Gillman, M.S. Hsiao, P. Buskohl, L.F. Drummy, et al., Deformation behavior of polystyrene-grafted nanoparticle assemblies with low grafting density, Macromolecules (2018), doi:10.1021/acs.macromol.8b01524.

[22] J. Wu, H. Cai, K. Xu, Z. Fu, X. Liu, M. Chen, A facile route to carboxylated carbon nanofibers and their functionalization by robust octasilsesquioxanes via conjugated linkage, J. Mater. Chem. C. (2013), doi:10.1039/c2tc00134a.

[23] M. Ahmad, J. Ren, Y. Zhang, H. Kou, M. ud din Naik, Q. Zhang, B. Zhang, Simple and facile preparation of tunable chitosan tubular nanocomposite microspheres for fast uranium removal from seawater, Chem. Eng. J. 427 (2022), doi:10.1016/j.cej.2021.130934.

[24] R.L. Jacobsen, T.M. Tritt, J.R. Guth, A.C. Ehrlich, D.J. Gillespie, Mechanical properties of vapor-grown carbon fiber, Carbon N. Y. (1995), doi:10.1016/0008-6223(95)00057-K.

[25] G.G. Tibbetts, M. Endo, C.P. Beetz, Carbon fibers grown from the vapor phase: a novel material, SAMPE J. (1986), doi:10.1016/0010-4361(87)90302-8.

[26] G.G. Tibbetts, G.L. Doll, D.W. Gorkiewicz, J.J. Moleski, T.A. Perry, C.J. Dasch, et al., Physical properties of vapor-grown carbon fibers, Carbon N. Y. (1993), doi:10.1016/0008-6223(93)90054-E.

[27] G.G. Tibbetts, Why are carbon filaments tubular? J. Cryst. Growth. (1984), doi:10.1016/0022-0248(84)90163-5.

[28] F. Benissad, P. Gadelle, M. Coulon, L. Bonnetain, Formation of carbon fibres from methane. II, Germination of Carbon and Fusion of the Catalyst Particles 26 (4) (1988) 425–432.

[29] A. Madroñero, E. Ariza, M. Verdú, W. Brandl, C. Barba, Some microstructural aspects of vapour-grown carbon fibres to disclose their failure mechanisms, J. Mater. Sci. (1996), doi:10.1007/bf00354437.

[30] A. Oberlin, M. Endo, T. Koyama, Filamentous growth of carbon through benzene decomposition, J. Cryst. Growth. (1976), doi:10.1016/0022-0248(76)90115-9.

[31] P.V. Lakshminarayanan, H. Toghiani, C.U. Pittman, Nitric acid oxidation of vapor grown carbon nanofibers, Carbon N. Y. (2004), doi:10.1016/j.carbon.2004.04.040.

[32] R.D. Patton, C.U. Pittman, L. Wang, J.R. Hill, Vapor grown carbon fiber composites with epoxy and poly(phenylene sulfide) matrices, Compos. Part A Appl. Sci. Manuf. (1999), doi:10.1016/S1359-835X(99)00018-4.

[33] R.D. Patton, C.U. Pittman, L. Wang, J.R. Hill, A. Day, Ablation, mechanical and thermal conductivity properties of vapor grown carbon fiber/phenolic matrix composites, Compos.-Part A Appl. Sci. Manuf. (2002), doi:10.1016/S1359-835X(01)00092-6.

[34] S. Kumar, H. Doshi, M. Srinivasarao, J.O. Park, D.A. Schiraldi, Fibers from polypropylene/nano carbon fiber composites, Polymer (Guildf) (2002), doi:10.1016/S0032-3861(01)00744-3.

[35] H. Darmstadt, C. Roy, S. Kaliaguine, J.M. Ting, R.L. Alig, Surface spectroscopic analysis of vapour grown carbon fibres prepared under various conditions, Carbon N. Y. (1998), doi:10.1016/S0008-6223(98)00096-7.

[36] H. Darmstadt, L. Sümmchen, J.M. Ting, U. Roland, S. Kaliaguine, C. Roy, Effects of surface treatment on the bulk chemistry and structure of vapor grown carbon fibers, Carbon N. Y. (1997), doi:10.1016/S0008-6223(97)00116-4.

[37] J. Zhang, X. Liu, R. Blume, A. Zhang, R. Schlögl, S.S. Dang, Surface-modified carbon nanotubes catalyze oxidative dehydrogenation of n-butane, Science (80-.) (2008), doi:10.1126/science.1161916.

[38] P.X. Hou, C. Liu, H.M. Cheng, Purification of carbon nanotubes, Carbon N. Y. (2008), doi:10.1016/j.carbon.2008.09.009.

[39] T.W. Ebbesen, P.M. Ajayan, H. Hiura, K. Tanigaki, Purification of nanotubes [6], Nature (1994), doi:10.1038/367519a0.

[40] S.C. Tsang, P.J.F. Harris, M.L.H. Green, Thinning and opening of carbon nanotubes by oxidation using carbon dioxide, Nature (1993), doi:10.1038/362520a0.

[41] P.M. Ajayan, T.W. Ebbesen, T. Ichihashi, S. Iijima, K. Tanigaki, H. Hiura, Opening carbon nanotubes with oxygen and implications for filling, Nature (1993), doi:10.1038/362522a0.

[42] H.C. Schniepp, J.L. Li, M.J. McAllister, H. Sai, M. Herrera-Alonson, D.H. Adamson, et al., Functionalized single graphene sheets derived from splitting graphite oxide, J. Phys. Chem. B. (2006), doi:10.1021/jp060936f.

[43] M.S.U. Sarwar, Z. Xiao, T. Saleh, A. Nojeh, K. Takahata, Micro glow plasma for localized nanostructural modification of carbon nanotube forest, Appl. Phys. Lett. (2016), doi:10.1063/1.4961629.

[44] C.H. Tseng, C.C. Wang, C.Y. Chen, Functionalizing carbon nanotubes by plasma modification for the preparation of covalent-integrated epoxy composites, Chem. Mater. (2007), doi:10.1021/cm062277p.

[45] R. Brukh, S. Mitra, Kinetics of carbon nanotube oxidation, J. Mater. Chem. (2007), doi:10.1039/b609218g.

[46] A. Hashimoto, K. Suenaga, A. Gloter, K. Urita, S. Iijima, Direct evidence for atomic defects in graphene layers, Nature (2004), doi:10.1038/nature02817.

[47] M. Ishigami, H.J. Choi, S. Aloni, S.G. Louie, M.L. Cohen, A. Zettl, Identifying defects in nanoscale materials, Phys. Rev. Lett. (2004), doi:10.1103/PhysRevLett.93.196803.

[48] L. Aballe, A. Barinov, A. Locatelli, S. Heun, M. Kiskinova, Tuning surface reactivity via electron quantum confinement, Phys. Rev. Lett. (2004), doi:10.1103/PhysRevLett.93.196103.

[49] A. Barinov, L. Gregoratti, P. Dudin, S. La Rosa, M. Kiskinova, Imaging and spectroscopy of multiwalled carbon nanotubes during oxidation: Defects and oxygen bonding, Adv. Mater. (2009), doi:10.1002/adma.200803003.

[50] W. Yang, O. Auciello, J.E. Butler, W. Cai, J.A. Carlisle, J. Gerbi, et al., DNA-modified nanocrystalline diamond thin-films as stable, biologically active substrates, Nat. Mater. (2002), doi:10.1038/nmat779.

[51] M. Lu, T. Knickerbocker, W. Cai, W. Yang, R.J. Hamers, L.M. Smith, Invasive Cleavage Reactions on DNA-Modified Diamond Surfaces, Biopolymers (2004), doi:10.1002/bip.20007.

[52] T.L. Lasseter, W. Cai, R.J. Hamers, Frequency-dependent electrical detection of protein binding events, Analyst (2004), doi:10.1039/b307591e.

[53] S.E. Baker, K.Y. Tse, E. Hindin, B.M. Nichols, T.L. Clare, R.J. Hamers, Covalent functionalization for biomolecular recognition on vertically aligned carbon nanofibers, Chem. Mater. (2005), doi:10.1021/cm051024d.

[54] C.A. Dyke, J.M. Tour, Solvent-free functionalization of carbon nanotubes, J. Am. Chem. Soc. (2003), doi:10.1021/ja0289806.

[55] Carbon nanotubes, Advanced topics in the synthesis, structure, properties and applications, Mater. Today (2008), doi:10.1016/s1369-7021(08)70021-x.

[56] X. Lou, C. Detrembleur, C. Pagnoulle, R. Jérôme, V. Bocharova, A. Kiriy, et al., Surface modification of multiwalled carbon nanotubes by poly(2-vinylpyridine): Dispersion, selective deposition, and decoration of the nanotubes, Adv. Mater. (2004), doi:10.1002/adma.200400298.

[57] V. Georgakilas, V. Tzitzios, D. Gournis, D. Petridis, Attachment of magnetic nanoparticles on carbon nanotubes and their soluble derivatives, Chem. Mater. (2005), doi:10.1021/cm0483590.

[58] F. Stoffelbach, A. Aqil, C. Jérôme, R. Jérôme, C. Detrembleur, An easy and economically viable route for the decoration of carbon nanotubes by magnetite nanoparticles, and their orientation in a magnetic field, Chem. Commun. (2005), doi:10.1039/b506758h.

[59] S. Munirasu, J. Albuerne, A. Boschetti-de-Fierro, V. Abetz, Functionalization of carbon materials using the diels-alder reaction a, Macromol. Rapid Commun. (2010), doi:10.1002/marc.200900751.

[60] Y.H. Liao, O. Marietta-Tondin, Z. Liang, C. Zhang, B. Wang, Investigation of the dispersion process of SWNTs/SC-15 epoxy resin nanocomposites, Mater. Sci. Eng. A. (2004), doi:10.1016/j.msea.2004.06.031.

[61] T. Villmow, P. Pötschke, S. Pegel, L. Häussler, B. Kretzschmar, Influence of twin-screw extrusion conditions on the dispersion of multi-walled carbon nanotubes in a poly(lactic acid) matrix, Polymer (Guildf) (2008), doi:10.1016/j.polymer.2008.06.010.

[62] G. Broza, K. Schulte, Melt processing and filler/matrix interphase in carbon nanotube reinforced poly(ether-ester) thermoplastic elastomer, Polym. Eng. Sci. (2008), doi:10.1002/pen.21075.

[63] M. Moniruzzaman, F. Du, N. Romero, K.I. Winey, Increased flexural modulus and strength in SWNT/epoxy composites by a new fabrication method, Polymer (Guildf) (2006), doi:10.1016/j.polymer.2005.11.011.

[64] F.H. Gojny, M.H.G. Wichmann, B. Fiedler, I.A. Kinloch, W. Bauhofer, A.H. Windle, et al., Evaluation and identification of electrical and thermal conduction mechanisms in carbon nanotube/epoxy composites, Polymer (Guildf) (2006), doi:10.1016/j.polymer.2006.01.029.

[65] X. Xin, G. Xu, T. Zhao, Y. Zhu, X. Shi, H. Gong, et al., Dispersing carbon nanotubes in aqueous solutions by a starlike block copolymer, J. Phys. Chem. C. (2008), doi:10.1021/jp8059344.

[66] X.H. Men, Z.Z. Zhang, H.J. Song, K. Wang, W. Jiang, Functionalization of carbon nanotubes to improve the tribological properties of poly(furfuryl alcohol) composite coatings, Compos. Sci. Technol. (2008), doi:10.1016/j.compscitech.2007.07.008.

[67] B.X. Yang, J.H. Shi, K.P. Pramoda, S.H. Goh, Enhancement of stiffness, strength, ductility and toughness of poly(ethylene oxide) using phenoxy-grafted multiwalled carbon nanotubes, Nanotechnology (2007), doi:10.1088/0957-4484/18/12/125606.

[68] D. Tasis, N. Tagmatarchis, A. Bianco, M. Prato, Chemistry of carbon nanotubes, Chem. Rev. (2006), doi:10.1021/cr050569o.

[69] Q.P. Feng, X.M. Xie, Y.T. Liu, W. Zhao, Y.F. Gao, Synthesis of hyperbranched aromatic polyamide-imide and its grafting onto multiwalled carbon nanotubes, J. Appl. Polym. Sci. (2007), doi:10.1002/app.26772.

[70] H. Kong, W. Li, C. Gao, D. Yan, Y. Jin, D.R.M. Walton, et al., Poly(N-isopropylacrylamide)-Coated carbon nanotubes: Temperature-sensitive molecular nanohybrids in water, Macromolecules (2004), doi:10.1021/ma048682o.

[71] X. Wang, Z. Du, C. Zhang, C. Li, X. Yang, H. Li, Multi-walled carbon nanotubes encapsulated with polyurethane and its nanocomposites, J. Polym. Sci. Part A Polym. Chem. (2008), doi:10.1002/pola.22818.

[72] C. Gao, C.D. Vo, Y.Z. Jin, W. Li, S.P. Armes, Multihydroxy polymer-functionalized carbon nanotubes: Synthesis, derivatization, and metal loading, Macromolecules (2005), doi:10.1021/ma050823e.

[73] M.R. Loos, D. Gomes, situ-polymerization of fluorinated polyoxadiazole with carbon nanotubes in poly(phosphoric acid), Mater. Lett. (2009), doi:10.1016/j.matlet.2008.11.042.

[74] X. Lu, F. Tian, N. Wang, Q. Zhang, Organic functionalization of the sidewalls of carbon nanotubes by Diels-Alder reactions: A theoretical prediction, Org. Lett. (2002), doi:10.1021/ol026956r.

[75] L. Zhang, J. Yang, C.L. Edwards, L.B. Alemany, V.N. Khabashesku, A.R. Barron, Diels-Alder addition to fluorinated single walled carbon nanotubes, Chem. Commun. (2005), doi:10.1039/b500125k.

[76] N.K. Gupta, A. Sengupta, A. Gupta, J.R. Sonawane, H. Sahoo, Biosorption-an alternative method for nuclear waste management: A critical review, J. Environ. Chem. Eng. (2018), doi:10.1016/j.jece.2018.03.021.

[77] M. Monier, D.A. Abdel-Latif, Synthesis and characterization of ion-imprinted resin based on carboxymethyl cellulose for selective removal of UO_2^{2+}, Carbohydr. Polym. (2013), doi:10.1016/j.carbpol.2013.05.062.

[78] R.A.A. Muzzarelli, Potential of chitin/chitosan-bearing materials for uranium recovery: An interdisciplinary review, Carbohydr. Polym. (2011), doi:10.1016/j.carbpol.2010.12.025.

[79] Z. Feng, T. Danjo, K. Odelius, M. Hakkarainen, T. Iwata, A.C. Albertsson, Recyclable fully biobased chitosan adsorbents spray-dried in one pot to microscopic size and enhanced adsorption capacity, Biomacromolecules (2019), doi:10.1021/acs.biomac.9b00186.

[80] Y. Zhou, D. Yang, X. Chen, Q. Xu, F. Lu, J. Nie, Electrospun water-soluble carboxyethyl chitosan/poly(vinyl alcohol) nanofibrous membrane as potential wound dressing for skin regeneration, Biomacromolecules (2008), doi:10.1021/bm7009015.
[81] L. Zhang, W. Xia, X. Liu, W. Zhang, Synthesis of titanium cross-linked chitosan composite for efficient adsorption and detoxification of hexavalent chromium from water, J. Mater. Chem. A. (2015), doi:10.1039/c4ta05194g.
[82] K.G.H. Desai, H.J. Park, Preparation of cross-linked chitosan microspheres by spray drying: Effect of cross-linking agent on the properties of spray dried microspheres, J. Microencapsul. (2005), doi:10.1080/02652040500100139.
[83] Y. Huang, Y. Lapitsky, Salt-assisted mechanistic analysis of chitosan/tripolyphosphate micro- and nanogel formation, Biomacromolecules (2012), doi:10.1021/bm3014236.
[84] I. Broder, P. Corey, P. Brasher, M. Lipa, P. Cole, Formaldehyde exposure and health status in households, Environ. Health Perspect. (1991), doi:10.1289/ehp.9195101.
[85] W. Zhao, R.W.N. Nugroho, K. Odelius, U. Edlund, C. Zhao, A.C. Albertsson, In situ cross-linking of stimuli-responsive hemicellulose microgels during spray drying, ACS Appl. Mater. Interfaces. (2015), doi:10.1021/am5084732.
[86] V. Arias, K. Odelius, A. Höglund, A.C. Albertsson, Homocomposites of Polylactide (PLA) with Induced Interfacial Stereocomplex Crystallites, ACS Sustain. Chem. Eng. (2015), doi:10.1021/acssuschemeng.5b00498.
[87] S. Koushkbaghi, A. Zakialamdari, M. Pishnamazi, H.F. Ramandi, M. Aliabadi, M. Irani, Aminated-Fe_3O_4 nanoparticles filled chitosan/PVA/PES dual layers nanofibrous membrane for the removal of Cr(VI) and Pb(II) ions from aqueous solutions in adsorption and membrane processes, Chem. Eng. J. (2018), doi:10.1016/j.cej.2017.12.075.
[88] N. Ghaemi, A new approach to copper ion removal from water by polymeric nanocomposite membrane embedded with γ-alumina nanoparticles, Appl. Surf. Sci. (2016), doi:10.1016/j.apsusc.2015.12.109.
[89] H. Azad, M. Mohsennia, A novel free-standing polyvinyl butyral-polyacrylonitrile/ZnAl-layered double hydroxide nanocomposite membrane for enhanced heavy metal removal from wastewater, J. Memb. Sci. (2020), doi:10.1016/j.memsci.2020.118487.
[90] M. Ahmad, F. Wu, Y. Cui, Q. Zhang, B. Zhang, Preparation of novel bifunctional magnetic tubular nanofibers and their application in efficient and irreversible uranium trap from aqueous solution, ACS Sustain. Chem. Eng. (2020), doi:10.1021/acssuschemeng.0c00332.
[91] K. Li, Y. Zhao, C. Song, X. Guo, Magnetic ordered mesoporous Fe_3O_4 /CeO_2 composites with synergy of adsorption and Fenton catalysis, Appl. Surf. Sci. (2017), doi:10.1016/j.apsusc.2017.07.041.

CHAPTER 7

Oil adsorption performance of tubular hypercrosslinked polymer and carbon nanofibers

Ke Yang[a], Yuhong Cui[a] and Baoliang Zhang[a,b]
[a]School of Chemistry and Chemical Engineering, Northwestern Polytechnical University, Xi'an, China
[b]Xi'an Key Laboratory of Functional Organic Porous Materials, Northwestern Polytechnical University, Xi'an, China

7.1 Introduction

In the past few decades, oil spills have occurred frequently, which not only caused the waste of oil resources, pollution of marine and freshwater ecosystem but also caused huge economic losses [1–3]. For example, the Gulf of Mexico oil spill in 2010 released a total of 5 million barrels of crude oil into the ocean in the last three months, posing a serious threat to the survival of a variety of marine organisms and becoming the most serious oil spill in the world [4–6]. Therefore, the development of efficient oil–absorbing materials to deal with such emergencies has always attracted attention. In recent years, scientists have continuously explored and tried from traditional inorganic oil–absorbing materials to synthetic polymer oil–absorbing materials.

Traditionally, there are three main methods for spilled oil treatment [7]: biological, chemical, and mechanical. The biological method [8] is to use microorganisms to degrade the spilled oil. Chemical cleanup uses in-situ combustion, dispersion, etc. Mechanical cleanup relies on manpower or machinery to skim the spilled oil and use oil–absorbing materials to absorb. Compared with other oil removal methods, the use of oil–absorbing materials greatly saves manpower and material resources. Many materials are used for oil adsorption, mainly inorganic and organic. Inorganic oil–absorbing materials [9] mainly include activated carbon, lime, zeolite, silica gel, etc. The above-mentioned materials are characterized by low cost, low oil adsorption capacity, and high transportation costs. Due to its lightweight and high specific surface area, new carbon materials such as carbon nanotubes [10] and

Fabrication and Functionalization of Advanced Tubular Nanofibers and their Applications.
DOI: https://doi.org/10.1016/B978-0-323-99039-4.00002-4
153

graphene [11] are also used for oil adsorption. Compared with inorganic materials, synthetic polymers [12–14] (polystyrene fibers, polymethacrylate fibers, polypropylene fibers, etc.) and natural polymers [15–18] (cotton, straw, wheat straw, etc.) show high oil adsorption and have obvious advantages in use.

It is not difficult to find that one-dimensional (1D) materials are favored by researchers in the field of oil adsorption due to their advantages of large aspect ratio and high specific surface area. Carbon nanotubes, polymer fibers, carbon fibers, and plant fibers are widely used. Compared with bulk materials, this type of material has the characteristics of lightweight, high oil adsorption rate, and good reusability. Meanwhile, nano-scale materials with cavities and large pore volumes are more suitable for oil adsorption. In this chapter, the adsorption force form, adsorption mechanism, and classification of adsorbents of oil-absorbing materials will be discussed. Meanwhile, the current research progress of 1D oil-absorbing materials with obvious advantages will be reviewed.

7.2 Introduction of oil-absorbing materials

Since the birth of human civilization, people have learned to use natural plant fibers such as cotton and linen for adsorption activities. The use of oil-absorbing materials has also undergone a process of transition from traditional to high-performance. At first, people used traditional oil-absorbing materials such as corn stalks, animal and plant fibers to absorb oil. These materials mainly rely on the capillary action of plant or animal fibers to complete adsorption, and the effect is not good. The main manifestations are (1) low oil adsorption capacity, (2) slow adsorption rate, (3) poor selective adsorption, which absorbs oil and water at the same time, (4) poor oil retention. In the face of emergencies such as large-scale oil spills at sea, traditional oil-absorbing materials obviously cannot meet the requirements of waste oil recovery and environmental governance. As a new type of functional material, high-performance oil-absorbing materials are different from traditional oil-absorbing materials. The obvious characteristics of this type of material are: (1) high oil adsorption capacity, and can handle different types of oil, (2) good water buoyancy, good oil-water selectivity, only absorbs oil but not water, (3) good performance in repeated use.

7.2.1 Adsorption force of oil-absorbing material

Judging from the research status at home and abroad, there are three stages in the generally accepted oil adsorption process of materials. First of all,

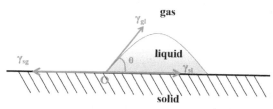

Figure 7.1 The cross-sectional view of the contact angle formed by the liquid on the solid surface.

relying on the Van der Waals force between the oil–absorbing material and oil molecules, a small amount of oil diffuses into the materials to achieve the effect of wetting. Wetting is a very common phenomenon. The stronger the interaction force between the liquid and the material, the easier the wetting phenomenon will occur. The degree of difficulty is usually expressed by the contact angle (CA). As shown in Fig. 7.1, θ is the contact angle between solid and liquid. Under ideal conditions, when the liquid reaches an equilibrium state on the solid surface, the solid–liquid, gas–liquid, and solid–gas interfacial tensions γ_{sl}, γ_{gl}, and γ_{sg} satisfy Young's equation [19–21]: $\gamma_{sg} = \gamma_{sl} + \gamma_{gl} \cos \theta$. θ is used to judge wettability. Generally speaking, when $\theta = 0°$, it means that the solid surface can be completely wetted and the liquid can spread on the solid surface. $\theta < 90°$, indicating the liquid can wet the solid surface. The smaller the θ, the better the wettability. $\theta > 90°$, the liquid cannot wet the solid surface. When $\theta = 180°$, the solid cannot be wetted at all, and the liquid condenses into small balls on the solid surface.

After the wetting process is completed, enter the second stage. At this time, the oil molecules diffuse into the material are solvated with the smallest unit inside the material, so that the material can be slowly stretched in the oil. However, in the initial stage, oil molecules cannot fully stretch the material, and the diffusion effect still dominates the adsorption at this time. When a steady stream of oil molecules continues to diffuse into the oil-absorbing material, the solvation is completed, and the material is fully unfolded in the oil, and saturated adsorption is reached. This process is also called the thermodynamic control stage. Eventually, the diffusion of oil into and out of the material reaches equilibrium, that is, reaches a state of thermodynamic equilibrium. From the initial wetting to the diffusion of oil molecules and then to the thermodynamic equilibrium state, it takes a long time. This process generally takes several hours, and the length of time varies depending on the material.

According to the theory of macro mechanics, any matter is always in a relatively stable state, that is, the force it receives is balanced, and the resultant force is zero. At the interface of the two phases, due to the difference between the material molecules, the interface layer will be in a state of an unbalanced force, and it is this unbalanced force that causes the adsorption phenomenon. Adsorption is the interaction between adsorbate and adsorbent. Adsorbents and adsorbents have different types and properties. This difference determines the adsorption force. Generally, the force between adsorbate and adsorbent mainly includes dispersion forces, dipole interaction, electrostatic force, and hydrogen bonding.

7.2.1.1 Dispersion force

As the main source of Van der Waals force, dispersion force exists between all molecules or atoms. Dispersion force is the force between the instantaneous dipoles of molecules. For the electrons in motion, the position of the instantaneous electrons is asymmetric to the nucleus, resulting in the instantaneous discard of the positive and negative charge centers, thus generating the instantaneous dipole. The dispersion force belongs to the intermolecular force, and its size is inversely proportional to the 6th power of the intermolecular distance. In addition, its energy also strongly depends on the number of electrons outside the nucleus, atomic weight, and atomic number.

7.2.1.2 Dipole interaction

Due to the difference in electronegativity of adsorbate molecules or atoms, when forming a chemical bond, the charge will be biased towards the more electronegative atom. If the center distance between the offset $+e$ and $-e$ is r, the electric distance generated by them is $\mu=er$, which is called the bond distance. Such surface dipole or bond distance with surface polar functional groups interacts with dipolar adsorbent molecules, and its force is inversely proportional to r^3, which is smaller than the dispersion force. When one of the adsorbed molecules or surface molecules has a dipole moment, it can induce the other to produce a dipole moment and weak interaction occurs.

7.2.1.3 Electrostatic force

When insulating solids or liquids are in contact with each other, electrostatic charges will be generated at their interfaces. Although their electricity is relatively small, they can generate a strong electric field of several thousand volts. Therefore, the solid interface is often charged. The electrostatic force is often generated between the adsorbate and the adsorbent or between the adsorbents, which is a long-distance interaction.

7.2.1.4 Hydrogen bonding

Hydrogen bonding and Van der Waals forces together constitute intermolecular forces. The formation of a hydrogen bond is mainly due to many polar groups containing hydrogen atoms on the surface of the material. When the hydrogen atoms in these groups meet the atoms containing high electronegative properties (such as O, S, N, F, Cl, etc.) in the adsorption molecule, they will interact with their lone pair electrons to form hydrogen bonds. The hydrogen bond is stronger than the other forces mentioned above. Therefore, the adsorbate adsorbed by the hydrogen bond is difficult to desorb at room temperature, and other external forces are required. When hydrogen bonding occurs, the surface of the material often contains other highly electronegative atoms, so the dipole interaction will also occur in the process.

The above forces are based on adsorption forces at the molecular level. In addition to these forces, there are also capillary adsorption, pore adsorption in porous bodies, and adsorption type complex between adsorbents and adsorbents. When 1D materials are used in the oil adsorption process, capillary adsorption and pore adsorption are significant. Through the analysis of these forces, it is clear that the adsorption performance of adsorbents to adsorbents not only depends on the interaction between them at the molecular level but also the effective construction of micro/nanochannels in adsorbents is more important, which has guiding significance for the design of new adsorbents.

7.2.2 The adsorption principle of oil-absorbing materials

Generally speaking, according to the size and strength of the force between the adsorbent and the adsorbate, adsorption can be divided into physical adsorption, chemical adsorption and hydrogen bond adsorption.

7.2.2.1 Physical adsorption

When the adsorbent is adsorbing, from the perspective of adsorption kinetics, the adsorbate is a process of continuous adsorption and desorption on the surface layer of the adsorbent. When the adsorption and desorption process is stable, the adsorption equilibrium is reached. This adsorption process is also called a reversible adsorption process. If reversible adsorption occurs between the adsorbate and the adsorbent, it means that there is only a relatively weak interaction between them, which is often referred to as Van der Waals force. Van der Waals force mainly includes dispersion force, orientation force, and inductive force. This adsorption phenomenon caused by Van der Waals force is called physical adsorption. The dispersion force generally plays a major role between the nonpolar adsorbent material and the adsorbate. However, when the polar molecules interact with the surface of the electrostatically charged adsorption material, the electronic structure of the two may change to produce a dipole moment. At this time, the orientation force and the inductive force also play an essential role. The adsorption selectivity of physical adsorption is poor, it is common among various materials, and it is easily affected by the external environment.

7.2.2.2 Chemical adsorption

When chemical bonding occurs between the adsorbent and the adsorbate, the adsorbate will undergo a chemical change after being adsorbed, which is the chemical adsorption process. The chemical adsorption process is often a strong interaction between the adsorbent and the adsorbate, which is generally an irreversible process. In addition, chemical adsorption generally only occurs in the monolayer adsorption on the surface of the adsorbent material. Therefore, its adsorption capacity is limited, but the chemical adsorption process is often accompanied by a physical adsorption process.

7.2.2.3 Hydrogen bond adsorption

Hydrogen bond adsorption is a form of adsorption between physical adsorption and chemical adsorption. Its strength is higher than physical adsorption based on Van der Waals forces. Therefore, the material needs to be eluted and regenerated with an organic solvent during repeated use. Hydrogen bond adsorption is also classified as physical adsorption because the adsorption force of hydrogen bond is close to Van der Waals force. However, some people believe that the retention, selectivity and desorption performance of the adsorbate by the adsorbent after the hydrogen bond adsorption is essentially different from that of physical adsorption, so they are classified as a separate adsorption phenomenon.

With the continuous advancement of science and technology, researchers have endowed it with various adsorption capabilities at the beginning of material design. Often the occurrence of the adsorption process is not just a function of an adsorption mechanism, but the result of a combination of multiple mechanisms. In this process, there are not only physical, chemical, and hydrogen bond adsorption, but also synergistic effects at the molecular level and pore effects at the material space level beyond these forces, and finally, achieve collaborative adsorption [22–24]. Therefore, at the beginning of the material design, the influence of various factors should be fully considered. Not only should it have enough lipophilic groups, but also the pore structure inside the material and the three-dimensional (3D) network of the material should be fully considered.

7.2.3 Classification of oil-absorbing materials

Oil-absorbing materials can be divided into traditional oil-absorbing materials and high-performance oil-absorbing materials [25]. Traditional oil-absorbing materials mainly include activated carbon, vermiculite [26], zeolite [27], animal and plant fibers, etc. Such materials mainly rely on the material surface or internal pores for simple physical adsorption of oil, the effect is not good. The following is a list of several common high oil adsorption materials.

7.2.3.1 Rubber-based material with high oil adsorption

The rubber-based high oil adsorption material is a type of high oil adsorption material formed by the introduction of lipophilic chemical functional groups or physical blending (mechanical blending and emulsion blending) on the traditional rubber matrix material through chemical grafting [28–30]. Among them, the chemical grafting method uses chemical reactions to directly introduce lipophilic group segments into the active sites of the original rubber matrix. The obtained material has good micro-compatibility, good strength, and repeated use performance in the process of repeated oil adsorption and deoiling. However, the reaction process is difficult and it is not easy to be prepared on a large scale. The blending method relies on physical blending to disperse oil-absorbing resins with better oil-absorbing properties in the rubber matrix. The blending method has a simple process, wide sources of raw materials and low cost. However, this method also has some problems, such as the interfacial compatibility problem caused by the introduction of lipophilic components, the precipitation of lipophilic components from the rubber matrix, and the problems of swelling and reduced tensile strength during recycling.

The rubber matrix that is often used as an oil-absorbing material mainly includes ethylene-propylene rubber, butyl rubber, styrene-butadiene rubber, neoprene rubber, and silicone rubber. The oil adsorption process of this type of oil-absorbing material is: the internal capillary action and surface adsorption of the rubber make the oil molecules quickly enter the rubber interior and combine with the internal lipophilic components. Meanwhile, the rubber swells, and finally reaches the swelling equilibrium, that is, the adsorption equilibrium.

7.2.3.2 Resin-based high oil-absorbing materials

Resin lipophilic material is a type of functional polymer material formed by the polymerization and cross-linking of lipophilic monomers. Generally in the form of particles, organic solvents (such as xylene, etc.) need to be added to make holes inside the material in the later stage of the polymerization to increase the oil adsorption rate. Therefore, it contains a 3D network cross-linked structure inside and forms a microporous structure. The monomers often used for polymerization are mainly styrene and acrylic esters. It has the characteristics of high oil adsorption performance and only oil adsorption without water adsorption and can absorb more than 10 times to dozens of times of oil. Such materials have structural designability, which significantly improves their performance.

7.2.3.3 Fibrous synthetic high oil-absorbing material

The original intention of the development of oil-absorbing materials is to deal with oil spills on the water surface, so there are certain requirements for the shape of the material. Although the above-mentioned resin-based oil-absorbing materials have high oil-absorbing properties, most of them are in granular form and cannot be used for immediate treatment of oil spills with large contaminated areas on the water surface and fast-spreading. The fibrous oil-absorbing material has a higher specific surface area, a faster oil adsorption rate, and can be processed into any shape, which is convenient for rapid oil spill recovery treatment. Some polymer materials can be prepared into polymer fibers by the melt spinning process, such as polypropylene and polystyrene. They have the characteristics of lightweight, good lipophilicity, low density, and good buoyancy in water. The common synthetic methods of fibrous oil-absorbing materials mainly include wet spinning, melt spinning, and electrostatic spinning. Among them, electrospinning is jet spinning of polymer solution or melting under a strong electric field. Under the action of an electric field, the droplet at the needle will change from a spherical shape to a conical shape and extend from the tip of the cone to obtain fibrous filaments. Electrospinning can produce nanometer-level superfine fibers. In recent years, electrospinning technology has developed rapidly. More than one hundred kinds of polymer materials can be prepared into ultra-fine nanofibers by electrospinning technology, which is widely used in the fields of filter materials and biomedical materials.

7.2.3.4 Natural material modified oil-absorbing material

Although polymer oil-absorbing materials exhibit excellent oil-absorbing properties, they generally have problems such as being difficult to degrade and easily causing secondary pollution. Therefore, people have been trying to use natural plant fibers to prepare high oil-absorbing materials through modification. Common methods mainly include chemical grafting of natural cellulose with polymer oil-absorbing materials and direct conversion of cellulose groups. Our group uses the Platanus fiber derived from biomass as the raw material to create pores on the surface and obtain an oil-absorbing material with ultra-high specific surface area, which will be described in detail later.

7.3 Research progress of one-dimensional oil-absorbing materials

One-dimensional oil-absorbing materials are favored by researchers and engineering applications due to the performance advantages brought by morphological characteristics. Commonly reported 1D materials used for oil spill recovery treatment mainly to include polymer fibers and carbon nanotubes. Because this type of material has a large aspect ratio, high specific surface area, fluffy structure, and good lipophilic and hydrophobic properties, it shows more obvious advantages than rubber matrix and resin matrix materials when dealing with large-scale oil spills. As polypropylene fiber, which has long been commercialized, is regarded as an ideal oil-absorbed material because it contains only carbon and hydrogen. Melt-blown polypropylene fiber has long been widely used in degreasing operations on seas, rivers, and other water surfaces. And oil-absorbing chains, oil-absorbing felts, and oil-absorbing wipes made of melt-blown polypropylene as raw materials have long been used in ports and oil transportation.

However, even polypropylene fibers that have long been commercialized show limited adsorption capacity. To further increase the adsorption capacity of 1D polymer fiber materials, it is imperative to study the design and preparation of new materials.

7.3.1 One-dimensional polymer nanofibers

Electrospinning, as a simple technology for preparing 1D fiber materials with nanometer or micrometer diameter, has achieved rapid development in recent years. It can not only control the diameter of the fiber but also manipulate the structure of a single fiber through controllable adjustment [31,32]. Electrospinning technology utilizes a special form of electrostatic

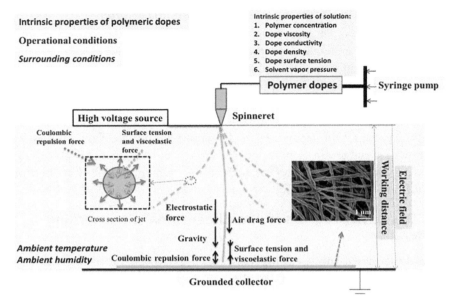

Figure 7.2 Schematic illustration of the nanofibers formation in electrospinning. The copyright of the figure obtained from Elsevier under the license No: 5272960986179, Mar 20, 2022 [38].

atomization of polymer fluids to produce tiny streams of polymer that can travel considerable distances and then stretch and stretch to produce fibers [33,34]. As shown in Fig. 7.2, its main components include four parts: a propulsion pump, an injector, a high-voltage power supply, and a receiving device. Under the action of an electric field, the droplet at the needle will change from a spherical shape to a conical shape (Taylor cone) and extend from the tip of the cone to obtain fiber filaments. When the jet is stretched to a finer diameter, it solidifies quickly, causing solid fibers to deposit on the grounded collector. Generally speaking, the electrospinning process can be divided into four consecutive steps [35,36]: (1) The electrification of droplets and the formation of cone-shaped jets. (2) The charged jet extends along a straight line. (3) The jet becomes thinner under the instability of the electric field and electric bending. (4) Collect the solidified fiber jet on the grounded collector [37,38].

7.3.1.1 One-dimensional polystyrene oil-absorbing material

Electrospinning technology has been applied to various types of materials to produce nanofibers, including organic polymers, small molecules, colloidal

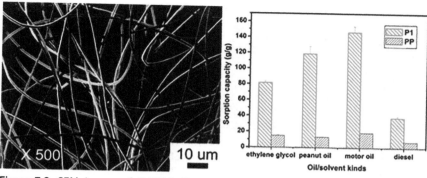

Figure 7.3 SEM image of PVC/PS fibers (left) and adsorption performance (right). Reprinted (adapted) with permission from Copyright 2022 American Chemical Society [39].

particles, and composite materials. The most common materials are polymers in the form of solutions or melts. By introducing nano–components of different sizes and morphologies into the polymer solution, the resulting mixture can also be used for electrospinning. At present, it has been proved that more than one hundred different types of organic polymers can be directly used for solution electrospinning to obtain nanofibers, including natural polymers and synthetic polymers. However, the report on the use of electrospun fibers for oil adsorption is relatively late. In 2011, Zhu et al. [39] of Qingdao University of Science and Technology reported for the first time the use of electrospun polyvinyl chloride/polystyrene (PVC/PS) fibers as oil–absorbing materials. Its highest adsorption capacity for motor oil, soybean oil, diesel and ethylene glycol reached 146, 119, 38, and 81 gg^{-1}. It was 5–9 times the adsorption capacity of commercial polypropylene nonwoven fabrics. Meanwhile, the prepared PVC/PS adsorbent also had excellent oil–water selectivity (about 1000 times) and good buoyancy in water. The voids between the fibers were the key factor that made the material have the ultra-high capacity (Fig. 7.3). Thus, the prelude to the study of electrospinning polymer fibers for oil adsorption was unveiled.

Afterward, Wu et al. [40] obtained an adsorbent with high oil adsorption capacity and oil-water selectivity by controlling the diameter and forming a porous morphology on superhydrophobic and super lipophilic electrostatic PS fibers. The results showed that the small diameter, the porous surface structure of PS fiber had an oil adsorption capacity of about 7.13, 81.40, 112.30, and 131.63 g g^{-1} for diesel, silicone oil, peanut oil and motor oil. Lin et al. [41] of Donghua University used a one-step electrospinning method to

Figure 7.4 SEM images (A and C) and optical profile images (B and D) of two polystyrene fibers with different surface roughness. The copyright of the figure obtained from Elsevier under the license No: 5272970624040, Mar 20, 2022 [41].

prepare two superhydrophobic and super lipophilic nano-porous PS fibers with different roughness (Fig. 7.4). The relationship between the surface fine structure and specific surface area of PS fibers and the amount of oil adsorption was explored. The highest adsorption capacity for motor oil, soybean oil, and sunflower oil reached 113.87, 111.80, 96.89 g g^{-1}. It was about 3–4 times that of natural adsorbents and polypropylene non-woven fiber mats. This was because electrospun polystyrene fibers had a smaller pore structure and a larger specific surface area than polypropylene fibers. Chen et al. [42] prepared nonsolvent-induced macroporous PS fibers by one-step electrostatic spinning and carried out adsorption tests on silicone oil, pump oil, sunflower oil, and diesel oil with different viscosity. They found that PS fibers with high porosity had excellent oil adsorption performance and could adsorb up to 900 g g^{-1}.

Since then, various polymer-based electrostatic spinning oil-absorbing materials have been reported one after another, mainly focusing on PS-based oil-absorbing materials. Li et al. [43] carried out electrostatic spinning on the blend of PS and polyacrylonitrile and obtained an oil-absorbing material with high mechanical strength and high capacity. It was found that the chemical composition of fiber and fiber diameter directly affect the oil adsorption capacity of fiber, and the oil adsorption capacity increased with the decrease of fiber diameter. The maximum adsorption capacities for pump oil, peanut oil, diesel oil, and gasoline were 194.85, 131.70, 66.75, and 43.38 g g^{-1}. The polystyrene/polyurethane (PS/PU) oil-absorbing pad prepared by The polystyrene/polyurethane (PS/PU) oil-absorbing pad prepared by Lin et al. [57] had an adsorption capacity of 64.40 and 47.48 g g^{-1} for engine oil and sunflower seed oil, which was 2–3 times that of conventional polypropylene fiber.

Although the oil-absorbing material based on PS shows high oil adsorption and fast adsorption rate in the initial use, it also faces the disadvantages of decreased adsorption capacity after oil adsorption regeneration and low repeated use rate. Based on this, Wu et al. [44] obtained carbon nanotubes (CNTs) with good dispersibility through covalent modification and fluorination and co-spinned them with PS to obtain a rigid oil-absorbing material PS-CNTs (Fig. 7.5). An interesting phenomenon was discovered in the research. There were two stages in the oil adsorption process of PS-CNTs. The first was that the oil quickly entered the holes formed by interconnected nanofibers, which was caused by the lipophilic adsorption of micron-scale voids. Later, due to the capillary action and Van der Waals force, the oil not only moved to the nano-internal space of the fiber but also remained on the surface of the fiber. The maximum adsorption capacity of PS-CNTs materials for sunflower oil, soybean oil, and motor oil reached 116, 123, and 112 g g^{-1}, respectively. Due to the addition of carbon nanotubes, the fiber had good mechanical strength. Although the adsorbent was repeatedly compressed and extracted to recover the adsorbed oil, its oil removal efficiency was still more than 80%, indicating the effectiveness and reusability of PS-CNTs composite adsorbent.

7.3.1.2 Other one-dimensional polymer oil-absorbing materials

In addition to the above PS matrix oil-absorbing materials, other polymer materials are also used to make oil-absorbing materials. Gao et al. [45] simultaneously prepared a flexible hybrid nanofiber membrane (FHNM) containing PVDF nanofibers and SiO_2/PVDF microspheres through electrospinning and point spraying. FHNM had superhydrophobicity, super lipophilicity and a unique porous structure. It can be used for oil adsorption and oil/water separation. Shu et al. [46] prepared polyethylene terephthalate nano-porous luminescent fiber (PNPLF) by electrostatic spinning technology, which had a good oil adsorption function, and the oil adsorption capacity was up to 135 g g^{-1}.

7.3.1.3 Magnetic one-dimensional oil-absorbing material

Magnetic 1D oil-absorbing materials are mainly prepared by adding magnetic particles (such as Fe_3O_4, etc.) into the polymer matrix during electrostatic spinning. It is beneficial to the recovery of materials after adsorption. Dorneanu et al. [47] added cobalt ferrite ($CoFe_2O_4$) to PVDF to obtain magnetic fiber composite materials and compared it with the adsorption capacity of pure PVDF fiber. Although the adsorption capacity of the fiber

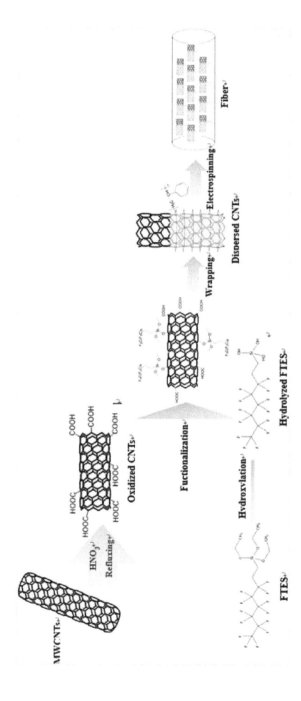

Figure 7.5 A schematic diagram of CNTs functionalization to fabricate an electrospun fibrous sorbent. The copyright of the figure obtained from Elsevier under the license No: 5272960257480, Mar 20, 2022 [44].

decreased slightly after adding $CoFe_2O_4$, the excellent magnetic properties were favorable for the recovery of the materials after oil adsorption by an external magnetic field. Song et al. [48] composited PS and Fe_3O_4 through electrospinning and obtained oil-absorbing fibers with a good magnetic response, special wettability, and long-range oil adsorption. The adsorption rate of edible oil, paraffin oil, and silicone oil can reach 87, 65, and 94 g g^{-1}.

7.3.2 Carbon nanotube oil-absorbing material

Carbon Nanotubes (CNTs), also known as Bucky tubes, are allotropes of carbon. They were accidentally discovered by Japanese physicist Iijima in 1991 when observing arc evaporation products with high-resolution transmission electron microscopy [49]. According to the number of graphite sheets in the tube, CNTs are divided into single-walled carbon nanotubes (SWCNTs) and multi-walled carbon nanotubes (MWCNTs). MWCNTs are formed by concentrically nesting multiple single-layer tubes, which can have two to dozens of layers, and the layer spacing is about 0.34 nm. The diameter is from a few nanometers to tens of nanometers, and the length can reach the order of micrometers or millimeters.

Initially, the preparation methods of CNTs mainly include the arc method and laser evaporation method. Although the products obtained by these two methods have high purity, the equipment is complicated and the cost is high, and it is difficult to realize industrialization. After learning from the preparation method of carbon fiber, chemical vapor deposition (CVD) is developed [50]. CVD method, also known as catalytic pyrolysis, uses transition metals such as iron, cobalt, and nickel or their compounds as catalysts to decompose hydrocarbons (ethylene, acetylene, and alcohol) under protective gas (argon, nitrogen) and ereduce gas (hydrogen) at high temperature, and then carbon atoms are regrown on metal particles to obtain CNTs. The equipment required for this method is simple and has already been industrialized. In 2002, Zhu et al. [51] used the CVD method to synthesize a carbon nanotube rope with a length of 40 cm, and related results were published in Science.

CNTs can be directly used as oil-absorbing materials, and many studies have further prepared them into 3D materials for oil adsorption. Fan et al. [52] prepared 3 mm long vertically arranged CNTs for oil adsorption, which had unique mechanical strength and excellent resilience under high strain. After 10 repeated oil adsorption tests, they still had good oil adsorption performance and could adsorb oil up to 69 g g^{-1}. Gui et al. [53] prepared

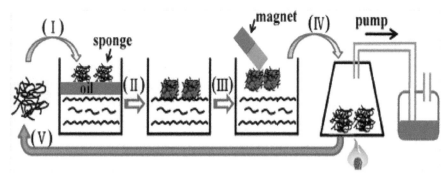

Figure 7.6 Schematic of the recycling of Me-CNT sponges used for spilled oil sorption. Reprinted (adapted) with permission from Copyright 2022 American Chemical Society [53].

a magnetic carbon nanotube sponge (Me-CNT), which was composed of abundant Fe encapsulated CNTs and had an adsorption capacity of 56 g g^{-1} for diesel oil, corresponding to 99% of the volumetric adsorption capacity. The sponge had good mechanical strength and can be recovered by a magnetic field and simply extruded to extrude the oil, which can be desorbed and recycled by heat treatment (Fig. 7.6) and reused up to 1000 times. Gui et al. [54] prepared porous carbon nanotube materials with porosity up to 98%. It had a unique 3D structure, large specific surface area and good hydrophobicity. Its adsorption capacity for diesel oil, vegetable oil, and mineral oil was more than 100 g g^{-1}. Meanwhile, CNTs can be reused after a single use through combustion or pressure oil removal.

7.3.3 Biomass-derived one-dimensional oil-absorbing materials

Natural biomass materials have been used directly as oil-absorbing materials for a long time. As early as the beginning of human society, they knew how to use cotton and linen for adsorption. In addition, animal fur or plant straws are also directly used as oil-absorbing materials. However, most biomass materials do not have a fixed form, causing mass transfer difficulties. Therefore, people have high hopes for the use of fixed-form biomass materials in a specific direction (adsorption separation).

The use of biomass-derived 1D oil-absorbing materials mainly includes three methods: direct use, use after surface hydrophobic modification, and preparation of 1D materials into 3D aerogels for oil adsorption. Li et al. [55] adopted carbon tube clusters with micron wall thickness and micron diameter from Sonchus Oleraceus L. Its maximum adsorption capacity for lubricating oil and waste oil can reach 103.76 g g^{-1} and 91.70 g g^{-1}, with low cost, simple and renewable, and can be used as an effective adsorbent for marine oil spill treatment. This was due to the rich pore structure, large specific surface area and hydrophobicity. Nine et al. [56] extracted super-wetting microfibers from a chestnut shell, which can be used for continuous oil-water separation and dye adsorption. Due to the presence of aliphatic and aromatic hydrocarbons on the surface, it had super-oil-philic characteristics and can absorb oil 94 times of its weight. The filter membrane designed with it as the raw material can carry out continuous oil-water separation of a series of organic solvents, including toluene, rapeseed oil, engine oil and so on.

The above work is to directly use biomass fiber for oil adsorption, but many biomass fibers will absorb water at the same time as oil adsorption. So, hydrophobic modification is a favorable strategy. Lee et al. [57] soaked SiO_2 particles coated with polydimethylsiloxane on the surface of cotton and kapok for hydrophobic modification. The contact angle of the material to water exceeded 150°, and the soaked cotton and kapok fibers can absorb 20–60 g g^{-1} of oil, and still showed stable oil adsorption ability after 10 cycles. In addition, the 3D aerogelation of 1D biomass materials is also a research hotspot. Song et al. [58] prepared carbon microtubule aerogel from kapok fiber with excellent stability and mechanical properties. It showed a high adsorption capacity for a variety of oils and organic solvents, up to 78–348 times of its weight. The recycling performance was tested by distillation, combustion, extrusion and other methods. After ten cycles, the adsorption capacity remained above 90%. Li et al. [59] reported a sustainable, flexible and superhydrophobic functionalized cellulose aerogel, namely cellulose aerogel coated with copper nanoparticles (Cu/CEA). The preparation process was shown in Fig. 7.7. The hydrophobicity of Cu/CEA was obtained by one-step deposition of Cu nanoparticle coating on cellulose fiber extracted from natural sisal. This method was economical, simple and environmentally friendly. Cu/CEA showed high oil adsorption capacity (67.8–164.5 g g^{-1}) and oil adsorption rate, as well as good recovery characteristics.

7.4 Tubular hypercrosslinked polymer

In Fig. 7.8, by adjusting the ratio between monomer 1, 4-p-dichlorobenzyl and catalyst $FeCl_3$, three kinds of tubular hypercrosslinked polymer nanofibers (THPFs) with different tube wall thickness and inner diameter

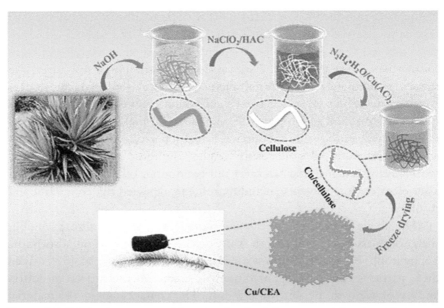

Figure 7.7 Fabrication process of functionalized cellulose aerogel. Reprinted (adapted) with permission from Copyright 2022 American Chemical Society [59].

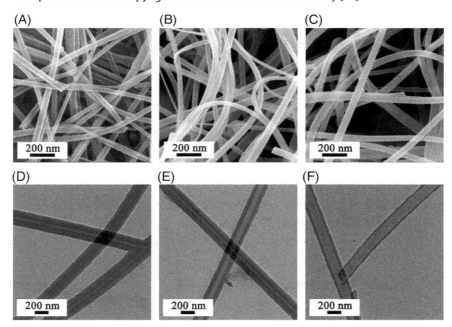

Figure 7.8 SEM and TEM images of THPFs-1 (A, B), THPFs-2 (C, D) and THPFs-3 (E, F).

Table 7.1 Statistics on the inner diameter and wall thickness of THPFs.

Samples	Internal diameter/nm	Wall thickness/nm
THPFs–1	33	87
THPFs–2	69	66
THPFs–3	107	36

sizes were obtained. SEM showed (Fig. 7.8A, C, and E) they all presented a uniform 1D structure, intertwined with each other. THPFs showed a large aspect ratio with a diameter between 120 and 150 nm. Meanwhile, TEM further showed the cavity structure (Fig. 7.8B, D, and F), and the inner diameter and tube wall thickness were calculated, as listed in Table 7.1. THPFs–1 had the thickest tube wall and the thinnest inner diameter, 87 nm and 33 nm, respectively. THPFs–3 was just the opposite, with the thickest inner diameter and the thinnest tube wall, 107 nm and 36 nm, respectively. THPFs–2 was between THPFs–1 and THPFs–3, and the tube wall value and inner diameter value were equivalent, 66 nm and 69 nm, respectively. The above results proved that the inner diameter and wall thickness of tubular polymer nanofibers can be controlled by adjusting the ratio of polymerized monomer, catalyst and solvent.

We took THPFs–1 as an example to test the relationship between adsorption capacity and adsorption time. Five oils including gasoline, diesel, motor oil, soybean oil and silicone oil were selected as adsorbed substances in the experiment. The entire adsorption process was completed within the first 10 min, that was, THPFs–1 can reach saturated adsorption within 10 min. Taking diesel as an example, the adsorption capacity of THPFS–1 at 10 min was 31.85 g g^{-1}, and 31.97 g g^{-1} at 2 h, which was the same, so it can be judged that the fiber has reached saturation adsorption. The adsorption mechanism can be attributed to the synergistic principle of physical adsorption and capillary adsorption. In subsequent experiments, to enable the fiber to fully achieve saturated adsorption, the adsorption time was determined to be 2 h.

Generally speaking, the ambient temperature also had an important influence on the oil adsorption capacity. The temperature will affect the speed of movement of oil molecules, thereby affecting the ease with which they diffuse into the adsorbent material. Meanwhile, the temperature will also affect the degree of spreading of the oil-absorbing material in the oil. In Fig. 7.9B, we further studied the adsorption performance of THPFs–1 on five oils at different ambient temperatures. It can be seen that as the temperature increased, the adsorption amount increased first and then stabilized. At

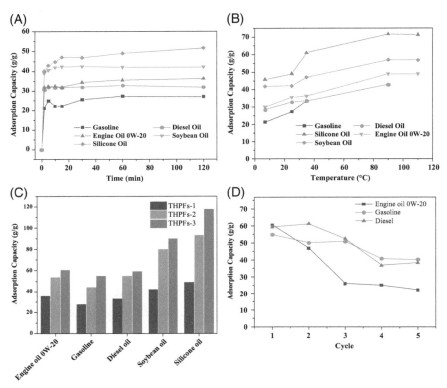

Figure 7.9 The adsorption rate of THPFs-1 on various oils at 25°C (A). The adsorption capacity of THPFs-1 for various oils as a function of temperature (B). Maximum adsorption capacity (C) of THPFs and recycling performance of THPFs-3 (D).

lower temperatures, oil molecules were not easy to diffuse into the fiber, so the adsorption capacity decreased. With the increase of temperature, the molecular movement intensified and the adsorption capacity increased. In particular, the maximum adsorption capacity of silicone oil increased from 45.86 g g^{-1} at 7°C to 71.79 g g^{-1} at 90°C. After 90°C, the adsorption capacity of THPFS-1 on various oils did not change significantly. After 90°C, the cavity structure of the fiber was covered with oil. At this time, if the temperature was increased, no more oil can be stored in the limited volume, so the adsorption capacity did not change much. Due to the volatility at high temperatures, the upper limit temperatures for gasoline and diesel testing were 35 and 90 °C, respectively.

After that, using 2 h as the adsorption time, the highest adsorption capacity of THPFs for various oils was tested. As shown in Fig. 7.9C, the adsorption capacity of THPFs-1, THPFs-2 and THPFs-3 for various oils showed an upward trend in turn. Among them, THPFs-3 had the highest adsorption capacity for various oils, and THPFs-1 was the lowest. The adsorption capacity increased with the increase of fiber diameter. It was understandable that the increase of the cavity would allow THPFs-3 to have a larger volume to hold the oil. THPFs-3 had the smallest specific surface area, only 30.81 m^2 g^{-1}. Therefore, the adsorption of oil was not only dependent on the specific surface area of the material but also related to the internal pore structure of the material. Finally, the highest adsorption capacity of THPFs-3 for engine oil 0W-20, gasoline, diesel, soybean oil and silicone oil were 60.39, 54.94, 59.41, 89.92, and 117.75 g g^{-1}, respectively. On this basis, engine oil 0W-20, gasoline and diesel oil were selected to test the repeated use performance of THPFs-3. As shown in Fig. 7.9D, although THPFs-3 showed a higher adsorption capacity in the first use, its adsorption capacity dropped rapidly in the second cycle. Even so, after 5 cycles, the adsorption capacity of THPFs-3 remained above 22 g g^{-1}.

7.5 Length controllable hydrophobically modified tubular carbon nanofibers

Carboxylated tubular carbon nanofibers (C-CTFs) with different aspect ratios were prepared by vacuum calcination, liquid-phase oxidation and ultrasonic treatment of tubular hypercrosslinked polymer nanofibers (TH-PFs) prepared by restricted condensation method (Fig. 7.10). Oil adsorption required the material to have a certain pore volume and appropriate surface hydrophobicity, so we modified the hydrophilic C-CTFs surface. Alkyl chains were introduced onto the surface of tubular carbon nanofibers (CTFs) by hydrolysis condensation using silane coupling agents with different saturated carbon chain lengths. C-CTFs modified by hexyl triethoxysilane (HTOS), octyl triethoxysilane (OTOS), dodecyltriethoxysilane (DTOS), and octadecyltriethoxysilane (OdTOS) were named h-CTFS-L, O-CTFS-L, D-CTFS-L, and Od-CTFS-L, respectively. Their morphology was shown in Fig. 7.11B–I. This was why CTFs were carboxylated. Meanwhile, nitric acid oxidation combined with ultrasonic treatment can achieve the cutting of long C-CTFs. Adjusting the centrifuge speed can obtain C-CTFs with lengths of 1.2, 2.1, 4.0, and >10 μm, respectively named C-CTFs-12, C-CTFs-21, C-CTFs-40, and C-CTFs-L. The length of the fiber determined its ability to physically overlap, which in turn affected its oil adsorption performance. After investigation and optimization, it was clear that the best oil-absorbing material was O-CTFs-L. The BET surface area of O-CTFs-L was 35.73 m^2 g^{-1}, and the average pore diameter was 33.99 nm.

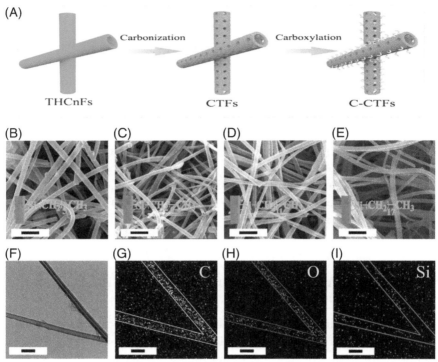

Figure 7.10 The schematic diagram of C-CTFS preparation process (A). SEM images of H-CTFs-L (A), O-CTFs-L (B), D-CTFs-L (C) and Od-CTFs-L (D). TEM image of O-CTFs-L (E). EDS analysis of O-CTFs-L: C element (F), O element (G), Si element (H). The scale bar in A, B, C, and D was 500 nm. The scale bar in E, F, G and H was 200 nm. The copyright of the figure obtained from Elsevier under the license No: 5274200043525, Mar 22, 2022.

H-CTFs-L, O-CTFs-L, D-CTFs-L, and Od-CTFs-L all showed compatibility with engine oil 0W-20 (density 0.8264 g cm$^{3-1}$, viscosity 103.899 mPa·s). The contact angle of O-CTFs-L was the smallest, with a value of 16.61°. The contact angle was shown in Fig. 7.11. We investigated the fiber's adsorption performance for different oils (Fig. 7.11B and C), O-CTFs-L for gasoline (density 0.7511 g cm$^{3-1}$, viscosity 1.132 mPa·s), diesel (density 0.8287 gcm$^{3-1}$, Viscosity 6.002 mPa·s), motor oil 0W-20, soybean oil (density 0.8993 g cm$^{3-1}$, viscosity 64.492 mPa·s) and silicone oil (density 0.9355 g cm$^{3-1}$, viscosity 157.751 mPa·s) had the highest adsorption capacities of 13.51, 26.33, 32.94, 33.17, 46.73 g g^{-1}. The adsorption speed and adsorption capacity were determined by the viscosity of the adsorbed oil, the length of the fiber, and the wettability of the oil to the fiber, as shown in Fig. 7.11C and D. The adsorption capacity of the fiber for hot oil and cold oil was evaluated

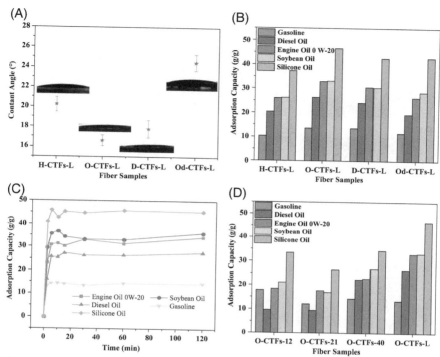

Figure 7.11 The contact angle between different silanized fibers with engine oil 0W-20 (A). The maximum adsorption capacity of different silanized fibers (B) for various oils under 25°C (error bars represent ± standard deviations, n = 3). Adsorption rate of O-CTFs-L for various oils at 25°C (C). The adsorption capacities of O-CTFs-12, O-CTFs-21, O-CTFs-40 and O-CTFs-L for various oils at 25°C (D). The copyright of the figure obtained from Elsevier under the license No: 5274200043525, Mar 22, 2022.

at 90°C and 10°C. O-CTFs–L showed high adsorption capacity for oil at different temperatures. It was a highly efficient material with the potential to deal with oil leakage under extreme conditions.

7.6 Porous biomass tubular carbon fibers from platanus orientalis

Biomass materials are widely used, which provides a feasible and easy way to obtain cheap carbon materials. Biomass after a reasonable and controllable carbonization process, can make it has rich pores and high specific surface area, showing the characteristics of lightweight, especially suitable for adsorption separation materials. Many biomass carbon materials have

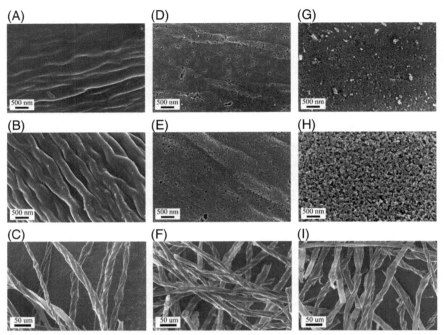

Figure 7.12 SEM images and surface structure of F700-5 (A), F700-10 (B, C), F800-5 (D), F800-10 (E, F), F900-5 (G) and F900-10 (H, I). The copyright of the figure obtained from Elsevier under the license No: 5274200471488, Mar 22, 2022.

been used in oil–water separation and oil adsorption. Previous studies on tubular polymer nanofibers have found that in addition to the oil adsorption characteristics of 1D materials, the physically cross-linked voids formed by the overlap between the fibers have a significant effect on oil enrichment. The fruit of the platanus orientalis appeared as a sphere with a rough surface, and microscopically it was assembled by countless bamboo-shaped fibers. The disassembled plant fibers (SnFs) were hydrophobic, with a contact angle of up to 151° with water, and can be completely wetted by most oils. This was due to the suitable roughness of the fiber surface and the crosslinked monosaccharide layer. After carbonization, the tubular carbon fibers (PCnFs) with porous tube walls had obvious pores, as shown in Fig. 7.12D, E, G, and H. These channels were passageways for oil products to enter and exit. Although the diameter of PCnFs was between 25 and 40 μm, each "bamboo node" was not connected to others, so oil adsorption from the end site was not possible. The sample was named after carbonization temperature and time, for example F900–5 represented vacuum carbonization at 900°C

Table 7.2 Pore performance data of the obtained products.

Samples	BET ($m^2 g^{-1}$)	Average pore size (nm)	Pore volume ($cm^3 g^{-1}$)
F700–5	12.16	6.50	0.02
F700–10	27.01	8.88	0.04
F800–5	38.25	7.74	0.04
F800–10	335.86	6.36	0.24
F900–5	342.70	8.90	0.31
F900–10	702.44	9.88	0.64

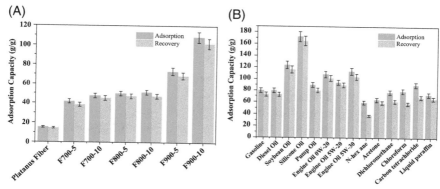

Figure 7.13 Maximum adsorption capacity of different fibers at 25°C (A). The maximum adsorption capacity of F900-10 (B) for various oils and organic solvents (error bars represent ± standard deviations, n = 3). The copyright of the figure obtained from Elsevier under the license No: 5274200471488, Mar 22, 2022.

for 5 h. The morphologies of different carbonation products were shown in Fig. 7.12A–H. The surface pores of PCnFs were visible, and the pore performance parameters of fibers were listed in Table 7.2.

The effect of carbonization temperature on adsorption properties was studied. As the carbonization temperature and time increased, the adsorption capacity of fiber to engine oil 0W–20 increased, as shown in Fig. 7.13A. The adsorption capacity of F900-10 was the highest, reaching 108.71 g g^{-1}. Meanwhile, F900-10 showed excellent adsorption capacity for different oils (Fig. 7.13B), and the adsorption process could be completed within 90 min. The desorbed oil can be simply squeezed physically, and the recovery of engine oil 0W–20 F900-10 can reach 93.87% of the adsorption capacity.

7.7 Summary and outlook

High oil adsorption fiber materials have entered the stage of rapid development, and various new materials and preparation methods keep emerging. At present, its development is not limited to the pursuit of a high oil adsorption rate but tends to be more functional. Using new materials, new preparation technology, new technology, cross-disciplinary fields, prepare new high-efficiency oil-absorbing materials. In the future, its development direction will mainly include the following two points.

(1)   Single material multifunctional design

After satisfying the pursuit of the basic oil adsorption rate and adsorption capacity, starting from the structural design, it will become inevitable to seek the multi-functionalization of the materials. For example, the introduction of some special groups in the material design process can not only meet the high adsorption capacity but also respond to conditions such as PH or light in the environment, so that the material can realize an intelligent cycle. In short, through the precise design of a molecular structure, the deepening function of high oil-absorbing materials can be given.

(2)   Multimaterial composite design

In practical applications, the composite use of high oil-absorbing materials with other excellent properties materials can combine the advantages of multiple materials, and it will also be an important trend in the development of high oil-absorbing materials in the future. For example, the high oil-absorbing material is compounded with magnetic nanoparticles to give the material magnetism, which is convenient for the separation and recovery of the material after the adsorption is completed. Or the combination of high oil-absorbing materials and high-efficiency catalytic degradation nanomaterials can improve the degradation of pollutants by catalytic degradation materials and improve the efficiency of pollution control. This will be a hot issue for research in the future.

References

[1] S. Kabiri, D.N.H. Tran, T. Altalhi, D. Losic, Outstanding adsorption performance of graphene–carbon nanotube aerogels for continuous oil removal, Carbon 80 (2014) 523–533.

[2] S. Songsaeng, P. Thamyongkit, S. Poompradub, Natural rubber/reduced-graphene oxide composite materials: morphological and oil adsorption properties for treatment of oil spills, J. Advanced Res. 15 (2019) 79–89.

[3] T. Liu, S. Chen, H. Liu, Oil adsorption and reuse performance of multi-walled carbon nanotubes, Procedia Eng. 102 (2015) 1896–1902.

[4] Y. Nishi, N. Iwashita, Y. Sawada, M. Inagaki, Sorption kinetics of heavy oil into porous carbons, Water Res. 36 (20) (2002) 5030–5036.

[5] A.A. Al-Majed, A.R. Adebayo, M.E. Hossain, A sustainable approach to controlling oil spills, J. Environ. Manag. 113 (2012) 213–227.

[6] J.L Schnoor, The gulf oil spill, Environ. Sci. Technol. 44 (2010) 4833.

[7] M.Z. Iqbal, A.A. Abdala, Oil spill cleanup using graphene, Env. Sci. Poll. Res. 20 (5) (2013) 3271–3279.

[8] R.P. Swannell, K. Lee, M. Mcdonagh, Field evaluations of mrine oil spill bioremediation., Microbiol. Rev. 60 (2) (1996) 342–365.

[9] M.V. Buzaeva, E.N. Kalyukova, E.S. Klimov, Sorption properties of AG-3 activated carbon in relation to oil products, Russ. J. Appl. Chem. 83 (10) (2010) 1883–1885.

[10] B. Zhao, L. Zhang, Y. Liang, H. Qiu, Efficient growth of millimeter-long few-walled carbon nanotube forests and their oil sorption, Appl. Phys. A 108 (2) (2012) 351–355.

[11] Y.P. Zheng, H.N. Wang, F.Y. Kang, Sorption capacity of exfoliated graphite for oils-sorption in and among worm-like particles, Carbon 42 (2004) 2603–2607.

[12] E. Ajenifuja, S.O. Alayande, O.A. Aromolaran, Equilibrium Kinetics Study of electrospun polystyrene and polystyrene-zeolite fibres for crude oil-water separation, J. Water Process Eng. 5 (2017) 253–259.

[13] S.J Choi, T. Kwon, H. Im, D. Moon, D. Baek, M. Seol, Y. Choi, A polydimethylsiloxane(PDMS) sponge for the selective adsorption of oil from water, ACS Appl. Mater. Interfaces 3 (12) (2011) 4552–4556.

[14] P. Li, Y. Qiao, L. Zhao, D. Yao, H. Sun, Electrospun PS/PAN fibers with improved mechanical property for removal of oil from water, Mar. Pollut. Bull. 93 (1) (2015) 75–80.

[15] V. Singh, R.J. Kendall, K. Hake, S. Ramkumar, Crude oil sorption by raw cotton, Ind. Eng. Chem. Res. 52 (18) (2013) 6277–6281.

[16] T. Annunciado, T. Sydenstricker, S. Amico, Experimental investigation of various vegetable fibers as sorbent materials for oil spills, Mar. Pollut. Bull. 50 (11) (2005) 1340–1346.

[17] G. Thilagavathi, C. Praba Karan, D. Das, Oil sorption and retention capacities of thermally-bonded hybrid nonwovens prepared from cotton, kapok, milkweed and polypropylene fibers, J. Environ. Manage. 219 (1) (2018) 340–349.

[18] Y. Xu, Q. Su, H. Shen, G. Xu, Physicochemical and sorption characteristics of poplar seed fiber as a natural oil sorbent, Text. Res. J. (2019) 4186–4194.

[19] F. Li, Z. Wang, S. Huang, Y. Pan, X. Zhao, Flexible, durable, and unconditioned superoleophobic/superhydrophilic surfaces for controllable transport and oil-water separation, Adv. Funct. Mater. (2018) 1706867.

[20] R. Wenzel, Resistance of solid surfaces to wetting by water, Industr. Eng. Chem. 28 (1936) 988–994.

[21] A. Cassie, S. Baxter, Wettability of porous surface, Trans. Faraday Soc. 40 (1994) 546–551.

[22] S. Svensson, A. Burke, D. Carrad, M. Leijnse, H. Linke, A. Micolich, Using polymer electrolyte gates to set-and-freeze threshold voltage and local potential in nanowire-based devices and thermoelectrics, Adv. Funct. Mater. 25 (2) (2014) 255–263.

[23] Lu J. Sponge-like polymeric adsorption material[P]. US 9427724B2. 2016 -08-30.

[24] Lu J. Method for preparing fibrous polymeric adsorption materials[P]. US 9427926B2. 2016 -08-30.

[25] F. Moura, R. Lago, Catalytic growth of carbon nanotubes and nanofibers on vermiculite to produce floatable hydrophobic "nanosponges" for oil spill remediation, Applied Catalysis B Environmental 90 (3–4) (2009) 436–440.

[26] V. Rajakovic, G. Aleksic, L. Rajakovic, Governing factors for motor oil removal from water with different sorption materials, J. Hazard. Mater. 154 (1-3) (2008) 558–563.

[27] A. Gunasekara, J. Donovan, B. Xing, Ground discarded tires remove naphthalene, toluene, and mercury from water, Chemosphere 41 (8) (2000) 1155–1160.

[28] C. Lin, Y. Hong, A. Hu, Using a composite material containing waste tire powder and polypropylene fiber cut end to recover spilled oil, Waste Manage. (Oxford) 30 (2) (2010) 263–267.

[29] D. Ceylan, S. Dogu, B. Karacik, S. Yakan, O. Okay, O. Okay, Evaluation of butyl rubber as sorbent material for the removal of oil and polycyclic aromatic hydrocarbons from seawater, Environ. Sci. Technol. 43 (10) (2009) 3846–3852.

[30] D. Ceylan, O. Okay, Macroporous polyisobutylene gels: A novel tough organogel with superfast responsivity, Macromolecules 40 (24) (2009) 8742–8749.

[31] D.H. Reneker, I. Chun, Nanometre diameter fibres of polymer, produced by electrospinning, Nanotechnology 7 (3) (1999) 216–223.

[32] D. Lxy, Y.N. Xia, Electrospinning of nanofibers: Reinventing the wheel? Adv. Mater. 16 (14) (2004) 1151–1170.

[33] J. Xue, J. Xie, W. Liu, Y. Xia, Electrospun nanofibers: new concepts, materials, and applications, Acc. Chem. Res. 50 (2017) 1976–1987.

[34] B. Sun, Y.Z. Long, H.D. Zhang, M.M. Li, Advances in three-dimensional nanofibrous macrostructures via electrospinning, Prog. Polym. Sci. 39 (5) (2014) 862–890.

[35] E. Zhmayev, D. Cho, LJ. Yong, Nanofibers from gas-assisted polymer melt electrospinning, Polymer 51 (18) (2010) 4140–4144.

[36] D. Yu, X. Li, X. Wang, J. Yang, S. Bligh, G. Williams, Nanofibers fabricated using triaxial electrospinning as zero order drug delivery systems, ACS Appl. Mater. Interfaces 7 (2015) 18891–18897.

[37] Reneker D., Fong H. Polymeric nanofibers: Introduction. 2006.

[38] Y. Liao, C. Loh, M. Tian, R. Wang, A. Fane, Progress in electrospun polymeric nanofibrous membranes for water treatment: Fabrication, modification and applications, Prog. Polym. Sci. 77 (2018) 69–94.

[39] H. Zhu, S. Qiu, J. Wei, Evaluation of electrospun polyvinyl chloride/polystyrene fibers as sorbent materials for oil spill cleanup, Environ. Sci. Technol. 45 (10) (2011) 4527–4531.

[40] J. Wu, N. Wang, L. Wang, H. Dong, Y. Zhao, L. Jiang, Electrospun porous structure fibrous film with high oil adsorption capacity, ACS Appl. Mater. Interfaces 4 (6) (2012) 3207–3212.

[41] J. Lin, Y. Shang, B. Ding, J. Yang, J. Yu, Nanoporous polystyrene fibers for oil spill cleanup, Mar. Pollut. Bull. 64 (2) (2012) 347–352.

[42] P.Y. Chen, S.H. Tung, One-step electrospinning to produce nonsolvent-induced macroporous fibers with ultrahigh oil adsorption capability, Macromolecules 50 (6) (2017) 2528–2534.

[43] P. Li, Y. Qiao, L. Zhao, D. Yao, H. Sun, Y. Hou, et al., Electrospun PS/PAN fibers with improved mechanical property for removal of oil from water, Mar. Pollut. Bull. 93 (1-2) (2015) 75–80.

[44] J. Wu, A.K. An, J. Guo, E. Lee, M. Farid, S. Jeong, CNTs reinforced super-hydrophobic-oleophilic electrospun polystyrene oil sorbent for enhanced sorption capacity and reusability, Chem. Eng. J. 314 (2017) 526–536.

[45] J. Gao, B. Li, L. Wang, X. Huang, H. Xue, Flexible membranes with a hierarchical nanofiber/microsphere structure for oil adsorption and oil/water separation, J. Ind. Eng. Chem. 68 (2018) 416–424.

[46] D. Shu, P. Xi, S. Li, C. Li, X. Wang, B. Chen, Morphologies and properties of PET nano porous luminescence fiber: Oil adsorption and fluorescence-indicating functions, ACS Appl. Mater. Interfaces 10 (2018) 2828–2836.

[47] P.P. Dorneanu, C. Cojocaru, N. Olaru, P. Samoila, A. Airinei, L. Sacaresau, Electrospun PVDF fibers and a novel PVDF/CoFe$_2$O$_4$ fibrous composite as nanostructured sorbent materials for oil spill cleanup, Appl. Surf. Sci. 424 (3) (2017) 389–396.

[48] B. Song, J. Zhu, N.H. Fa, Magnetic fibrous sorbent for remote and efficient oil adsorption, Mar. Pollut. Bull. 120 (1-2) (2017) 159–164.

[49] S. Iijima, Helical microtubules of graphitic carbon, Nature 354 (1991) 56–58.

[50] K. Hata, D.N. Futaba, K. Mizuno, Water assisted highly efficient synthesis of impurity free single walled carbon nanotubes, Science 306 (5700) (2004) 1362–1364.

[51] H.W. Zhu, C.L. Xu, D.H. Wu, B.Q. Wei, R. Vajtai, P.M. Ajayan, Direct synthesis of long single-walled carbon nanotube strands, Science 296 (5569) (2002) 884–886.

[52] Z. Fan, J. Yan, G. Ning, T. Wei, W. Qian, S. Zhang, et al., Oil sorption and recovery by using vertically aligned carbon nanotubes, Carbon 48 (14) (2010) 4197–4200.

[53] X. Gui, Z. Zeng, Z. Lin, Q. Gan, R. Xiang, Y. Zhu, et al., Magnetic and highly recyclable macroporous carbon nanotubes for spilled oil sorption and separation, ACS Appl. Mater. Interfaces 5 (12) (2013) 5845–5850.

[54] X. Gui, J. Wei, K. Wang, A. Cao, H. Zhu, Y. Jia, et al., Carbon nanotube sponges, Adv. Mater. 22 (5) (2010) 617–621.

[55] H. Li, Y. Li, Y. Chen, M. Yin, T. Jia, S. He, et al., Carbon tube clusters with nanometer walls thickness, micrometer diameter from biomass and its adsorption property as bioadsorbent, ACS Sustain. Chem. Eng. 7 (1) (2019) 858–866.

[56] J. Nine, S. Kabiri, A.K. Sumona, T. Tung, M. Moussa, D. Losic, Superhydrophobic/superoleophilic natural fibres for continuous oil-water separation and interfacial dye-adsorption, Sep. Purif. Technol. 233 (2020) 116062.

[57] J. Lee, D. Kim, S. Han, B. Kim, E. Park, M. Jeong, et al., Fabrication of superhydrophobic fibre and its application to selective oil spill removal, Chem. Eng. J. 289 (2016) 1–6.

[58] P. Song, J. Cui, J. Di, D. Liu, M. Xu, B. Tang, et al., Carbon microtube aerogel derived from kapok fiber: An efficient and recyclable sorbent for oils and organic solvents, ACS Nano 14 (2020) 595–602.

[59] Z. Li, L. Zhong, T. Zhang, F. Qiu, X. Yue, D. Yang, Sustainable, flexible and superhydrophobic functionalized cellulose aerogel for selective and versatile oil/water separation, ACS Sustain. Chem. Eng. 7 (2019) 9984–9994.

CHAPTER 8

Fabrication of carboxylated tubular carbon nanofibers as anode electrodes for high-performance lithium-ion batteries

Yu Huyan[a], Junjie Chen[a] and Baoliang Zhang[a,b]
[a]School of Chemistry and Chemical Engineering, Northwestern Polytechnical University, Xi'an, China
[b]Xi'an Key Laboratory of Functional Organic Porous Materials, Northwestern Polytechnical University, Xi'an, China

8.1 Introduction

With the increase in energy demand and aggravation of environmental pollution, much attention has been paid to the development of new energy conversion and storage devices. Rechargeable lithium–ion batteries (LIBs) have become an important energy storage device due to their excellent cycle stability, low self-discharge, and no memory effect [1]. To satisfy the requirements of the development of modern high-tech products such as hybrid vehicles and artificial satellites, LIBs with higher reversible capacity, cycle stability, and rate performance have been proposed. Graphite is mostly employed in LIBs as anode material. However, it is restricted due to its low theoretical capacity and poor overcharge/discharge resistance. These make graphite unable to meet the requirements of high energy density and power density of LIBs [2,3]. Thus, one of the key factors to high–performance LIBs lies in the designation and preparation of carbon-based anode with large reversible capacity and stable cycling.

Compared with zero-dimensional (0D) and two-dimensional (2D) carbon nanomaterials, one-dimensional (1D) carbon nanomaterials (e.g., nanorods, nanowires, and nanotubes) show special properties in mechanics, electricity, and optics. They possess several excellent performances such as large aspect ratio and dimensional confinement effect. These structural characteristics make 1D carbon nanomaterials display great application prospects in energy storage [4,5], catalysis [6–8], and sensing [9,10]. Among

Fabrication and Functionalization of Advanced Tubular Nanofibers and their Applications.
DOI: https://doi.org/10.1016/B978-0-323-99039-4.00005-X
183

these materials, porous carbon nanofibers (PCnF) exhibit several unique advantages in LIBs electrode materials because of their large aspect ratio ($\alpha > 1000$), high specific surface area, abundant pore channels, and small-fiber diameter [11,12]. On the one hand, the diffusion path of lithium ions in PCnF is short. It facilitates the rapid insertion and extraction of lithium ions. On the other, the large specific surface area and abundant pore structure of PCnF help to increase the contact between electrode and electrolyte. It is beneficial for the transport of lithium ions. Unfortunately, the low theoretical specific capacity and surface chemical inertness result in poor energy density and power density. Meantime, PCnF and their derivatives used in LIBs anode materials are mainly obtained by electrospinning and post-treatment. The preparation process of the methods is complicated [13,14]. Thus, it is necessary to design a new functional PCnF anode and develop simple synthesis methods.

8.2 Introduction of anode materials for lithium-ion batteries

In the 1980s, the Canadian Moli Energy Company designed commercial lithium secondary batteries with lithium metal as the anode and MoS_2 as the cathode [13]. However, due to the uneven dissolution and deposition of Li+ on the electrode surface during the electrochemical reactions, specific capacity attenuation and safety hazards are inevitable. Therefore, the application of secondary batteries with lithium as the anode is limited. To solve the above problems, in 1980, a "rocking-chair" battery was proposed by Armand [14], the interlayer compound $Li_yM_nY_m$ with the low lithium intercalation potential was as the anode and A_zB_w with the high potential was as the cathode, which inhibits the formation of "dead" Li. Although the rocking chair battery solves the safety problem, the low working voltage and slow ion transport still exist. In 1990, Japan's Sony Energy Company innovatively adopted petroleum coke as the anode material of rocking chair battery, which effectively improved the working voltage and energy density, the battery was named LIBs [15]. Fig. 8.1 shows the schematic diagram of the working principle for LIBs. LIBs are mainly composed of four parts: anode, cathode, separator, and electrolyte. During the charging process, Li^+ are extracted from the cathode, and electrons are released, then Li^+ are embedded in the lattices of anode materials through the electrolyte. Meanwhile, the electrons are transferred from cathode to anode through external circuits, the electrical energy is converted into chemical energy, the discharging process is vice versa. Among them, anode,

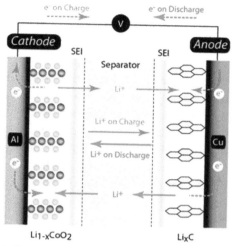

Figure 8.1 Schematic illustrating the mechanism of LIBs charging and discharging process [16]. The copyright of the Figure obtained from Elsevier under the license No: 5234210908509, Jan 22, 2022.

as one of the core structural units of LIBs, should mainly meet the four principles:

(1) Environmentally friendly and low cost; (2) Superior mechanical and structural stability, electronic and ionic conductivity; (3) Low and stable intercalation/extraction potentials; (4) Reversible insertion and de-insertion, etc. According to the working mechanism, anode materials of LIBs can be divided into three types: intercalation, alloy, and conversion type.

8.2.1 Intercalation type anode materials

Intercalation anode materials include carbon and titanium oxides. Among them, carbon materials include graphite, acetylene black, microbead carbon, petroleum coke, and carbon fiber. Carbon materials as anode materials for LIBs exhibit the advantages of high working voltage, good safety, and low cost, but the high hysteresis voltage and irreversible capacity still exist, which limit the application of LIBs with high energy density and power density. The high hysteresis voltage can be explained by the microporous lithium storage mechanism in amorphous carbon, Li^+ is inserted into the graphite crystallites and the micropores to form lithium clusters or molecules during the insertion process. During the extraction process, Li^+ is firstly extracted from outer graphite crystallites, then lithium clusters in the micropores are extracted from graphite crystallites. Since the strong interaction between free radical carbon atoms and Li^+, the extraction of Li^+ requires some force, resulting in voltage hysteresis. In recent years, graphene, as a derivative of graphite, shows a large specific surface area, good electronic conductivity, chemical stability, and excellent mechanical properties, which may be the promising anode material for LIBs [17–19]. However, graphene cannot be directly used as anode material for LIBs due to low initial coulombic efficiency and rapid capacity fading. Extending the interlayer spacing of graphene is one of the effective ways to solve the above problems. Yoo et al. [18] introduced macromolecules to extend the interlayer distance of graphene to increase the lithium storage capacity. By introducing CNT or C60 into graphene, the specific capacity increases from 540 to 730 and 784 mAh g^{-1}. Therefore, the electrochemical performance of graphene nanosheets can be effectively improved by adjusting the interlayer spacing. Secondly, the energy band and electronic structure of graphene can also be changed via heteroatom doping. The specific capacities of N-doped graphene and B-doped graphene prepared by Wu et al. [20] are about 199 and 235 mAh g^{-1} at 25 A g^{-1}. The N or B doping effectively increases the defects on the surface of graphene nanosheets and extends the interlayer distance, improves the electronic conductivity and thermal stability, enhances the adsorption of Li^+ and the electron transport rate. Reddy et al. [21] directly grew N-doped graphene on Cu collectors via chemical vapor deposition and studied the lithium storage properties. Electrochemical tests show that due to nitrogen doping, abundant defects are generated on the surface, which makes the reversible discharge capacity of nitrogen-doped graphene almost twice that of pristine graphene. In addition, graphene can also be used as the matrix to support other anode nanomaterials on the surface, such as silicon, tin, transition metal oxides or sulfides, etc. [22–26]. Graphene can not only provide sufficient loading sites for other anode materials to avoid the agglomeration of metals, transition metal oxides, or sulfides but also effectively alleviate the volume change during charging and discharging process.

Titanium oxides mainly include $Li_4Ti_5O_{12}$ and TiO_2. Among them, spinel-structure $Li_4Ti_5O_{12}$ is a "zero-strain" intercalation anode material with excellent structural stability and safety, however, the low electronic and ionic conductivity and theoretical capacity (175 mAh g^{-1}) limit the application in high-power LIBs. In recent years, to improve the conductivity and rate performance of $Li_4Ti_5O_{12}$, the measures mainly include the following three aspects: (1) Preparing $Li_4Ti_5O_{12}$ nanomaterials with different morphologies [27,28]. The large specific surface area of nanomaterials can effectively increase the lithium storage capacity. The small size is beneficial to shorten the ion transport path so that the electrode materials exhibit rapid charging performance. (2) Doping modification. Appropriate doping of Na^+, Cu^{2+}, Mg^{2+}, and Al^{3+} can effectively improve the electronic and ionic conductivity of $Li_4Ti_5O_{12}$, rapid charging performance is also improved [29,30]. (3) Surface modification. $Li_4Ti_5O_{12}$ is coated with nanomaterials to improve the electronic conductivity and cycle stability, including carbon materials, metals, and oxides, etc. [31–34].

8.2.2 Alloy-type anode materials

The charge-discharge process of alloy-type anode materials is realized via alloying and dealloying reactions between active materials and Li^+. The lithium storage process can usually be expressed as $xLi^+ + xe^- + M \rightarrow Li_xM$, where M includes Si, Ge, Sn et al. Among them, Si is the anode material with the highest theoretical capacity. During the lithiation process, Si reacts with Li^+ to form $Li_{12}Si_7$, Li_7Si_3, $Li_{13}Si_4$, $Li_{15}Si_4$, $Li_{22}Si_{15,}$ and other alloys, the theoretical capacity up to 4200 mAh g^{-1} is about 11 times larger than the graphite anode. Moreover, the delithiation voltage of Si is 0.45 V, which is within the reasonable voltage range. However, in the lithiation process, with the continuous insertion of Li^+, the volume expansion rate of Si is as high as 300%–400%, while in the delithiation process, the exfoliation of active materials in electrodes and the structural collapse lead to rapid capacity fading. In addition, during the initial charge-discharge process, Si reacts with the organic electrolyte at solid-liquid interfaces to form the solid electrolyte interface (SEI) films on the surface of Si electrodes. The initial SEI films are destroyed by the volume expansion/contraction of Si during lithiation/delithiation process, while new SEI films are formed during the subsequent cycles, which will inevitably lead to poor cycle stability of anode materials.

The preparation of nano-Si is an effective way to solve the above problems. Si nano-materials can not only effectively alleviate the volume change and improve the cycle performance, but also shorten the Li^+ diffusion distance and accelerate electrochemical reactions. For the structural design, the preparation of nano-Si includes three types: (1) Design and preparation of solid Si nanostructures with different morphologies, such as nanoparticles, nanowires, etc. Wang et al. [35] first adopted aqueous NaCl particles as the sacrificial template, tetraethyl orthosilicate as the silicon source, Si nanosheets were prepared through the crystal formation

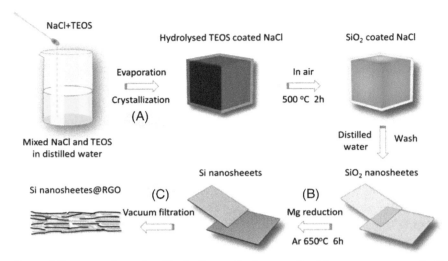

Figure 8.2 The schematic illustration of the synthesizing processes of Si nanosheets@rGO [35] (The copyright of the Figure obtained from Elsevier under the license No: 5279210289245 Mar 31, 2022).

and chemical reduction process, then Si nanosheets@rGO electrodes were obtained via the vacuum filtration, the preparation process is shown in Fig. 8.2. The electrode delivers a high reversible capacity of 2250 mAh g^{-1} after 50 cycles at 0.2 A g^{-1}. (2) Design of hollow Si nanostructures. The internal space in a hollow structure can effectively relieve the stress and strain due to the volume changes during the cycling process, the structural stability of Si-based electrodes is maintained. Yao et al. [36] combined the improved Stöber method, CVD, and HF etching strategy to prepare interconnected hollow Si nanospheres. The capacity attenuation was only 8% after 700 cycles at 0.5 C, showing excellent cycling stability. (3) Design of confined hollow Si structures. Compared with hollow Si nanostructures, the structure can buffer the volume changes of Si during the cycling process, the surface modification of hollow Si electrodes is beneficial to improve the stability of SEI films. Hu et al. [37] first prepared Si@SiO$_2$@RF@SiO$_2$@PDA composites via improved Stöber method and carbon coating, then combining carbonization and HF etching to obtain core-shell structured Si/SiO with double carbon layers. The electrodes exhibit excellent structural stability and maintain the specific capacity of 1113 mAh g^{-1} at 100 mA g^{-1} after 200 cycles. Firstly, hollow porous pollen microspheres (HPP) were obtained by Chen et al. [38] via etching biological pollen with HCl solution, HPP was used as templates to accommodate the silane coupling agent modified nano-Si. Then,

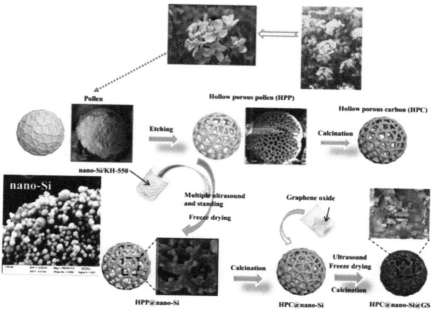

Figure 8.3 Schematic illustration of the synthesis procedure of HPC@nano-Si@GS [38] (The copyright of the Figure obtained from Elsevier under the license No: 5279211314570 Mar 31, 2022).

nano-Si-coated hollow porous carbon microspheres (HPC@nano-Si) were obtained by carbonization. Finally, graphene nanosheet-coated HPC@nano-Si microspheres (HPC@nano-Si@GS) were prepared by combining ultrasonic, freeze-drying, and carbonization techniques. The preparation process is shown in Fig. 8.3. HPC@nano-Si@GS was used as anode materials for LIBs, the reversible capacity reached 1043 mAh g^{-1} with the capacity retention of 67% after 200 cycles at 0.1 A g^{-1}.

8.2.3 Conversion type anode materials

Conversion-type anode materials include oxides, phosphides, sulfides, and nitrides of Fe, Co, Ni, Mn, Cu, Cr, etc. The charging and discharging process is realized by the reactions between active materials and Li$^+$ to form metal clusters, the lithium storage process is generally expressed as $M_xO_y + 2yLi^+ + 2ye^- \rightarrow yLi_2O + xM$. The conversion-type materials exhibit high theoretical capacities of 700–1200 mAh g^{-1} and volume capacities between 4000 and 5500 mAh cm^{-3}. However, like alloy-type materials, conversion-type materials also suffer from problems such as the large volume change, unstable SEI films, low initial coulombic efficiency, and poor cycling stability. In addition, due to the interconversion among multiple solid phases (MO_x, Li_2O, and M) of different structures during the (de)intercalation of Li^+, which requires the breaking of chemical bonds, conversion-type materials generally show larger lagging voltages. In recent years, to solve the problems, researchers usually adopt measures such as nanometerization or matching with buffer substrates [39–44].

8.3 Research status of carbon nanotube-based anode materials

As the allomorph of graphite, CNTs with the length-diameter ratio of > 1000 can be regarded as rolled by graphene sheets. According to the number of graphene layers, CNTs can be divided into single-walled (SWCNTs) and multiwalled carbon nanotubes (MWCNTs), the preparation methods include the catalytic cracking method, chemical vapor deposition method, arc method, and laser etching method. Compared with traditional graphite electrodes, CNTs have been widely used as anode materials for LIBs due to their high tensile strength and excellent electrical conductivity. The lithium storage mechanism [45–47] of CNTs suggests that Both the inside and outside of CNTs can serve as active sites for the insertion and adsorption of Li^+, the intercalation capacity is related to the morphology of nanotubes. The size of pores depends on the number of removed carbon atoms. When three carbon atoms are removed, Li^+ can easily diffuse into the interior of CNTs through defect sites and accumulate at the outside, the reversible capacity is also increased. In addition, the length of CNTs shows a significant effect on the diffusion efficiency of Li^+ [48]. When inserting into CNTs, Li^+ will be randomly diffused inside the nanotubes. If the length of CNTs is too long, it is not conducive to the extraction of Li^+, resulting in inefficient diffusion. Pure CNTs only exhibit a slight increase in capacity and cycling performance compared to graphite. In recent years, researchers have combined CNTs with other materials to prepare LIBs anodes with high energy and power density. Among them, CNTs can provide active sites for lithium storage and serve as scaffolds for other high-capacity materials, the electronic conductivity and structural stability of electrode materials are improved.

The electrochemical performance of CNTs electrodes can be improved via matching with other carbon materials, such as graphene, graphene oxide, carbon nanofibers, etc. The research mainly focuses on the design and preparation of CNTs-based composite electrodes, generally including blending, coating, in-situ chemical vapor deposition, etc. Zhang et al. [49] prepared graphene oxide/graphite/CNTs self-supporting electrodes by the blending method. Graphite particles and CNTs were uniformly dispersed on graphene oxide flakes, which effectively prevented the aggregation of graphene oxide flakes and promoted electron transfer in electrodes. The electrode delivers a high specific capacity of 1050.3 mAh g^{-1} with the capacity retention rate of 89.6% after 60 cycles at 0.5 C. Sahoo et al. [50] coated graphene nanosheets on the surface of MWNTs by the simple chemical vapor deposition. MWNTs exhibit a wrinkled surface, which increases the porosity of electrodes and the adsorption sites for Li$^+$, the specific capacity of 373 mAh g^{-1} is achieved over 150 cycles at 100 mA g^{-1}. Polyacrylonitrile/nickel acetate/polymethylmethacrylate composite nanofibers were prepared by Chen et al. [51] via the coaxial electrospinning technology. Nitrogen-doped hollow CNT-CNF hybrids were obtained combining the pyrolysis, KOH activation with HNO$_3$ treatment, the capacity loss was only 20% after 3500 cycles at 8 A g^{-1}.

8.4 Preparation of carboxyl modified carbon nanotube anode materials

To improve the electrochemical performance of PCnF electrodes, the most commonly used methods are oxidation treatment [52], surface coating [52,53] and heteroatom doping [7,54]. The oxidation treatment of PCnF is an obvious approach for obtaining functionalized materials. By this approach, the performance improvement of the PCnF-based electrode is realized.

Wang et al. [55] reported a carboxyl functional carbon fiber (PFCF) prepared via chemical oxidation by H$_3$PO$_4$/H$_2$SO$_4$/HNO$_3$. Compared with unmodified carbon fiber, PFCF exhibited excellent reversible, low charge transfer resistance and higher stable capacity during charge and discharge. Wang et al. [56] prepared carbon nanotube-based Li-S batteries cathodes by CO$_2$ oxidation, sonication-assist and vacuum filtration. The high reversible capacity of 459.6 mAh g^{-1} at 5 C was achieved. Oxidation treatment can generate nanopores in PCnF. These pores can effectively increase the storage sites and migration channels for lithium ions. Meanwhile, the oxygen-containing functional groups on the surface of materials can form passivation film during lithium ion insertion. It helps to improve the cycling performance of electrode. The gas phase oxidation strategy is difficult to ensure the homogeneity of products. It cannot meet the demands of industrial production. However, the materials modified by liquid phase oxidation process have single product and excellent electrochemical performance.

8.5 Fabrication of acidified tubular carbon nanofibers

In previous work, we have fabricated polymer nanofibers with tubular structure using benzyl halide as a monomer by restricted self-condensation approach. Then, porous tubular carbon nanofibers are developed through subsequent carbonization [25,26]. The successful development of these materials has provided new electrodes for LIBs. Herein, the tubular carbon nanofibers (TCFs) developed by our group are chosen as a substrate. A novel acidified tubular carbon nanofibers (CMTCFs) are fabricated via surface modification of TCFs by an oxidant (H_2O/HNO_3). The acidification time is adjusted to achieve the products with different carboxyl content. Furthermore, the electrochemical property of the obtained CMTCFs is studied and compared. The structural advantages of CMTCFs anode are responsible for outstanding electrochemical performance. Such as (1) CMTCFs with large aspect ratios are interconnected to provide continuous electron channels for an electrode. (2) The mesoporous of CMTCFs not only provide sufficient channels for lithium ions transmission but also effectively increase the lithium ions storage sites of the electrode. (3) The proper carboxyl content on CMTCFs can effectively balance the relationship between the capacity and reaction kinetics. Both the specific capacity and rate performance of the electrode have been improved. The study of the influence of carboxyl content on electrochemical performance provided a reference for the development of other functional groups modified electrodes.

8.5.1 Design and characterization of CMTCFs

Fig. 8.4 showed the whole strategy for the preparation of CMTCFs. First, TPFs were fabricated by Friedel-Crafts alkylation reaction in dual oil phase system and subsequent soxhlet extraction techniques. The obtained TPFs exhibited regular tubular structure and a large aspect ratio (Fig. 8.5A and B). After carbonization, the morphology of TCFs did not change significantly. However, the mass thickness contrast of the tube wall was reduced. It signified that abundant channels were successfully created on the tube wall of TCFs (Fig. 8.6C and D). Then, the liquid phase oxidation of TCFs was carried out under reflux conditions by HNO_3/H_2O. In this process,

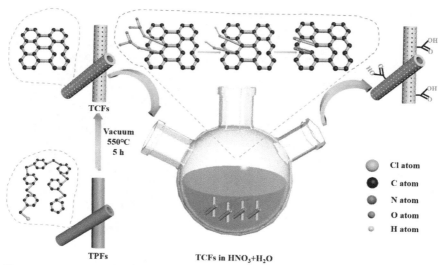

Figure 8.4 Schematic illustration of the synthesis of CMTCFs.

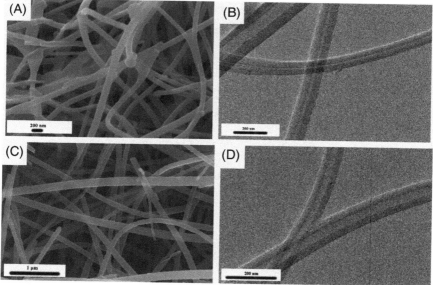

Figure 8.5 SEM (A and C) and TEM (B and D) images of the nanofibers: TPFs (A and B); TCFs (C and D).

functional groups including nitro and oxide were formed on TCFs as intermediates for carboxyl modification. Subsequently, their groups were hydrolyzed. They were transformed into O–H bonds and further into C=O bonds. It was accompanied by a fracture of the C–C bonds. Finally, CMTCFs

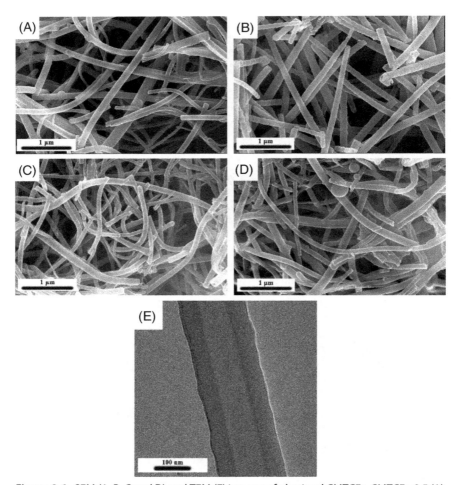

Figure 8.6 SEM (A, B, C and D) and TEM (E) images of obtained CMTCFs: CMTCFs-0.5 (A); CMTCFs-1 (B and E); CMTCFs-3 (C); CMTCFs-5 (D).

were obtained by further oxidizing the O-H bond and C=O bond to – COOH [55].

To clarify the effect of liquid-phase oxidation on the morphology of TCFs, SEM and TEM were used to observe the morphology of CMTCFs at different reaction times. As shown in Fig. 8.6, the morphology of CMTCFs did not change obviously with the increase of acidification time. The fibers were intertwined with each other to form a 3D network structure. It was beneficial to electronic transmission. Compared with TCFs, the diameter of CMTCFs was unchanged. The average diameter was about 120 nm

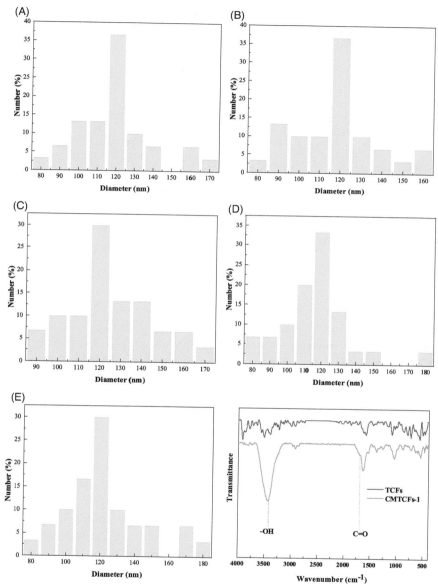

Figure 8.7 Diameter distribution curves of the nanofibers: TCFs (A); CMTCFs-0.5 (B); CMTCFs-1 (C); CMTCFs-3 (D); CMTCFs-5 (E); FTIR of TCFs and CMTCF-1 (F).

(Fig. 8.7A–E). Meantime, the tubular structure of CMTCFs was well maintained. The above results showed that the obtained TCFs had excellent acid resistance. In addition, from the pore size distribution curve of CMTCFs-1

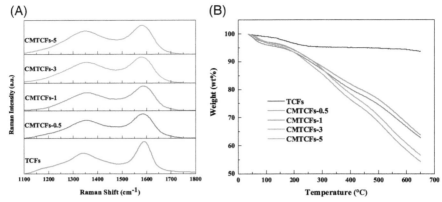

Figure 8.8 Raman spectra (A) and TGA curves (B) of the TCFs and CMTCFs.

Table 8.1 I_D/I_G intensity ratios of TCFs and CMTCFs.

Samples	TCFs	CMTCFs-0.5	CMTCFs-1	CMTCFs-3	CMTCFs-5
I_D/I_G	0.708	0.718	0.734	0.773	0.793

(Fig. 8.7F), pores at 4–6 nm were observed. This could be attributed to the mesoporous on the tube wall of CMTCFs-1. The size of lithium ions before and after solvation were 0.08 and 0.4 nm [57]. It was smaller than 4 nm. Therefore, the mesoporous on the tube wall of CMTCFs-1 not only facilitated ion transmission but also increased the lithium storage site of the electrode.

Raman was used to ingevaluate the effect of acidification on carbon in TCFs. As depicted in Fig. 8.8A, TCFs and CMTCFs had two distinct broad peaks between 1100 and 1800 cm^{-1}. The characteristic peaks located at ~1348 and ~1580 cm^{-1} corresponded to amorphous carbon (D band) and ordered graphitic carbon (G band) [58,59]. The intensity ratio of these two peaks could directly reflect the content of defects in carbon materials. Compared with TCFs, the I_D/I_G value increased after liquid-phase treatment. The I_D/I_G value increased from 0.718 to 0.793 as the acidification time prolonged (Table 8.1). It indicated that acidification increased the content of amorphous carbon in TCFs. Fig. 8.8B showed the TGA curves of TCFs and CMTCFs. The slight weight loss below 150°C was attributed to the evaporation of adsorbed water on samples. The loss between 150 and 500°C could be explained by the removal of functional groups on materials. Thermal degradation above 500°C was related to the oxidation process of disordered carbon. The TGA curve of TCFs reached a steady-state after 150°C. However, significant weight loss of CMTCFs was detected

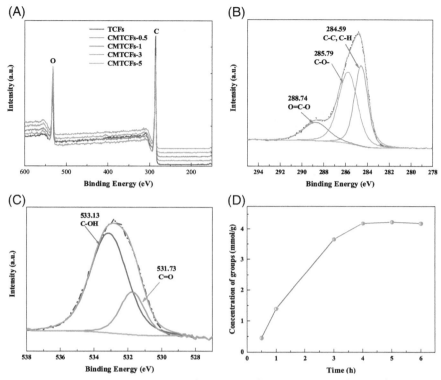

Figure 8.9 XPS survey scan spectra of CMTCFs and TCFs (A); C 1s and O 1s of CMTCFs-1 (B and C); acid-base titration results of CMTCFs at different acidification times (D).

after 150°C. This signified the increase of defect content and successful modification of oxygenous groups on TCFs after acidification. Besides, the degree of thermal degradation of CMTCFs-0.5 to CMTCFs-5 increased with the extension of reaction time. It was consistent with the Raman results.

Acidification not only changed the state of carbon but also introduced oxygenous groups on TCFs. To investigate the chemical composition of CMTCFs, we performed XPS and acid–base titration analysis. As shown in Fig. 8.9A, both C and O elements were detected in TCFs and CMTCFs. The O element in TCFs could be attributed to air oxidation during sample storage. Table 8.2 showed the O/C content ratio of TCFs and CMTCFs. The O/C content ratio in CMTCFs increased from 22.13% to 24.67% as the acidification time was extended from 0.5 to 5 h. It indicated that the oxygenous groups increase with the extension of oxidation time. In Fig. 8.9B, the C 1s spectrum of CMTCFs-1 can be divided into three peaks.

Table 8.2 O/C content ratio of CMTCFs and TCFs.

Samples	TCFs	CMTCFs-0.5	CMTCFs-1	CMTCFs-3	CMTCFs-5
O/C	12.87%	22.13%	23.53%	24.64%	24.67%

These peaks located at 284.59, 285.79, and 288.74 eV corresponded to C-C/C-H, C-O-, and O=C-O [60,61]. The peaks at 531.73 and 533.13 eV can be assigned to C=O and C-OH (Fig. 8.9C) [62]. Besides, the infrared spectral peaks at 1710 and 3456 cm^{-1} were attributed to the vibrational absorption peaks of C=O and O-H (Fig. 8.7F) [63,64]. The above results showed that the carboxyl groups were successfully modified on TCFs after oxidation. Acid-base titration was used to quantify the content of carboxyl groups on CMTCFs. As shown in Fig. 8.9D, the carboxyl group content on CMTCFs increased with prolonged oxidation time. The content reached 4.232 mmol g^{-1} and remained stable when the acidification time exceeded 4 h.

8.5.2 Electrochemical performance of carboxyl-modified tubular carbon nanofibers

The electrochemical performances of CMTCFs-0.5, CMTCFs-1, CMTCFs-3 and CMTCFs-5 were compared by CV, galvanostatic charge/discharge and EIS. The influence of carboxyl group content on electrochemical performance of CMTCFs anodes was clarified.

CV was used to analyze the electrochemical reactions of CMTCFs anode, the results were summarized in Fig. 8.10A–D. In the first negative scan, CMTCFs-0.5, CMTCFs-1, CMTCFs-3 and CMTCFs-5 all showed broad reduction peak at 1-2 V. It was related to the decomposition of electrolyte and the formation of solid electrolyte interface (SEI) film [65,66]. This peak disappeared in subsequent cycles. It signifies that the formation of SEI film mainly occurred in the first circle. The sharp reduction peak around 0.03 V corresponded to the insertion of lithium ions in CMTCFs electrodes. The CV curves were overlapped well in the last four cycles, indicating that CMTCFs anodes had excellent reversibility. There was no significant change in the CV profiles of CMTCFs electrodes compared to TCFs (Fig. 8.10). Thus, the modification of carboxyl groups on CMTCFs did not affect the electrochemical reaction of electrodes. Fig. 8.9E–H showed the galvanostatic charge/discharge curves of CMTCFs at 1000 mA g^{-1}. The curves all appeared as inclined lines. No obvious platform was observed. Meanwhile, The charge/discharge curves of CMTCFs were overlapped well

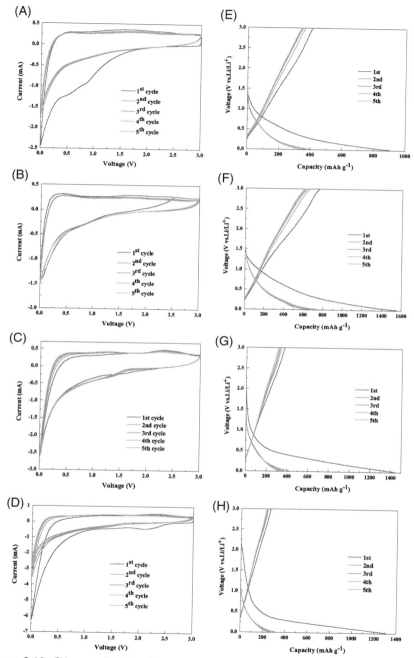

Figure 8.10 CV curves (A–D) and discharge/charge curves (E–H) of CMTCFs: CMTCFs-0.5 (A and E); CMTCFs-1 (B and F); CMTCFs-1 (C and G); CMTCFs (D and H).

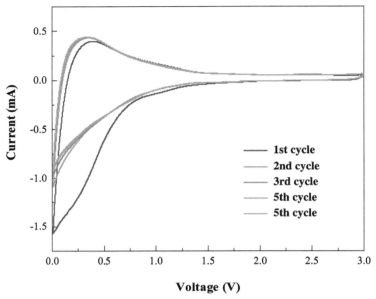

Figure 8.11 CV curves of TCFs.

after the first cycle. It was consistent with the CV results. The charge and discharge specific capacities of CMTCFs-1 in the first cycle were 752.7 mAh g^{-1} and 1554.5 mAh g^{-1}. The initial coulombic efficiency was 48.4%. The large irreversible capacity was mainly attributed to the formation of SEI film, decomposition of electrolyte and irreversible insertion of lithium ions (Fig. 8.11) [67,68].

The cycling performance of CMTCFs with different carboxyl contents at 1000 mA g^{-1} was compared, as shown in Fig. 8.12A. Compared with TCFs, the specific capacities of CMTCFs electrodes were all improved after acidification at 0.5, 1, and 3 h. The cycling performance of CMTCFs-5 was similar to that of TCFs. The specific capacity of CMTCFs-3 increased slowly during the first 200 cycles and then decreased continuously. CMTCFs-1 electrode underwent capacity decay within 25 cycles and remained stable in subsequent cycles. The decrease in capacity was related to the formation of SEI film and the deposition of electrolytes. The reversible specific capacity of CMTCFs-1 was 588.5 mAh g^{-1} after 500 cycles. It was only 10% lower than the second cycle. Although the stability of CMTCFs-0.5 and CMTCFs-5 anodes was maintained well during 500 cycles. The specific capacity was much lower than that of CMTCFs-1. Thus, carboxyl groups content on CMTCFs had a significant effect on electrochemical performance.

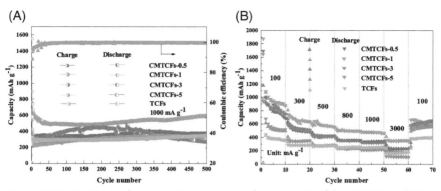

Figure 8.12 Cycling performance (A) and rate performance (B) of the CMTCFs and TCFs.

Appropriate carboxyl functional groups (CMTCFs-1) could introduce a pseudocapacitive reaction to increasing the specific capacity. However, excessive carboxyl groups (CMTCFs-3 and CMTCFs-5) hindered the migration and diffusion of ions. The electrochemical property of the electrode was reduced [57].

The rate performance of CMTCFs in the range of 100–3000 mA g^{-1} was investigated and compared. As shown in Fig. 8.12B, CMTCFs-1 showed the best rate performance. The average reversible specific capacities at different current rates of 100, 300, 500, 800, 1000, and 3000 mA g^{-1} were 932.53, 673.21, 582.51, 500.46, 474.51, and 354.73 mAh g^{-1}. When the current rate returned to 100 mA g^{-1}, the average reversible capacity was maintained at 615.29 mAh g^{-1}. Moderate carboxyl groups content, stable framework structure, and mesoporous on tube wall made CMTCFs-1 anode achieve outstanding rate performance. The cycling performance of CMTCFs-3 and CMTCFs-5 was better or similar to that of TCFs. However, the rate performance was poor at high current density (3000 mAh g^{-1}). It indicated that the specific capacity of the electrode could be improved by carboxyl modification. Excessive carboxyl groups could reduce the rate of performance of materials. This was similar to the reported results [57].

Fig. 8.13 showed EIS curves of CMTCFs anodes after 500 cycles at 1000 mA g^{-1}. The inset was the corresponding equivalent circuit. All Nyquist curves consisted of two concave semicircles and an inclined line. The concave semicircle at the high-frequency region corresponded to R_S, C_{SEI} and R_{SEI}. R_S represented the electrolyte resistance. C_{SEI} and R_{SEI} were resistance and capacitance of solid interface layer formed on electrode [69]. The concave semicircle located at the middle-high frequency region

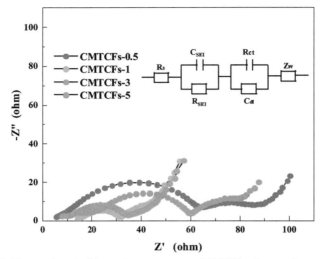

Figure 8.13 Electrochemical impedance spectra of CMTCFs after cycling performance test. The inset is the equivalent circuit model.

Table 8.3 Simulation results of the kinetic parameters of CMTCFs after cycling performance test.

Samples	R_S (ohm)	R_{SEI} (ohm)	R_{ct} (ohm)
CMTCFs-0.5	1.77	59.27	26.12
CMTCFs-1	11.28	14.42	7.34
CMTCFs-3	18.59	42.73	11.51
CMTCFs-5	5.61	19.63	10.73

corresponded to R_{ct} and C_{dl}. They were electric double-layer capacitors and charge transfer resistors [70]. The inclined line at the low-frequency region corresponded to Z_W. It was related Warburg impedance of the lithium ions diffusion process. CMTCFs-1 anode had the smallest R_{ct} value (Table 8.3). It signified that the electrode had the lowest charge transfer resistance. Thus, optimal electrochemical performance was obtained.

SEM was used to detect the morphological changes of CMTCFs-1 before and after cycling. As shown in Fig. 8.14A, uniform dispersion of CMTCFs-1 in electrode material was observed. The fibers were intertwined with each other to form 3D conductive network. This facilitated the transfer of electrons (Fig. 8.14B). CMTCFs-1 still had a regular fiber structure and smooth surface after electrode preparation. It showed that the mixing and coating did not affect CMTCFs-1. The fibrous structure remained well after

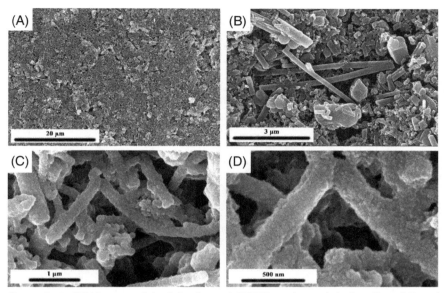

Figure 8.14 SEM images of the fully charged CMTCFs-1 electrode before (A and B) and after (C and D) cycling performance test.

500 cycles at 1000 mA g^{-1}, which indicated that CMTCFs-1 had excellent structure stability.

The lithium ions storage mechanism of CMTCFs-1 anode (500 cycles at 1000 mA g^{-1}) was studied by CV curves at different scan rates, as shown in Fig. 8.15. The following relationship existed between current (i) and scan rate (υ):

$$i = a\upsilon^b$$

Where i was peak current value, υ was scan rate, a and b were constants [71]. When b value was 1, the peak current value was proportional to the scan rate. The electrode appeared as a capacitive process. When b value was 0.5, the peak current value was proportional to the square root of the scan rate. It meant the diffusion control process [72,73]. In Fig. 8.15B, the b value of Peak1 was 0.77, which was close to 1. It represented the surface control process. The k$_1$ value was calculated by the following equation:

$$i(V)/\upsilon^{1/2}=k_1\upsilon^{1/2}+k_2$$

Where k$_1$ and k$_2$ were constants, k$_1\upsilon$ represented the capacitive controlled contribution [74]. As shown in Fig. 8.15D, with the scan rate increased from

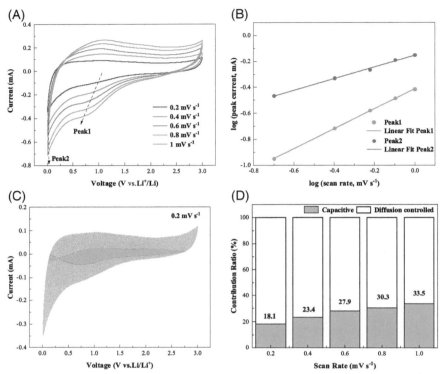

Figure 8.15 CV curves of the CMTCFs-1 electrode (after 500 cycles at 1000 mA g^{-1}) at different scan rates (A); determination of the b value at specific peak current (B); capacitive and diffusion controlled contribution at a scan rate of 0.2 mV s^{-1}, the estimated capacitive contribution is shown in the shaded area (C); contribution ratios of surface controlled at various scan rates (D).

0.2 to 1 mV s^{-1}, the pseudocapacitive contribution rate increased from 18.1% to 33.5% (Fig. 8.16). The interfacial lithium ions storage behavior effectively increased the specific capacity of CMTCFs.

8.6 Summary and prospect

In summary, one-dimensional carbon materials, especially with porous/hollow or tubular structures, exhibit more significant advantages than other carbon materials with the same composition but different morphology as anodes for batteries. Typically, carbon nanotubes mainly show high cost and poor dispersibility in the matrix, which affects their industrial application. Polymer precursors with one-dimensional structure

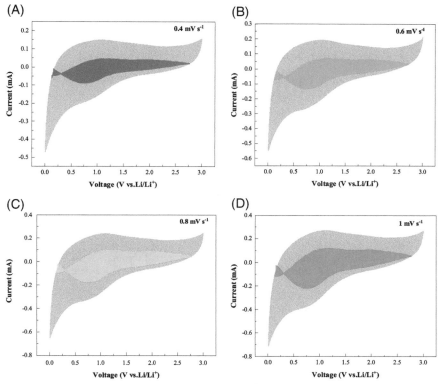

Figure 8.16 CV curve with the pseudocapacitive fraction shown by the shaded area at different scan rate: 0.4 mV s^{-1} (A); 0.6 mV s^{-1} (B); 0.8 mV s^{-1} (C); 1 mV s^{-1} (D).

or carbon nanotube-like morphology are prepared via conventional techniques such as electrospinning and liquid-phase polymerization. Simply carbonizing porous nanofibers to prepare electrode materials is a feasible, low-cost and easy-to-industrial approach. However, pure carbon materials are still limited by their inherent capacity. The main research directions of one-dimensional carbon materials as battery anodes include:

(1) Controllable construction of porous polymer fibers precursor, the development of pore-building and atomic doping technologies for efficient solid polymer fibers during the carbonization.

(2) Volume shrinkage and deformation will occur during the carbonization process of polymers, the loss of organic components creates new pores, the original channels of polymers will also change, these can lead to the uncertainty of pore performance for as-obtained carbon fibers. Suitable pore size control methods need to be further developed.

(3) The mild functional group modification method on the surface of carbon fibers should be carried out. Strong zoxidizing agents show a destructive effect on the materials, which will reduce the strength of the carbon fibers.

(4) There are many types of one-dimensional carbon material for battery anodes, many exhibits the potential for industrial application, but "the last mile" still needs a lot of systematic work.

References

[1] M. Li, J. Lu, Z. Chen, K. Amine, 30 Years of Lithium-Ion Batteries, Adv. Mater. 30 (33) (2018) e1800561.

[2] S. Hossain, Y.-K. Kim, Y. Saleh, R. Loutfy, Comparative studies of MCMB and C-C composite as anodes for lithium-ion battery systems, J. Power Sources. 114 (2) (2003) 264–276.

[3] T.K. Yoshio Idota, Akihiro Matsufuji, Yukio Maekawa, Tsutomu Miyasaka, Tin-based amorphous oxide: A high-capacity lithium-ion–storage material, Science. 276 (1997) 1395–1397.

[4] C.J. Poonam Sehrawat, S.S. Islam, Carbon nanotubes in Li-ion batteries: A review, Mat. Sci. Engin. B 213 (2016) 12–40.

[5] M.J.G. Brian, J. Landi, Cory D. Cress, †Roberta A. DiLeo, Ryne P. Raffaelle, Carbon nanotubes for lithium ion batteries, Energy Environ. Sci. 2 (2009) 638–654.

[6] Y.P. Zhu, Y. Jing, A. Vasileff, T. Heine, S.-Z. Qiao, 3D synergistically active carbon nanofibers for improved oxygen evolution, Adv. Energy Mat. 7 (14) (2017) 1602928.

[7] Y.-E. Miao, J. Yan, Y. Ouyang, H. Lu, F. Lai, Y. Wu, T. Liu, A bio-inspired N-doped porous carbon electrocatalyst with hierarchical superstructure for efficient oxygen reduction reaction, Appl. Surf. Sci. 443 (2018) 266–273.

[8] Z. Liu, Z. Zhao, Y. Wang, S. Dou, D. Yan, D. Liu, Z. Xia, S. Wang, Situ exfoliated, edge-rich, oxygen-functionalized graphene from carbon fibers for oxygen electrocatalysis, Adv. Mater. 7 (18) (2017) 1602928.

[9] X. Peng, K. Wu, Y. Hu, H. Zhuo, Z. Chen, S. Jing, Q. Liu, C. Liu, L. Zhong, A mechanically strong and sensitive CNT/rGO–CNF carbon aerogel for piezoresistive sensors, J. Mat. Chem. A. 6 (46) (2018) 23550–23559.

[10] A.S.M.I. Uddin, P.S. Kumar, K. Hassan, H.C. Kim, Enhanced sensing performance of bimetallic Al/Ag-CNF network and porous PDMS-based triboelectric acetylene gas sensors in a high humidity atmosphere, Sens. Actuators B: Chem. 258 (2018) 857–869.

[11] D. Nan, Z.-H. Huang, R. Lv, L. Yang, J.-G. Wang, W. Shen, Y. Lin, X. Yu, L. Ye, H. Sun, F. Kang, Nitrogen-enriched electrospun porous carbon nanofiber networks as high-performance free-standing electrode materials, J. Mater. Chem. A 2 (46) (2014) 19678–19684.

[12] L.P.W. Yi Zhao, Moulay Tahar Sougrati, Zhenxing Feng, Yann Leconte, Adrian Fisher, Madhavi Srinivasan, Zhichuan Xu*, A review on design strategies for carbon based metal oxides and sulfides nanocomposites for high performance li and na ion battery anodes, Adv. Energy Mater. 7 (9) (2017) 1601424.

[13] L. Zhang, T. Wei, Z. Jiang, C. Liu, H. Jiang, J. Chang, L. Sheng, Q. Zhou, L. Yuan, Z. Fan, Electrostatic interaction in electrospun nanofibers: Double-layer carbon protection of $CoFe_2O_4$ nanosheets enabling ultralong-life and ultrahigh-rate lithium ion storage, Nano Energy. 48 (2018) 238–247.

[14] L. Tao, Y. Huang, Y. Zheng, X. Yang, C. Liu, M. Di, S. Larpkiattaworn, M.R. Nimlos, Z. Zheng, Porous carbon nanofiber derived from a waste biomass as anode material in lithium-ion batteries, J. Taiwan Inst. Chem. Eng. 95 (2019) 217–226.

[15] V. Datsyuk, M. Kalyva, K. Papagelis, J. Parthenios, D. Tasis, A. Siokou, I. Kallitsis, C. Galiotis, Chemical oxidation of multiwalled carbon nanotubes, Carbon 46 (6) (2008) 833–840.

[16] X. Huang, X. Cai, D. Xu, W. Chen, S. Wang, W. Zhou, Y. Meng, Y. Fang, X. Yu, Hierarchical Fe$_2$O$_3$@CNF fabric decorated with MoS2 nanosheets as a robust anode for flexible lithium-ion batteries exhibiting ultrahigh areal capacity, J. Mater. Chem. A 6 (35) (2018) 16890–16899.

[17] D. Wang, K. Wang, L. Sun, H. Wu, J. Wang, Y. Zhao, L. Yan, Y. Luo, K. Jiang, Q. Li, S. Fan, J. Li, J. Wang, MnO2 nanoparticles anchored on carbon nanotubes with hybrid supercapacitor-battery behavior for ultrafast lithium storage, Carbon 139 (2018) 145–155.

[18] S. Choi, M.-C. Kim, S.-H. Moon, J.-E. Lee, Y.-K. Shin, E.-S. Kim, K.-W. Park, 3D yolk–shell Si@void@CNF nanostructured electrodes with improved electrochemical performance for lithium-ion batteries, J. Ind. Eng. Chem. 64 (2018) 344–351.

[19] Y. Xu, T. Yuan, Z. Bian, H. Sun, Y. Pang, C. Peng, J. Yang, S. Zheng, Electrospun flexible Si/C@CNF nonwoven anode for high capacity and durable lithium-ion battery, Composites Communications 11 (2019) 1–5.

[20] M. Feng, S. Wang, Y. Yu, Q. Feng, J. Yang, B. Zhang, Carboxyl functionalized carbon fibers with preserved tensile strength and electrochemical performance used as anodes of structural lithium-ion batteries, Appl. Surf. Sci. 392 (2017) 27–35.

[21] D. Wang, K. Wang, H. Wu, Y. Luo, L. Sun, Y. Zhao, J. Wang, L. Jia, K. Jiang, Q. Li, S. Fan, J. Wang, CO2 oxidation of carbon nanotubes for lithium-sulfur batteries with improved electrochemical performance, Carbon 132 (2018) 370–379.

[22] B. Li, X. Huang, L. Liang, B. Tan, Synthesis of uniform microporous polymer nanoparticles and their applications for hydrogen storage, J. Mater. Chem. 20 (35) (2010).

[23] B. Li, F. Su, H.-K. Luo, L. Liang, B. Tan, Hypercrosslinked microporous polymer networks for effective removal of toxic metal ions from water, Microporous Mesoporous Mater. 138 (1-3) (2011) 207–214.

[24] X. Wang, P. Mu, C. Zhang, Y. Chen, J. Zeng, F. Wang, J.X. Jiang, Control synthesis of tubular hyper-cross-linked polymers for highly porous carbon nanotubes, ACS Appl. Mater. Interfaces 9 (24) (2017) 20779–20786.

[25] J. Wang, Z. Yang, M. Ahmad, H. Zhang, Q. Zhang, B. Zhang, Novel synthetic method for tubular nanofibers, Polym. Chem. 10 (31) (2019) 4239–4245.

[26] J. Wang, Y. Huyan, Z. Yang, A. Zhang, Q. Zhang, B. Zhang, Tubular carbon nanofibers: Synthesis, characterization and applications in microwave absorption, Carbon 152 (2019) 255–266.

[27] Y. Huyan, J. Wang, J. Chen, Q. Zhang, B. Zhang, Magnetic tubular carbon nanofibers as anode electrodes for high-performance lithium–ion batteries, Int. J. Energy Res. 43 (2019) 8242–8256.

[28] Z. Ding, V. Trouillet, S. Dsoke, Are functional groups beneficial or harmful on the electrochemical performance of activated carbon electrodes? J. Electrochem. Soc. 166 (6) (2019) A1004–A1014.

[29] K. Zhou, M. Hu, Y.-b. He, L. Yang, C. Han, R. Lv, F. Kang, B. Li, Transition metal assisted synthesis of tunable pore structure carbon with high performance as sodium/lithium ion battery anode, Carbon 129 (2018) 667–673.

[30] F. Zheng, Z. Yin, H. Xia, G. Bai, Y. Zhang, Porous MnO@C nanocomposite derived from metal-organic frameworks as anode materials for long-life lithium-ion batteries, Chem. Eng. J. 327 (2017) 474–480.

[31] Y. Song, Y. Chen, J. Wu, Y. Fu, R. Zhou, S. Chen, L. Wang, Hollow metal organic frameworks-derived porous ZnO/C nanocages as anode materials for lithium-ion batteries, J. Alloys Compd. 694 (2017) 1246–1253.

[32] B. Rajagopalan, B. Kim, S.H. Hur, I.-K. Yoo, J.S. Chung, Redox synthesis of poly (p–phenylenediamine)–reduced graphene oxide for the improvement of electrochemical performance of lithium titanate in lithium–ion battery anode, J. Alloys Compd. 709 (2017) 248–259.

[33] B. Zhang, Y. Huyan, J. Wang, X. Chen, H. Zhang, Q. Zhang, Fe3O4@SiO2@CCS porous magnetic microspheres as adsorbent for removal of organic dyes in aqueous phase, J. Alloys Compd. 735 (2018) 1986–1996.

[34] Z. Yang, J. Xu, J. Wang, Q. Zhang, B. Zhang, Design and preparation of self-driven BSA surface imprinted tubular carbon nanofibers and their specific adsorption performance, Chem. Eng. J. 373 (2019) 923–934.

[35] P.P. Wang, Y.X. Zhang, X.Y. Fan, J.X. Zhong, K. Huang, Synthesis of Si nanosheets by using Sodium Chloride as template for high-performance lithium-ion battery anode material[J], J. Power Sources 379 (2018) 20–25.

[36] H. Kim, N. Venugopal, J. Yoon, W.-S. Yoon, A facile and surfactant-free synthesis of porous hollow λ-MnO_2 3D nanoarchitectures for lithium-ion batteries with superior performance, J. Alloys Compd. 778 (2019) 37–46.

[37] R. Jin, Y. Cui, S. Gao, S. Zhang, L. Yang, G. Li, CNTs@NC@$CuCo_2S_4$ nanocomposites: An advanced electrode for high performance lithium-ion batteries and supercapacitors, Electrochim. Acta 273 (2018) 43–52.

[38] H. Chen, S. He, X. Hou, S. Wang, F. Chen, H. Qin, Y. Xia, G. Zhou, Nano-Si/C microsphere with hollow double spherical interlayer and submicron porous structure to enhance performance for lithium-ion battery anode[J], Electrochim. Acta 312 (2019) 242–250.

[39] Y. Jiang, D. Zhang, Y. Li, T. Yuan, N. Bahlawane, C. Liang, W. Sun, Y. Lu, M. Yan, Amorphous Fe_2O_3 as a high-capacity, high-rate and long-life anode material for lithium ion batteries, Nano Energy 4 (2014) 23–30.

[40] Z.X. Yang Xia, Xiao Dou, Hui Huang, Xianghong Lu, Rongjun Yan, Yongping Gan, Wenjun Zhu, Jiangping Tu, Wenkui Zhang, Xinyong Tao, Green and facile fabrication of hollow porous MnO/C microspheres from microalgaes for lithium-ion batteries, ACS NANO 7 (2013) 7083–7092.

[41] S. Zhang, L. Zhu, H. Song, X. Chen, J. Zhou, Enhanced electrochemical performance of MnO nanowire/graphene composite during cycling as the anode material for lithium-ion batteries, Nano Energy 10 (2014) 172–180.

[42] T. Yuan, Y. Jiang, W. Sun, B. Xiang, Y. Li, M. Yan, B. Xu, S. Dou, Ever-increasing pseudocapacitance in RGO-MnO-RGO sandwich nanostructures for ultrahigh-rate lithium storage, Adv. Funct. Mater. 26 (13) (2016) 2198–2206.

[43] Y. Yang, X. Zhao, H.-E. Wang, M. Li, C. Hao, M. Ji, S. Ren, G. Cao, Phosphorized SnO_2/graphene heterostructures for highly reversible lithium-ion storage with enhanced pseudocapacitance, J. Mater. Chem. A 6 (8) (2018) 3479–3487.

[44] B. Long, M.-S. Balogun, L. Luo, W. Qiu, Y. Luo, S. Song, Y. Tong, Phase boundary derived pseudocapacitance enhanced nickel-based composites for electrochemical energy storage devices, Adv. Energy Mater. 8 (5) (2018) 1701681.

[45] W.X. Chen, J.Y. Lee, Z. Liu, The nanocomposites of carbon nanotube with Sb and $SnSb_{0.5}$ as Li-ion battery anodes[J], Carbon 41 (5) (2003) 959–966.

[46] K. Nishidate, M. Hasegawa, Energetics of lithium ion adsorption on defective carbon nanotubes[J], Phys. Rev. B 71 (24) (2005) 245418.

[47] H. Shimoda, B. Gao, X.P. Tang, A. Kleinhammes, L. Fleming, Y. Wu, O. Zhou, Lithium intercalation in opened single-wall carbon nanotubes: Storage capacity and electronic properties[J], Phys. Rev. Lett. 88 (1) (2002) 015502.

[48] V. Meunier, J. Kephart, C. Roland, J. Bernholc, Ab initio investigations of lithium diffusion in carbon nanotube systems[J], Phys. Rev. Lett. 88 (7) (2002) 075506.

[49] J. Zhang, Z. Xie, W. Li, S. Dong, M. Qu, High-capacity graphene oxide/graphite/carbon nanotube composites for use in Li-ion battery anodes[J], Carbon 74 (2014) 153–162.

[50] M. Sahoo, S. Ramaprabhu, Effect of wrinkles on electrochemical performance of multiwalled carbon nanotubes as anode material for Li ion battery[J], Electrochim. Acta 186 (2015) 142–150.

[51] Y. Chen, X. Li, K. Park, J. Song, J. Hong, L. Zhou, Y.-W. Mai, H. Huang, a.J. B. Hollow carbon-nanotube/carbon-nanofiber hybrid anodes for Li-ion batteries[J], J. Am. Chem. Soc. 135 (44) (2013) 16280–16283.

[52] V. Datsyuk, M. Kalyva, K. Papagelis, J. Parthenios, D. Tasis, A. Siokou, I. Kallitsis, C. Galiotis, Chemical oxidation of multiwalled carbon nanotubes[J], Carbon 46 (6) (2008) 833–840.

[53] B. Li, X. Huang, L. Liang, B. Tan, Synthesis of uniform microporous polymer nanoparticles and their applications for hydrogen storage[J], J. Mater. Chem. 20 (35) (2010) 7444–7450.

[54] S. Choi, M.-C. Kim, S.-H. Moon, J.-E. Lee, Y.-K. Shin, E.-S. Kim, K.-W. Park, 3D yolk-shell Si@void@CNF nanostructured electrodes with improved electrochemical performance for lithium-ion batteries[J], J. Ind. Eng. Chem. 64 (2018) 344–351.

[55] M. Feng, S. Wang, Y. Yu, Q. Feng, J. Yang, B. Zhang, Carboxyl functionalized carbon fibers with preserved tensile strength and electrochemical performance used as anodes of structural lithium-ion batteries[J], Appl. Surf. Sci. 392 (2017) 27–35.

[56] D. Wang, Y. Wang, Q. Li, W. Guo, F. Zhang, S. Niu, Urchin-like α-Fe_2O_3/MnO_2 hierarchical hollow composite microspheres as lithium-ion battery anodes[J], J. Power Sources 393 (2018) 186–192.

[57] Z. Ding, V. Trouillet, S. Dsoke, Are functional groups beneficial or harmful on the electrochemical performance of activated carbon electrodes?[J], J. Electrochem. Soc. 166 (6) (2019) A1004–A1014.

[58] K. Zhou, M. Hu, Y.-b. He, L. Yang, C. Han, R. Lv, F. Kang, B. Li, Transition metal assisted synthesis of tunable pore structure carbon with high performance as sodium/lithium ion battery anode[J], Carbon 129 (2018) 667–673.

[59] F. Zheng, Z. Yin, H. Xia, G. Bai, Y. Zhang, Porous MnO@C nanocomposite derived from metal-organic frameworks as anode materials for long-life lithium-ion batteries[J], Chem. Eng. J. 327 (2017) 474–480.

[60] Y. Song, Y. Chen, J. Wu, Y. Fu, R. Zhou, S. Chen, L. Wang, Hollow metal organic frameworks-derived porous ZnO/C nanocages as anode materials for lithium-ion batteries[J], J. Alloys Compd. 694 (2017) 1246–1253.

[61] B. Rajagopalan, B. Kim, S.H. Hur, I.-K. Yoo, J.S. Chung, Redox synthesis of poly (p-phenylenediamine)-reduced graphene oxide for the improvement of electrochemical performance of lithium titanate in lithium-ion battery anode[J], J. Alloys Compd. 709 (2017) 248–259.

[62] B. Zhang, Y. Huyan, J. Wang, X. Chen, H. Zhang, Q. Zhang, Fe_3O_4@SiO_2@CCS porous magnetic microspheres as adsorbent for removal of organic dyes in aqueous phase[J], J. Alloys Compd. 735 (2018) 1986–1996.

[63] Z. Yang, J. Xu, J. Wang, Q. Zhang, B. Zhang, Design and preparation of self-driven BSA surface imprinted tubular carbon nanofibers and their specific adsorption performance[J], Chem. Eng. J. 373 (2019) 923–934.

[64] X. Li, X. Tian, T. Yang, W. Wang, Y. Song, Q. Guo, Z. Liu, Silylated functionalized silicon-based composite as anode with excellent cyclic performance for lithium-ion battery[J], J. Power Sources 385 (2018) 84–90.

[65] H. Kim, N. Venugopal, J. Yoon, W.-S. Yoon, A facile and surfactant-free synthesis of porous hollow λ-MnO_2 3D nanoarchitectures for lithium ion batteries with superior performance[J], J. Alloys Compd. 778 (2019) 37–46.

[66] R. Jin, Y. Cui, S. Gao, S. Zhang, L. Yang, G. Li, CNTs@NC@CuCo$_2$S$_4$ nanocomposites: An advanced electrode for high performance lithium-ion batteries and supercapacitors[J], Electrochim. Acta 273 (2018) 43–52.
[67] C. Jiang, J. Wang, Z. Chen, Z. Yu, Z. Lin, Z. Zou, Nitrogen-doped hierarchical carbon spheres derived from MnO$_2$-templated spherical polypyrrole as excellent high rate anode of Li-ion batteries[J], Electrochim. Acta 245 (2017) 279–286.
[68] Y. Jiang, D. Zhang, Y. Li, T. Yuan, N. Bahlawane, C. Liang, W. Sun, Y. Lu, M. Yan, Amorphous Fe$_2$O$_3$ as a high-capacity, high-rate and long-life anode material for lithium ion batteries[J], Nano Energy 4 (2014) 23–30.
[69] Z.X. Yang Xia, Xiao Dou, Hui Huang, Xianghong Lu, Rongjun Yan, Yongping Gan, Wenjun Zhu, Jiangping Tu, Wenkui Zhang, Xinyong Tao, Green and facile fabrication of hollow porous MnO/C microspheres from microalgaes for lithium-ion batteries[J], ACS Nano 7 (8) (2013) 7083–7092.
[70] S. Zhang, L. Zhu, H. Song, X. Chen, J. Zhou, Enhanced electrochemical performance of MnO nanowire/graphene composite during cycling as the anode material for lithium-ion batteries[J], Nano Energy 10 (2014) 172–180.
[71] T. Yuan, Y. Jiang, W. Sun, B. Xiang, Y. Li, M. Yan, B. Xu, S. Dou, Ever-increasing pseudocapacitance in RGO-MnO-RGO sandwich nanostructures for ultrahigh-rate lithium storage[J], Adv. Funct. Mater. 26 (13) (2016) 2198–2206.
[72] Y. Yang, X. Zhao, H.-E. Wang, M. Li, C. Hao, M. Ji, S. Ren, G. Cao, Phosphorized SnO$_2$/graphene heterostructures for highly reversible lithium-ion storage with enhanced pseudocapacitance[J], Journal of Materials Chemistry A, 6 (8) (2018) 3479–3487.
[73] B. Long, M.-S. Balogun, L. Luo, W. Qiu, Y. Luo, S. Song, Y. Tong, Phase boundary derived pseudocapacitance enhanced nickel-based composites for electrochemical energy storage devices[J], Adv. Energy Mater. 8 (5) (2018) 1701681.
[74] F. Ren, Z. Lu, H. Zhang, L. Huai, X. Chen, S. Wu, Z. Peng, D. Wang, J. Ye, Pseudocapacitance induced uniform platin/stripping of li metal anode in vertical graphene nanowalls[J], Adv. Funct. Mater. 28 (50) (2018) 1805638.

CHAPTER 9

Tubular carbon nanofibers loaded with different MnO$_2$: Preparation and electrochemical performance

Yu Huyan[a], Mengmeng Wei[a] and Baoliang Zhang[a,b]
[a]School of Chemistry and Chemical Engineering, Northwestern Polytechnical University, Xi'an, China
[b]Xi'an Key Laboratory of Functional Organic Porous Materials, Northwestern Polytechnical University, Xi'an, China

9.1 Introduction

Lithium–ion batteries (LIBs) have been recognized as ideal choice for next generation large-scale grids store and hybrid vehicles due to their no memory effect, excellent safety, and environmental friendliness. Currently, the energy density, power density, and cycle life of commercial LIBs are insufficient to meet the development requirements of modern society [1]. As the core components of LIBs, the characteristics of electrode materials directly determine their application performance. Therefore, the key point to LIBs lies in the development of new electrode materials, especially anode materials with long-term cycle stability and outstanding rate performance.

In recent years, transition metal oxides (TMOs, e.g., MnO [2,3], Fe$_3$O$_4$ [4,5], and Co$_3$O$_4$ [6–8]) have received considerable attention owing to their cost-effective, abundant resources, and easy preparation. In view of various TMOs, MnO$_2$ is touted by battery scientists as rechargeable LIBs anode materials due to its high theoretical capacity ($\sim$1230 mAh g^{-1}), low toxicity, and environmental friendliness [9,10]. In spite of the high theoretical capacity and practical application potential, MnO$_2$ electrodes suffer from poor rate capability and fast capacity fading due to low electrical conductivity and large volume changes during continuous cycling. In order to improve the electrochemical performance of MnO$_2$ as anode material of LIBs, intensive effort has been devoted to the composite of nanostructured MnO$_2$ and conductive buffer substrate. Successful examples are δ-MnO$_2$-CNTs-G-NF [11], α-Fe$_2$O$_3$/MnO$_2$ [12], PPy/MnO$_2$-rGO-CNTs [13] and MnO$_2$/ZIF-67-COOH [14].

Fabrication and Functionalization of Advanced Tubular Nanofibers and their Applications.
DOI: https://doi.org/10.1016/B978-0-323-99039-4.00010-3

211

9.2 Research progress of MnO$_2$-based anode materials

MnO$_2$ has been widely studied as an anode material for rechargeable LIBs. However, like other transition metal oxide anode materials, MnO$_2$ anode materials have poor cycle properties and magnification properties. On the one hand, in the process of charging and discharging, the volume of the MnO$_2$ anode material will expand and contract. At the same time, MnO$_2$ itself macroscopic mechanical energy makes it unable to resist the resulting stress and the electrode will deform and rupture. Consequently, it gradually gets powdered, collapsed, until failure. On the other hand, MnO$_2$ possesses low conductivity (10^{-5}-10^{-6} S cm^{-1}). It leads to rapid capacity decay during multiplier rate charge and discharge. Up till now, in order to enhance the electrochemical properties of MnO$_2$ as a LIBs anode material, recent research has aimed at material nanosizing and compounding with conductive buffer substrates.

9.2.1 MnO$_2$-based nanostructured electrodes

The objectives of nanosizing for MnO$_2$ electrode materials are as follows: (1) Increase the specific surface area of electrode material and expand the contact interface between the active material and the electrolyte to obtain a higher specific capacity. But in fact, the high specific surface area also leads to the side effect of increasing the electrolyte decomposition reaction on the electrode surface; (2) Shorten the diffusion path of lithium ions. It is conducive to improve kinetic performance and the multiplicative performance of the electrode material. Nevertheless, too small particle size is bad for the vibration density of the electrode material; (3) The stress relaxation time of the electrode material is increased to adapt to the volume change during charging and discharging. Furthermore, enhance the structural stability of the electrode material, and increase the cycling performance of the electrode material. The means of MnO$_2$ nanosizing can be divided into template methods and non-template methods. And its structure is mainly manifested as α-, β-, λ- and δ-MnO$_2$, and its morphology mainly includes nanorods, nanowires, nanoflowers, nanospheres, nanofibers, and nanosheets, and so on [15–17].

(1) Preparation of MnO$_2$-based nanostructured electrodes by nontemplate method

   Nontemplate methods for the preparation of MnO$_2$ mainly consists of redox reaction [18], thermal deposition [19], hydrothermal methods [20], solvothermal methods, sol-gel methods [21], etc. The morphology and crystalline shape of the obtained MnO$_2$ are closely related to the reaction system, while the electrochemical properties of MnO$_2$ electrode materials strongly depend on the crystalline structure, morphology, and size of MnO$_2$. Nulu et al. [22] prepared α-MnO$_2$ nanorods with concentrated distribution of diameter and length between 15 and 50 nm and 1–3 μm, respectively. One-dimensional rod-like structure and continuous electron transport path allowed them to exhibit high reversible capacity, good long-term cycling stability, and excellent multiplicative performance as LIBs anode materials. After 1000 cycles at 100 mA g^{-1}, the specific capacity was stabilized at 780 mA h g^{-1}. To explore the effect of pH on the MnO$_2$ morphology generated by the reduction of KMnO$_4$ by isopropanol and the effect of MnO$_2$ morphology on the electrochemical properties of electrode materials, Ma et al. [23] used the same method to control the MnO$_2$ morphology by varying the pH of the system and obtained nanosheet (pH2-MnO$_2$), flower-like (pH7-MnO$_2$) and onion-like (pH13-MnO$_2$). And electrochemical studies revealed that the overall electrochemical performance of the three different morphologies of MnO$_2$ electrodes tended to be: nanosheet > onion > flower, which reflected the dependence of the electrochemical performance of the electrode materials on the MnO$_2$ morphology. In addition, the electrochemical performance of MnO$_2$ electrodes depended on the crystalline structure of MnO$_2$. Lots of studies have shown that δ-MnO$_2$, among the many crystalline structures of MnO$_2$, is the most widely used class of MnO$_2$ electrode materials, mainly because the monoclinic structure of δ-MnO$_2$ is similar to graphene, which is more convenient for the rapid transport of lithium ions in the electrode material. Zhang et al. [24] prepared β-MnO$_2$ with a near-hexagonal nanosheet morphology and δ-MnO$_2$ with a spherical morphology composed of nanosheets by using a hydrothermal method and by regulating the reaction time and the pH of the system. The electrochemical test results showed that δ-MnO$_2$ had a reversible capacity of 154 mA h g^{-1} after 50 cycles at a current density of 100 mA g^{-1}, which was better than the cycling performance of β-MnO$_2$. In recent years, it has been found in the study of Ti-based and W-based [25–27] electrode material systems that the bound water or hydration groups embedded in the metal oxide lattice can keep the crystal in an open lamellar structure, increase the ionic conductivity of the electrode material, promote the electrochemical redox activity, and thus enhance the electrochemical performance of the electrode material, especially in terms of capacity and multiplicity performance [28]. Zhang et al. [29] prepared novel hierarchical hydrated manganese

Figure 9.1 Schematics of morphology evolution of nanostructured MnO_2.

oxide (HMO) aggregated from δ-MnO_2 nanosheets by combining liquid precipitation reaction and low–temperature heat treatment techniques. Compared with MnO after high-temperature pyrolysis, HMO electrodes exhibited superior electrochemical properties due to the presence of bound water and high specific surface area. After 300 cycles at a current density of 1 A g^{-1}, the specific capacity reached 508.9 mA h g^{-1}. After 300 cycles at a high current density of 4 A g^{-1}, the specific capacity reached 394.4 mA h g^{-1}.

(2) Preparation of MnO_2-based nanostructured electrodes by template method

With the continuous increase of template types and the maturation of the preparation process, the template method is widely used as a synthetic method with strong controllability and simple operation to prepare MnO_2 nanomaterials. The template method is to use substances with nanostructure, easily controlled shape, and inexpensive and easily available as templates, deposit related materials into the pores or surface of the templates by physical or chemical methods, and then use HCl, NaOH, calcination in air, etc. By removing the template, nanomaterials with template-specific morphology and size are obtained. Compared with direct synthesis methods, template synthesis methods can precisely control the size and shape of nanomaterials and solve the problem of dispersion stability of nanomaterials. Guo et al. [30] used graphene oxide (GO) as the morphology control agent and template to prepare MnO_2 by a simple sol–gel method. The morphology of MnO_2 can be effectively controlled by adjusting the content of GO in the system. With the increase of GO content, the morphology of GO changes from the initial nanorods to nanoparticles. The evolution process of MnO_2 morphology is shown in Fig. 9.1. After GO was removed, the MnO_2 nanoparticles remained uniformly dispersed. By investigating the electrochemical performance of this MnO_2 nanoparticle as a negative electrode material for LIBs, the uniformly dispersed MnO_2 nanoparticle exhibits an excellent initial Coulombic efficiency of 94.5%.

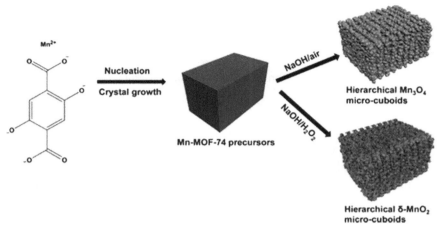

Figure 9.2 Schematic of the fabrication processes of the 3D hierarchical MnO_x micro-cuboids.

In recent years, the preparation of micro/nano-sized MnO_2 electrode materials with hierarchical mesoporous structure by template method has attracted extensive attention from researchers. Its structural advantages mainly include (1) preventing the self-aggregation of micro/nanoparticles and the destruction of the electrode structure; (2) effectively enhancing the penetration ability of the electrolyte in the electrode material; (3) the mesoporous channel can moderate the volume change of the electrode material to a certain extent during charge-discharge cycling. Hu et al. [31] used the solvothermal synthesized Mn-MOF-74 as a self-sacrificing template and dissolved it in H_2O_2-containing and H_2O_2-free alkaline solutions. By the controlled reaction of metal-organic skeleton ligands and OH-ligands and subsequent in situ oxidation of manganese hydroxide intermediates, two three-dimensional layered mesoporous microcubes were prepared, namely Mn_3O_4 microcubes assembled from nanoparticles and δ-MnO_2 microcubes assembled from nanosheets (Fig. 9.2). The ordered hierarchical mesoporous structure and micro/submicron size allowed the δ-MnO_2 and Mn_3O_4 microcubes to exhibit excellent cycle stability. After 400 cycles at 2 A g^{-1}, the specific capacities of the δ-MnO_2 and Mn_3O_4 electrodes were maintained at 362.1 and 437.1 mAh g^{-1} respectively. Zhang et al. [32] used eggshells as biological templates to synthesize villi-like nanospheres composed of δ-MnO_2 nanosheets by hydrothermal method. Electrochemical studies show that the δ-MnO_2 nanosphere electrode exhibits a reversible capacity of 1022.1 mAh g^{-1} after 100 cycles at 100 mA g^{-1}. Ramesha et al. [33] selected

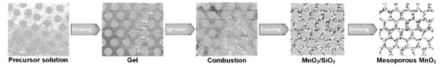

Figure 9.3 The schematic of fabricating 3D mesoporous δ-MnO$_2$.

SBA-15 and MCM-48 as templates to synthesize ordered mesoporous β-MnO$_2$ with interconnected nanorod morphology (SBA-MnO$_2$) and open cage/bowl morphology (MCM-MnO$_2$), respectively, by a two-solvent nano casting method. The specific capacities of SBA-MnO$_2$ and MCM-MnO$_2$ were about 245 and 345 mAh g^{-1}, respectively, when they were cycled 100 times at 250 mA g^{-1} as negative electrode materials for LIBs. In general, mesoporous materials prepared using soft structure-directing agents (SDAs), mesoporous silica, or carbon (SBA-15, CMK-3, and KIT-6) have pore sizes essentially less than 12 nm, while materials prepared using polymeric microspheres such as polystyrene or polymethyl methacrylate as colloidal templates have macroporous (>50 nm) structures. Recently, a low-cost, simple and efficient colloidal solution combustion synthesis (CSCS) method was developed by Chan et al. The main purpose is to generate crystalline mesoporous materials with controlled composition and adjustable porosity by using colloids in combustion synthesis, which enriches the synthesis method of mesoporous materials while extending the pore size distribution of mesoporous materials prepared by traditional template method. They successfully synthesized a novel 3D mesoporous δ-MnO$_2$ using the CSCS method and subsequent alkali etching (Fig. 9.3). The prepared δ-MnO$_2$ has a large specific surface area (128 m^2 g^{-1}), small crystal size (5–10 nm), and uniformly oversized mesopores (29.5 nm) [34]. In addition, porous hollow structured MnO$_2$ electrode materials prepared using the template method have attracted much attention for their unique internal porosity, high specific surface area, and structural stability [35,36]. Wang et al. used hollow SiO$_2$ cubes as sacrificial templates to synthesize hollow cubes composed of MnO$_2$ nanosheets by hydrothermal method. Compared with untemplated MnO$_2$ layered microflora, the MnO$_2$ hollow cubes showed better electrochemical performance [37]. Yoon et al. used Li$_2$O$_2$ as an inorganic template to obtain hollow λ-MnO$_2$ aggregated from nanosheets. This hollow λ-MnO$_2$ exhibited excellent electrochemical performance when used as an anode material for LIBs, maintaining a specific capacity of 810 mAh g^{-1} after 400 cycles at a current density of 1000 mA g^{-1}. It showed excellent multiplicative performance up to 508 mAh g^{-1} at a high current density of 2000 mA g^{-1} [38].

9.2.2 MnO_2-based composite electrode

One of the main limitations of using MnO_2 as anode material for LIBs stems from its microstructure that cannot form a good electron transport network between cells, resulting in electrode materials exhibiting low electronic conductivity, which makes nanostructured MnO_2 with conductive substrates composite electrodes an attractive research area. The purpose of compounding with the conductive substrates is mainly manifested in the following four aspects: (1) The electronic conductivity of MnO_2-based anodes are enhanced to increase the electron transfer rate of electrode materials; (2) the volume change of MnO_2 is buffered during lithium ion intercalation and deintercalation process, the stability of the electrode structure is maintained and the cycle stability of the electrode materials is improved; (3) as a substrate material, the MnO_2-based active substance is guaranteed to be uniformly dispersed in the electrode material; (4) as an active material for storing lithium, the energy density of the electrode material is enhanced. At present, the research hotspots of MnO_2-based composite electrodes mainly focus on structural design and the selection of conductive substrates. Among them, the conductive substrates used mainly include metal oxides and carbon materials.

(1) MnO_2-metal oxide composite electrode

Metal oxides include Fe_2O_3, NiO, MoO_3, TiO_2, etc. As another type of conductive composite substrates, they are widely used in the design and preparation of MnO_2 composite electrodes. By adjusting the morphology and composition of two metal oxides, the advantages of the two active materials are maximized to obtain the best battery performance. The composite forms are generally manifested as hierarchical structures, core-shell structures, hollow structures, and so on. Among them, hollow structures with high specific surface area and short ion diffusion paths can often significantly improve the electrochemical performance of MnO_2-based electrode materials. Its large specific surface area can provide sufficient active sites for electrode materials and reduce the diffusion paths of lithium ions. The hollow structure can provide additional space to accommodate large volume changes of metal oxides during charge-discharge and relieve the stress–strain associated with Lithium-ion intercalation/deintercalation

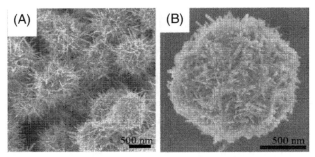

Figure 9.4 Urchin-likeα-Fe$_2$O$_3$/MnO$_2$ hierarchical hollow composite microspheres prepared by using carbon sphere as sacrificial templates: SEM image (A); high magnification SEM image (B).

processes. Wang and co-workers prepared urchin-like α-Fe$_2$O$_3$/MnO$_2$ hierarchical composite microspheres (Fig. 9.4) by using carbon spheres as sacrificial templates, combining the high electrical conductivity of Fe$_2$O$_3$ and high theoretical capacity characteristics of MnO$_2$, and used them as LIBs anode materials. Electrochemical analysis showed that this electrode was cycled 500 times at 0.5 A g^{-1} with a discharge specific capacity of 494 mAh g^{-1}. Hao and co-workers introduced oxygen vacancies through the nanoscale ionic mixing of Fe and Mn ions during the pyrolysis process, then oxygen-rich hollow Fe$_2$O$_3$/MnO$_2$ nanorod electrode was prepared. The electrode was cycled 2000 times at 5 C with a reversible capacity of 706 mAh g^{-1} [39]. Wang et al. took carbon spheres as sacrificial templates and prepared metal oxide/MnO$_2$ hierarchical hollow hybrid spheres through penetration, carbonization, redox and pyrolysis, where the metal oxides included NiO, Co$_3$O$_4$ and CuO. Electrochemical analysis showed that the NiO/MnO$_2$ electrode had the best electrochemical properties, cycled 100 times at 0.5 A g^{-1} with a reversible specific capacity of 670.3 mAh g^{-1} [40]. Su et al. successfully fabricated a double-shelled hollow-structured MnO$_2$@TiO$_2$ electrode by a facile two-step template method using hydrothermally synthesized monodisperse carbon spheres as a template. The electrode was cycled 200 times at 200 mA g^{-1} with specific capacity of 802 mAh g^{-1}, and the specific capacity at 1 A g^{-1} is 400 mAh g^{-1} [41]. In addition, another strategy is to prepare MnO$_2$-based composite electrodes by using both metal oxides and carbon materials as conductive composite substrates at the same time. For example: Fe$_2$O$_3$@C@MnO$_2$@C [42], MnO$_2$@Fe$_3$O$_4$/CNT [43],

SnO$_2$@C@MnO$_2$ [44], TiO$_2$@C@MnO$_2$ [45], C@Fe$_3$O$_4$@MnO$_2$ [46], etc.

(2) MnO$_2$-carbon composite electrode

Now, carbon materials are the most widely used category of conductive composite substrates, mainly including CNT, CNF, graphene, and its derivatives. Graphene is a two-dimensional carbon nanomaterial consisting of carbon atoms in a hexagonal shape with sp^2 hybridized orbitals in a honeycomb lattice. By virtue of excellent electronic conductivity and mechanical properties, it is considered an ideal conductive additive for the preparation of composite MnO$_2$ nanoelectrodes. However, graphene flakes are prone to irreversibly aggregate because of strong π-π interactions and their superhydrophobicity, which leads to a small specific surface area and low diffusion rate of Li$^+$. To solve these problems, it is always constructed graphene oxide (GO) nanosheets by Hummers' method. The modification by oxygen-containing functional groups not only improves the hydrophilicity of graphene but also acts as active sites to react with metal ions, which can promote the growth of MnO$_2$ crystals and evenly disperse on the GO surface. Using KMnO$_4$ and 2-morpholinoethanesulfonic acid as reaction monomers, Jiang et al. [47] grew MnO$_2$ nanorods on the surface of reduced graphene oxide (MnO$_2$-NR/rGO) by hydrothermal reaction in the presence of GO, and used them as anode materials of LIBs. Due to the change in the morphology and crystalline shape of the electrode material during charging and discharging, the MnO2-NR/rGO electrode exhibited enhanced electrochemical performance, whose specific capacity higher than 1500 mAh g^{-1} in 300 cycles at 1 A g^{-1}. Zhao et al. [48] prepared MnO$_2$ nanorods with a diameter of about 100 nm and a length of about 750 nm by hydrothermal method at first. Then, (3-aminopropyl) triethoxysilane was used for surface modification of MnO$_2$ nanorods to make them evenly dispersed on the rGO surface. Finally, MnO$_2$ nanorods anchored reduced graphene oxide (MnO$_2$/rGO) composite electrode was prepared by hydrothermal method. Compared with the electrode unmodified with (3-aminopropyl) triethoxysilane, the MnO$_2$/rGO electrode exhibited better electrochemical performance. It has a reversible capacity of 600.3 mAh g^{-1} in 650 cycles at 0.5 A g^{-1} and still maintained 168.2 mAh g^{-1} at a higher current density of 5 A g^{-1}. Weng et al. [49] prepared MnO$_2$ nanopins/reduced graphene oxide (MnO$_2$/rGO) composite electrode with a reversible capacity of 660.9 mAh g^{-1} in 50 cycles at 123 mAh g^{-1}. Liu et al. [50] reduced

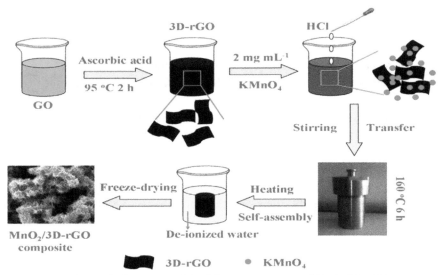

Figure 9.5 Schematic illustration of the fabrication process of MnO$_2$/3D-rGO.

GO with ascorbic acid to prepare reduced graphene oxide (rGO), and then prepare δ-MnO$_2$ nanoconvolution/rGO composite electrode by hydrothermal method, which had a discharge capacity of 528 mAh g^{-1} in 50 cycles at 100 mA g^{-1}. Subsequently, in a similar way, Liu et al. [51] prepared MnO$_2$ nanorods/three-dimensional reduced graphene oxide (MnO$_2$/3D-rGO) composite electrode, and the preparation process is shown in Fig. 9.5. Compared to the δ-MnO$_2$ nanorods/rGO composite electrode, this electrode exhibited improved electrochemical performance due to the presence of the 3D porous framework, and it has a stable specific capacity of 595 mAh g^{-1} at 100 mA g^{-1} in 60 cycles. In recent years, nitrogen-doped graphene has received much attention due to its excellent electrochemical properties, outstanding structural flexibility and large specific surface area. Nitrogen-doping can introduce defects on the surface of graphene, which is beneficial to the combination with Li$^+$ and increase the lithium storage capacity. At the same time, the doping of nitrogen atoms can obtain n-type semiconductor materials, which enhance the electronic conductivity of the substrate material. Consequently, the electrochemical performance of the electrode material is improved. Jiang et al. [52] successfully loaded MnO$_2$ nanowires on the surface of nitrogen-doped porous graphene (MNPGs) by hydrothermal method and applied them to the negative electrode material of LIBs. Electrochemical tests showed that the MNPGs electrode has a specific capacity of 300 mAh g^{-1} in 2400 cycles at 1 A g^{-1}, and even at a high current density of 10 A g^{-1}, it can still maintain at 248.5 mAh g^{-1}. Using polypyrrole as the nitrogen source, Yang et al. [53] prepared nitrogen-doped graphene nanosheets and then produced nitrogen-doped graphene/MnO$_2$ nanorods (MnO$_2$/NG) composite electrode by hydrothermal method. After 3000 cycles at 2500 mA g^{-1}, the capacity retention of the electrode was close to 100%.

Carbon nanofibers (CNFs) or carbon nanotubes (CNTs) have the advantages of large aspect ratio, high strength and toughness, large specific surface area, and good electrical conductivity, which are widely used in MnO$_2$ electrode conductive substrate. Currently, most of the constructed structure are composite substrates with CNF, CNT, and their derivatives, where MnO$_2$ is loaded or coated on the surface of the substrate material in the form of nanoparticles or nanosheets. Wang et al. [54] synthesized air-oxidized carbon nanotubes coated with MnO$_2$ nanoparticles (MnO$_2$/aCNT) flexible electrodes through the redox reaction between KMnO$_4$ and air-oxidized carbon nanotubes under room temperature. The specific capacity of the MnO$_2$/aCNT electrode is 395.8 mAh g^{-1} at 10 A g^{-1}. Zhang et al. [55] synthesized manganese dioxide/carbon nanotubes (MnO$_2$/CNTs) composite electrodes by a simple room temperature solution method. Electrochemical studies show that the load of MnO$_2$ on the surface of CNTs has a significant effect on the electrochemical performance of the electrode material. When the MnO$_2$ load reached 42%, the MnO$_2$/CNTs electrode showed the best electrochemical performance. Shen et al. [56] mixed colloidal MnO$_2$ nanosheets with CNTs to design and prepare a new layered MnO$_2$/CNTs composite. When the mass ratio of MnO$_2$ to CNTs is 1:1, the MnO$_2$/CNTs composite shows the best electrochemical performance with a specific capacity of 796 mAh g^{-1} after 200 cycles at 500 mA g^{-1}. CNT, CNF, and rGO mixtures are also frequently used as conductive carbon substrates for MnO$_2$ electrodes, such as δ-MnO$_2$-CNTs-G-NF, PPy/MnO$_2$-rGO-CNTs, rGO-coated CNT/rGO@MnO$_2$.

9.3 One-dimensional carbon nanomaterials and MnO$_2$ composite electrode materials

Among the composite substrate with MnO$_2$, one-dimensional carbon materials such as carbon nanofibers (CNFs) and carbon nanotubes (CNTs) possesses large aspect ratios, high electrical conductivity and specific surface area. These advantages of CNFs and CNTs make them considered as ideal conductive buffer substrates.

Dong and co-workers have synthesized layered $MnO_2/CNTs$ by one-step mixing method. The material possessed reversible capacity of 796 mAh g^{-1} after 200 charge-discharge cycles at 500 mA g^{-1}. Core-shell $MnO_2/CNFs$ composite film is used as anode for LIBs and displays reversible capacity of 778 mAh g^{-1} after 200 cycles at 50 mA g^{-1} [57]. In another report, Wang and co-workers demonstrated MnO_2 nanoparticles grown on CNTs. This anode material shows high specific capacity of 395.8 mA h^{-1} at 10000 mA g^{-1}. In recent years, tremendous success has been made in improving the electrochemical performance of one-dimensional MnO_2/C electrode. However, there is still a huge gap with the target of the next generation LIBs. Moreover, the research on rate performance is mostly focused on single cycle. This is not consistent with the practical application [58,59]. Thus, design and synthesis one-dimensional MnO_2/C anode materials with high specific capacity and outstanding rate performance is still a grand challenge.

9.4 Tubular carbon nanofibers loaded with worm-like MnO_2

Herein, we prepared carboxyl modified tubular carbon nanofibers (CMTCFs) using tubular carbon nanofibers (TCFs) and HNO_3/H_2O mixed solution as conductive buffer substrates and oxidant, respectively. Then, worm-like MnO_2 coated CMTCFs (CMTCFs@MnO_2) are obtained via simple one-step in-situ redox reaction at room temperature. The ratio of potassium permanganate and CMTCFs is adjusted to achieve the composites with different worm-like MnO_2 spatial distribution density. Meantime, the electrochemical performance of four CMTCFs@MnO_2 are systematically researched and compared. The obtained CMTCFs@MnO_2 electrode materials not only have the characteristics of two materials, but also show the synergistic effect of CMTCFs and worm-like MnO_2. Such as: (1) The CMTCFs provide continuous transport channels for electron migration. Moreover, using CMTCFs as substrate is beneficial to maintain the stability of electrode structure. (2) The interconnected worm-like MnO_2 form porous structure on CMTCFs. This can effectively buffer the large volume change of MnO_2 during Li^+ insertion and extraction process. (3) The moderate worm-like MnO_2 spatial density on CMTCFs is beneficial to the electron transfer between conductive carbon and CMTCFs. (4) The change of the structure of MnO_2 on CMTCFs during the electrochemical

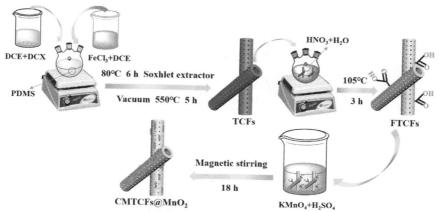

Figure 9.6 Schematic illustration of the fabrication process of CMTCFs@MnO$_2$.

test is conducive to enhancing the electrochemical performance of electrode material. Because of these structural advantages, CMTCFs@MnO$_2$ anode exhibit high reversible capacity and enhanced rate performance.

9.4.1 Characterization of CMTCFs@MnO$_2$

The fabrication procedure of worm-like MnO$_2$ coated CMTCFs was illustrated in Fig. 9.6. First, TCFs were prepared using DCX as monomer and FeCl$_3$ as catalyst via Friedel-Crafts alkylation reaction and subsequent carbonization. The TCFs exhibited regular tubular nanofiber morphology and smooth surface (Fig. 9.7A and C). Then, acidification on TCFs was carried out in a mixed solution with HNO$_3$ and H$_2$O. The tubular nanofiber morphology of CMTCFs can be well retained after liquid phase oxidation, as shown in Fig. 9.7B and D. Meantime, the characteristic peak at ~3433 and ~1760 cm^{-1} in FTIR signified the presence of carboxyl groups on CMTCFs [60] (Fig. 9.7E). The presence of carboxyl groups on CMTCFs facilitates coordination with metal ions. It was beneficial to promote the nucleation and growth of metal oxides on CMTCFs. Moreover, the binding force between CMTCFs and metal oxides was enhanced [61–63]. Finally, CMTCFs@MnO$_2$ were fabricated through in-situ redox process of CMTCFs and KMnO$_4$ in acidic environment.

Different MnO$_2$ spatial distribution density on CMTCFs were achieved by adjusting the ratio of potassium permanganate and CMTCFs. The SEM images of obtained CMTCFs@MnO$_2$ with CMTCFs/KMnO$_4$ weight ratio from 1:1 to 1:4 were shown in Fig. 9.8A trace amount of worm-like MnO$_2$ were coated on CMTCFs when the CMTCFs/KMnO$_4$ weight

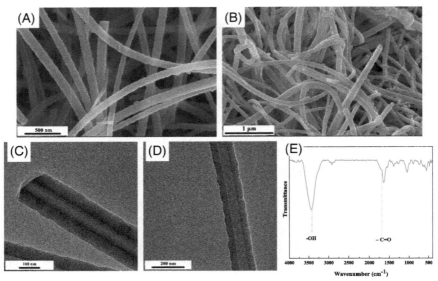

Figure 9.7 SEM (A and B) and TEM (C and D) of nanofibers: TCFs (A and C); CMTCFs (B and D); FTIR spectra of the CMTCFs (E).

ratio was 1:1. Moreover, pure CMTCFs were found in CMTCFs@MnO$_2$-1 (Fig. 9.8A and B). This was due to the small amount of KMnO$_4$ added in the reaction system. With the increase of KMnO$_4$ content (1:2, 1:3 and 1:4), the worm–like MnO$_2$ on CMTCFs was uniformly distributed and the spatial density of MnO$_2$ gradually increased, as displayed in Fig. 9.8C–H. This can be confirmed by the subsequent TGA data. Meantime, worm–like MnO$_2$ on CMTCFs were intertwined with each other to form porous network. It was not only beneficial to accommodate the volume change of MnO$_2$ during Li$^+$ insertion and extraction process, but also can effectively increase the lithium storage site. The specific surface areas of CMTCFs@MnO$_2$-1 to CMTCFs@MnO$_2$-4, determined by Nitrogen adsorption, were 35.42, 59.04, 37.71 and 14.13 m^2 g^{-1} respectively (Fig. 9.9A and B, Table 9.1). It should be mentioned that addition of H$_2$SO$_4$ in the reaction system was critical to the worm–like MnO$_2$ formation. Without H$_2$SO$_4$, MnO$_2$ were coated on CMTCFs in the form of nanoparticles, as shown in Fig. 9.9C–E.

TEM was used to further observe the microstructure of CMTCFs@MnO$_2$-3. It can be clearly seen from Fig. 9.10A, worm–like MnO$_2$ were uniformly coated on CMTCFs to construct one-dimensional coaxial layered structure. The high-magnification TEM image (Fig. 9.10B) shows worm–like MnO$_2$ were intertwined with each other and closely

Figure 9.8 SEM images of the obtained CMTCFs@MnO$_2$ with different MnO$_2$ density: CMTCFs@MnO$_2$-1 (A and B); CMTCFs@MnO$_2$-2 (C and D); CMTCFs@MnO$_2$-3 (E and F); CMTCFs@MnO$_2$-4 (G and H).

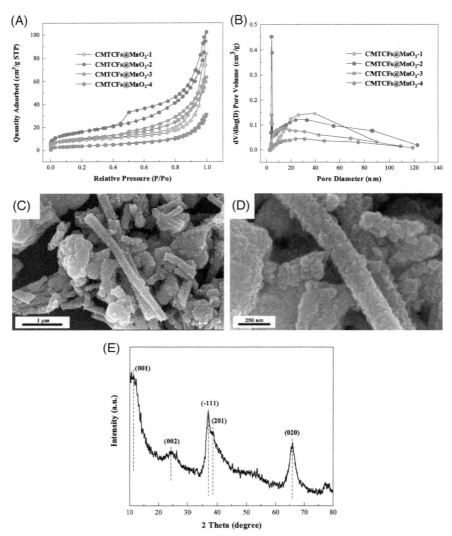

Figure 9.9 N_2 adsorption/desorption isotherms (A) and pore size distribution (B) of the CMTCFs@MnO_2; CMTCFs@MnO_2 without sulfuric acid: SEM (C and D); XRD (E).

Table 9.1 The pore performance of the as-prepared CMTCFs@MnO_2.

Sample information	Surface area ($m^2\ g^{-1}$)	Average pore size (nm)	Pore volume ($m^3\ g^{-1}$)
CMTCFs@MnO_2-1	35.42	19.27	0.13
CMTCFs@MnO_2-2	59.04	9.34	0.16
CMTCFs@MnO_2-3	37.71	11.1	0.10
CMTCFs@MnO_2-4	14.13	13.6	0.05

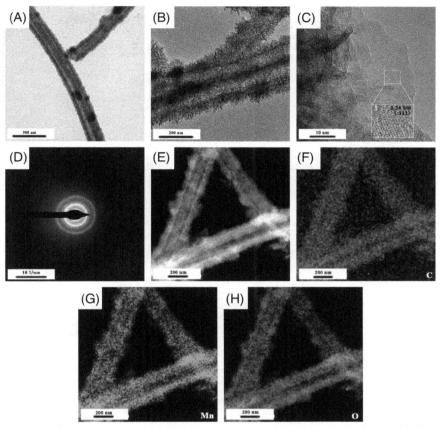

Figure 9.10 TEM images (A and B), high resolution TEM image (C), electron diffraction pattern (D), STEM image (E) and elemental mapping (F–H) of the obtained CMTCFs@MnO$_2$-3.

anchored on CMTCFs. It was consistent with the SEM results (Fig. 9.10E and F). To observe the crystallographic characteristics of CMTCFs@MnO$_2$-3, high resolution TEM (HRTEM) measurement was carried out, the result was shown in Fig. 9.10C. The lattice spacing at 0.24 nm corresponds to the (-111) crystal plane of δ-MnO$_2$. Meantime, the select area electron diffraction (SAED) image (Fig. 9.10D) shows multiple diffraction rings. It signified the poor crystallinity of MnO$_2$. The presence of amorphous regions can buffer mechanical strain of MnO$_2$ during discharge and charge process. It was beneficial for rapid insertion of lithium ions [64]. As can be observed in Elemental mapping (Fig. 9.10E–H), CMTCFs@MnO$_2$-3 had clear layered structure. The uniform distribution of Mn and O elements further proves the uniform coating of MnO$_2$ on CMTCFs.

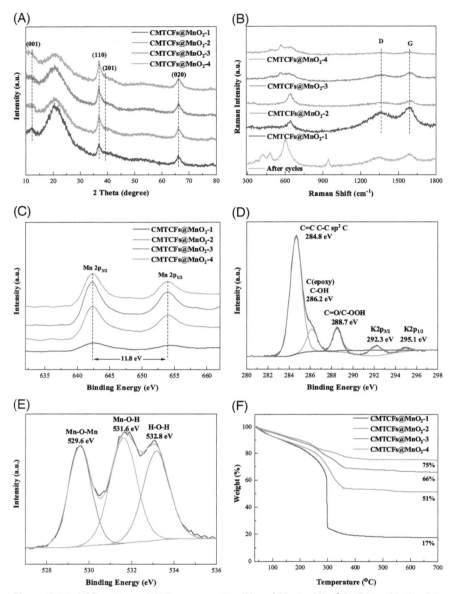

Figure 9.11 XRD patterns (A), Raman spectra (B) and Mn 2p (C) of CMTCFs@MnO$_2$; C 1s (D) and O 1s (E) of CMTCFs@MnO$_2$-3; TGA curves (F) of CMTCFs@MnO$_2$.

The XRD characterization was used to research the crystal structure of CMTCFs@MnO$_2$. As can be found in Fig. 9.11A, the broad peaks of four CMTCFs@MnO$_2$ located at 22° were assigned to the diffraction peaks of CMTCFs. Meantime, the crystal diffraction peaks of the four composites

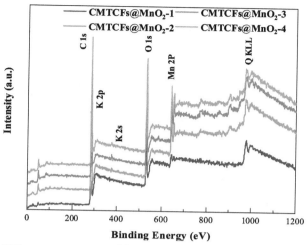

Figure 9.12 XPS survey spectrum of the CMTCFs@MnO_2.

in the range of 30–70° were clearly observed. The relatively broad diffraction peaks and low diffraction intensity indicated the poor crystallinity of CMTCFs@MnO_2 [65,66]. It was consistent with the SAED results (Fig. 9.10D). All the diffraction peaks can be well indexed to birnessite-type MnO_2 (JCPDS card No. 80-1098). No other impure diffraction peaks were observed. The diffraction peaks located at 11.8°, 36.82°, 39.02° and 65.94° were consistent with (001), (110), (201) and (020) planes of δ-MnO_2. According to previous reports, the birnessite-type MnO_2 had two-dimensional layered structure similar to graphene. It was more convenient for the rapid transport of lithium ions in electrode material [67].

Raman spectroscopy was used to further study the structure of CMTCFs@MnO_2. As depicted in Fig. 9.11B, the broad diffraction peaks at ~1349 and ~1582 cm^{-1} were assigned to the disordered carbon (D-bond) and the ordered graphitic carbon (G-bond) [68,69]. Meantime, the diffraction peaks of four composites located at 540–570 cm^{-1} can be attributed to the F_{2g} symmetrical υ_3 (Mn-O) stretching vibration. This proves the presence of Mn^{4+}. The peak at 639 cm^{-1} was related to υ_2 (Mn-O) symmetrical stretching vibration of $[MnO_6]$ octahedron.

XPS measurements were performed to research the chemical composition and surface electronic state of obtained samples. The overall XPS spectrum in Fig. 9.12 revealed the only exist of C, O, K, and Mn elements in four CMTCFs@MnO_2. In Fig. 9.11C, the peaks at 642.3 and 654.1 eV correspond to the characteristic peaks of Mn $2p_{3/2}$ and Mn $2p_{1/2}$. The splitting width between the two peaks was 11.8 eV. It confirmed that the manganese ions exist as Mn^{4+} [70]. The high-resolution C 1s spectrum of CMTCFs@MnO_2-3 (Fig. 9.11D) could be divided into three peaks: C=C/C-C (284.8 eV), C-OH (286.2 eV) and C=O/COOH (288.7 eV). In addition, the peaks located at 292.3 and 295.1 eV could be assigned to K $2p_{3/2}$ and K $2p_{1/2}$, which were derived from the interlayer hydration K^+ of δ-MnO_2. As shown in Fig. 9.11E, the O 1s spectrum of CMTCFs@MnO_2-3 displayed three characteristic peaks at 529.6, 531.6 and 531.6 eV. These peaks correspond to Mn-O-Mn, Mn-O-H and external water [10].

TGA was used to determine the MnO_2 content in composites, the results were summarized in Fig. 9.11F. The four CMTCFs@MnO_2 all exhibited two main weight loss temperature ranges. The final weight loss rate from CMTCFs@MnO_2-1 to CMTCFs@MnO_2-4 gradually decreased. It was consistent with the SEM results (Fig. 9.8). The decomposition temperature below 230°C could be attributed to water molecules adsorbed on the samples. The significant weight loss between 250 and 370°C was due to the decomposition of CMTCFs. The weight loss rate from CMTCFs@MnO_2-1 to CMTCFs@MnO_2-4 was 83%, 49%, 34% and 25%. Therefore, the MnO_2 content in CMTCFs@MnO_2-1, CMTCFs@MnO_2-2, CMTCFs@MnO_2-3 and CMTCFs@MnO_2-4 was 17%, 51%, 66% and 75%.

9.4.2 Effect of worm-like MnO_2 density on the cycle performance

The electrochemical properties of CMTCFs@MnO_2 anodes were evaluated using lithium metal as the counter electrode and the reference electrode. The electrochemical properties of CMTCFs@MnO_2-1, CMTCFs@MnO_2-2, CMTCFs@MnO_2-3 and CMTCFs@MnO_2-4 were measured and compared to clarify the effect of the MnO_2 spatial distribution density.

Fig. 9.13A present the CV curves of CMTCFs@MnO_2-3 for the initial five cycles at 0.5 mV s^{-1} in between 0.01 and 3 V. Significant differences in the initial CV curve were observed. Two reduction peaks could be clearly seen at around 1 and 0.25 V in the initial CV curve. One was due to the deposition of electrolyte on electrode and the formation of solid electrolyte interface (SEI) film. The other was the reduction of Mn^{4+} to Mn^{2+} and further reduction to Mn during lithium ions insertion. The reduction peak shifted to ~0.8 V in the following cycles. This was due to the structural restructuring caused by the formation of Li_2O and Metal Mn [71]. In addition, the sharp reduction peak at 0.01 V was related to the insertion of lithium ions in CMTCFs. Two oxidation peaks located at ~1.4 and ~2.2 V

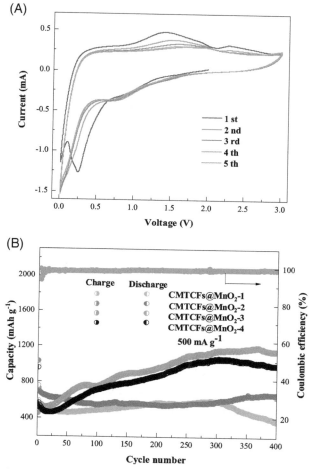

Figure 9.13 CV curves of CMTCFs@MnO₂-3 at a scan rate of 0.5 mV s⁻¹ (A); cycling performance of CMTCFs@MnO₂ at a current density of 500 mA g⁻¹ (B).

were Mn oxidation to Mn^{2+} and further oxidation to Mn^{4+}. The CV profiles of CMTCFs@MnO₂-3 were overlapped well after the first cycle. It indicated the electrode had good structure stability and electrochemical reversibility.

The cycling performances of CMTCFs@MnO₂ anodes were assessed at 500 mA g⁻¹ in between 0.01 and 3 V (Fig. 9.13B). The specific capacities of CMTCFs@MnO₂-3 and CMTCFs@MnO₂-4 electrode decreased slightly during initial 25 cycles. It could be ascribed to the deposition of electrolyte and the formation of SEI film. Then, the specific capacities of the electrodes

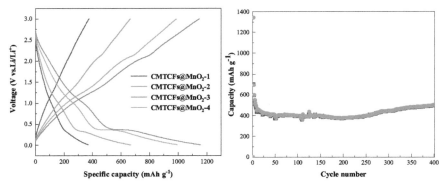

Figure 9.14 Discharge/charge curves of CMTCFs@MnO$_2$ at the 400th cycle; cycling performance of the CMTCFs at 500 mA g^{-1}.

increased slowly with the number of cycles. Similar trend was previously reported for nano metal oxide composites electrode [72,73]. One was due to the reversible growth of gel-like polymer film on the electrode. The other was attributed to the activation of the electrode during cycling [74]. The reversible specific capacities of CMTCFs@MnO$_2$-3 and CMTCFs@MnO$_2$-4 anode stabilized after 300 cycles and reached 1147.4 and 986.9 mAh g^{-1} at 400 cycles (Fig. 9.14). Compared with CMTCFs@MnO$_2$-3 and CMTCFs@MnO$_2$-4, CMTCFs@MnO$_2$-1 and CMTCFs@MnO$_2$-2 exhibited good cycle stability. But the reversible specific capacity was lower. Thus, the spatial distribution density of worm-like MnO$_2$ on CMTCFs had significant effect on electrochemical performance of electrode materials. Moderate worm-like MnO$_2$ loading density (CMTCFs@MnO$_2$-3) not only facilitated the electronic contact between CMTCFs and conductive carbon, but also maximizes the specific capacity of electrode material. In addition, CMTCFs@MnO$_2$-3 anode delivered discharge specific capacity of 1155.9 mAh g^{-1} after 400 cycles. From TGA curve of CMTCFs@MnO$_2$-3 in Fig. 9.11F, the weight ratio of CMTCFs, MnO$_2$ and H$_2$O was 20:66:14. The discharge specific capacity of CMTCFs was 506 mAh g^{-1} after 400 cycles at 500 mA g^{-1} (Fig. 9.14). The theoretical capacity of MnO$_2$ electrode was 1233 mAh g^{-1}. Therefore, the theoretical maximum discharge specific capacity of CMTCFs@MnO$_2$-3 electrode was 914.98 mAh g^{-1} (506 mAh g^{-1} × 20 wt% + 1233 mAh g^{-1} × 66 wt% = 914.98 mAh g^{-1}). This data was lower than the actual discharge specific capacity. The extra specific capacity was due to the change of electrode structure during cycling and capacitive lithium storage behavior at CMTCFs @MnO$_2$-3 electrode interface.

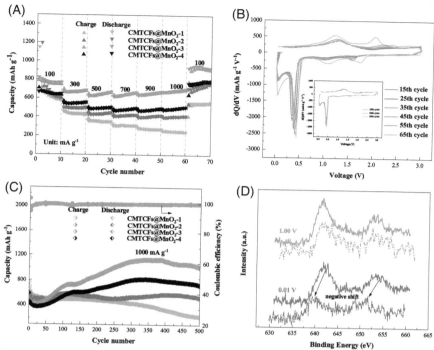

Figure 9.15 Electrochemical properties of CMTCFs@MnO$_2$ electrodes: rate performance (A); corresponding differential capacities (dQ/dV) versus voltages plots of the charge/discharge curves at different current density of CMTCFs@MnO$_2$-3 (B); cycling performance and coulombic efficiency at 1000 mA g^{-1} (C); Mn 2p XPS spectra of CMTCFs@MnO$_2$-3 electrode at various lithiated stages (D).

9.4.3 Enhanced rate performance of CMTCFs@MnO$_2$

Fig. 9.15A showed the rate performance of CMTCFs@MnO$_2$ anodes at 100, 300, 500, 700, 900, and 1000 mA g^{-1}. The rate performance of CMTCFs@MnO$_2$-3 electrode was obviously better than CMTCFs@MnO$_2$-4, CMTCFs@MnO$_2$-1 and CMTCFs@MnO$_2$-2 electrode. This was because the moderate MnO$_2$ spatial density could effectively balance the relationship between capacity and electronic contact. The average reversible capacity of CMTCFs@MnO$_2$-3 electrode was 680.24 mAh g^{-1} at 1000 mA g^{-1}. It was obviously higher than the average reversible capacity at low current densities of 300 (668.93 mAh g^{-1}), 500 (641.03 mAh g^{-1}), 700 (638.03 mAh g^{-1}), and 900 mA g^{-1} (651.05 mAh g^{-1}). The dQ/dV plots corresponding to the charge-discharge curves at different current densities were shown in Fig. 9.15B. The positions of characteristic peaks were basically unchanged in dQ/dV plots at 15th

(300 mA g^{-1}), 25th (500 mA g^{-1}) and 35th (700 mA g^{-1}) cycle. However, the characteristic peaks located at 1.25 and 2.26 V shifted in dQ/dV plots at 45th (900 mA g^{-1}) and 55th (1000 mA g^{-1}) cycle. It indicated that the electrochemical environment changed during cycling process. The electrode was activated. Thus, the capacity increase phenomenon was observed at high current density. More remarkable, when the current density returned to 100 mA g^{-1}, the average reversible capacities of CMTCFs@MnO$_2$-2, CMTCFs@MnO$_2$-3 and CMTCFs@MnO$_2$-4 anode reached 704.6, 908.22 and 756.51 mAh g^{-1}. These data significantly higher the average reversible capacities of 641.75, 759.78 and 651.43 mAh g^{-1} at the initial 100 mA g^{-1}. Moreover, the characteristic peaks in dQ/dV plots at 65th (100 mA g^{-1}) cycle showed more obvious shift than 45th (900 mA g^{-1}) and 55th (1000 mA g^{-1}) cycle. Therefore, the enhancement of lithium storage capacity was related to the activation of electrode.

In order to clarify the effect of large current density cycling on rate performance of electrode. The cycling performance at 1000 mA g^{-1} and subsequent rate performance of CMTCFs@MnO$_2$ anodes were carried out. Fig. 9.15C showed the cyclic performance curves of four CMTCFs@MnO$_2$ anodes at 1000 mA g^{-1} in between 0.01 and 3 V. The cycling curve trends of CMTCFs@MnO$_2$ electrodes were similar to those at 500 mA g^{-1}. The reversible specific capacity of CMTCFs@MnO$_2$-3 anode was maintained at 1013.5 mAh g^{-1} after 500 cycles. The Mn 2p spectra of CMTCFs@MnO$_2$-3 electrode at various lithiated stages at the first cycle and 75th cycles were examined (Fig. 9.15D). The position of the characteristic peaks moved to a low bind energy as the number of cycles increased, especially at low voltage. It further demonstrated the presence of changes in electrochemical environment and the activation behavior of electrode during cycling [75].

The rate performance of CMTCFs@MnO$_2$ anodes after 500 cycles at the current density of 1000 mA g^{-1} was evaluated, as shown in Fig. 9.16. Compared with CMTCFs@MnO$_2$-1 and CMTCFs@MnO$_2$-2, the rate performance of CMTCFs@MnO$_2$-3 and CMTCFs@MnO$_2$-4 was more outstanding after long-term cycling at large current density. In addition, the average reversible specific capacities of CMTCFs@MnO$_2$-3 and CMTCFs@MnO$_2$-4 at current densities of 100, 300, 500, 700, 900, and 1000 mA g^{-1} were (1422.7, 1098.233), (1149.33, 1021.79), (1037.33, 908.5), (959.41, 839.89), (881.29, 774.2), and (768.1, 718.22) mAh g^{-1}. Compared with the results in Fig. 9.15A, the average reversible specific capacity increased by (87%, 69%), (72%, 87%), (62%, 84%), (50%, 75%), (35%, 62%), and (13%, 46%) (Fig. 9.17). Thus, long-term cycling at high current density

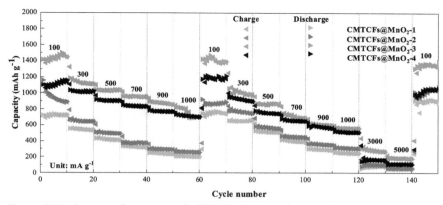

Figure 9.16 Rate performance of CMTCFs@MnO₂ after cycling performance at 1000 mA g⁻¹.

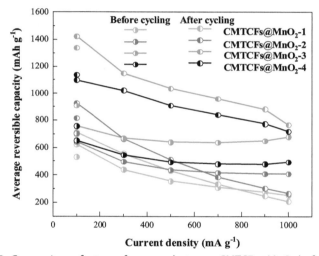

Figure 9.17 Comparison of rate performance between CMTCFs@MnO₂ before and after cycling.

had enhanced significantly on the rate performance of CMTCFs@MnO₂ electrode with moderate MnO₂ spatial density.

To explore the rate performance at large current density, the rate performance test of CMTCFs@MnO₂ anodes after cycling and rate performance was carried out. As shown in Fig. 9.16, CMTCFs@MnO₂-3 anode displayed the best rate performance. The electrode delivered high average specific capacities of 303.81 and 190.28 mAh g⁻¹ at 3000 and 5000 mA g⁻¹. When the current density returned to 100 mA g⁻¹, the average reversible capacities of CMTCFs@MnO₂-3 anode remained 1242.64 mAh g⁻¹. It was close to

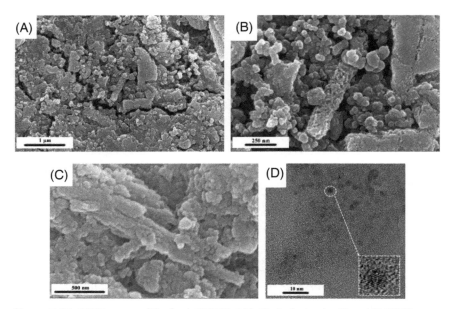

Figure 9.18 SEM images of the fresh CMTCFs@MnO$_2$-3 electrodes (A and B); SEM images of the fully charged CMTCFs@MnO$_2$-3 electrode after cycling performance and rate performance test (C); high resolution TEM image of the fully charged CMTCFs@MnO$_2$-3 electrode after cycling performance and rate performance test (D).

the average reversible capacity at the initial 100 mA g^{-1} (1422.7 mAh g^{-1}). Thus, CMTCFs@MnO$_2$-3 anode had outstanding rate performance.

9.4.4 Lithium storage mechanism of CMTCFs@MnO$_2$

To clarify the structure change of electrode material after electro-chemical performance test, SEM, TEM, and Raman were collected for CMTCFs@MnO$_2$-3 anode (Fig. 9.18). As shown in Fig. 9.18A and B, the fibrous structure of CMTCFs remained well. The CMTCFs were still coated with worm-like MnO$_2$. It indicated that the electrode preparation process had no significant effect on CMTCFs@MnO$_2$-3 anode. However, obvious attachments were observed on electrode material after electrochemical test (Fig. 9.18C). This might be related to additives in electrode material and by-products in chemical cycle. As can be seen from Fig. 9.18D, worn-like MnO$_2$ turned into nanoparticles after cycling and rate performance test. This change not only increased the electronic connection between conductive carbon and CMTCFs, but also facilitated the capacitive lithium storage behavior at electrode material interface. In addition, the lattice spacing of

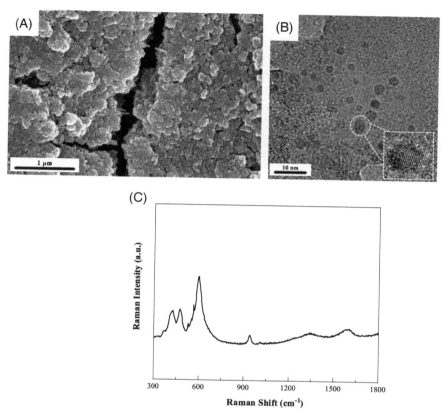

Figure 9.19 Structure of CMTCFs@MnO_2-3 after 400 cycles at 500 mA g^{-1}: SEM (A); high resolution TEM image (B); Raman spectra (C).

0.20 nm in the TEM image corresponds to the (400) plane of λ-MnO_2. The Raman peaks at 366.87, 417.81, and 479.64 cm^{-1} in Raman spectrum (Fig. 9.11B) appeared after cycling and rate performance test. This corresponds to λ-MnO_2. It illustrated that the crystal form of MnO_2 changed from δ-MnO_2 to λ-MnO_2 during electrochemical test. Meantime, the structure and crystal form of CMTCFs@MnO_2-3 anode had similar changes after 400 cycles at 500 mA g^{-1} (Fig. 9.19).

EIS was used to investigated the mechanism of enhanced electrochemical performance of CMTCFs@MnO_2-3 anode, as shown in Fig. 9.20. Both Nyquist curves consisted of concave semicircle at high frequency region and sloped line at low frequency region. The inset was the corresponding equivalent circuit. R_s, R_{ct}, and Z_w associated with the solution resistance, Faradaic impedance and ion transfer impedance [76,77]. The R_{ct} of

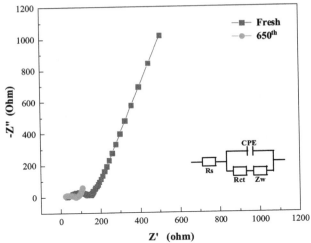

Figure 9.20 Electrochemical impedance spectra of MTCFs@MnO$_2$-3 before and after cycling performance and rate performance test. The inset is the equivalent circuit model for plot fitting.

Table 9.2 Simulation results of the kinetic parameters of MTCFs@MnO$_2$-3 before and after cycling performance and rate performance test.

Samples	Rs (ohm)	Rct (ohm)
Fresh	31.49	153.3
650 th	10.33	64.54

CMTCFs@MnO$_2$-3 anode was reduced from 153.3 Ω to 64.54 Ω after electrochemical performance test (Table 9.2). It indicated that the change of morphology and crystal form of CMTCFs@MnO$_2$-3 anode during electrochemical test effectively improved the electron transfer rate of the electrode.

CV curves at different voltage scan rates were used to investigate the lithium storage mechanism of CMTCFs@MnO$_2$-3 anode (after cycle and rate performance test), as shown in Fig. 9.21. The b values corresponding following formula:

$$i = a v^b$$

Where i was peak current value, v was scan rate, a, and b were constants [78]. When the b value was approaches 0.5 and 1, it meant the diffusion control process and the capacitance process, respectively [79]. The b values of Peak1 to Peak5 were 0.68, 0.78, 0.54, 0.93, and 0.55 (Fig. 9.21B). Thus,

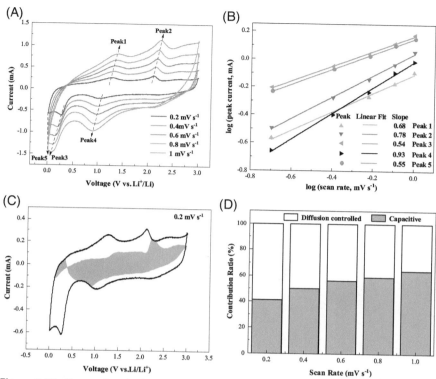

Figure 9.21 Electrochemical properties of the MTCFs@MnO$_2$-3 electrode: CV curves of the MTCFs@MnO$_2$-3 electrode (after cycling performance and rate performance test) at different scan rates (A); determination of the b value at specific peak current (B); capacitive and diffusion controlled contribution at a scan rate of 0.2 mV s^{-1}, the estimated capacitive contribution is shown in the shaded area (C); contribution ratios of surface controlled at various scan rates (D).

the lithium storage mechanism of CMTCFs@MnO$_2$-3 was controlled by both the diffusion process and the capacitance process. Further calculate the pseudocapacitive contribution rate by following formula:

$$i(V) = k_1 \upsilon + k_2 \upsilon^{1/2}$$

Where k_1 and k_2 were constants. The $k_1 \upsilon$ represented the capacitive controlled contribution [80]. As can be seen from Fig. 9.21C–D and Fig. 9.22, pseudocapacitive contribution rates were 41%, 50%, 56%, 59%, and 64% at scan rates of 0.2, 0.4, 0.6, 0.8, and 1 mV s^{-1}. Therefore, the interfacial lithium storage behavior existed in CMTCFs@MnO$_2$-3. It enhanced the electrochemical performance of electrode. The design and

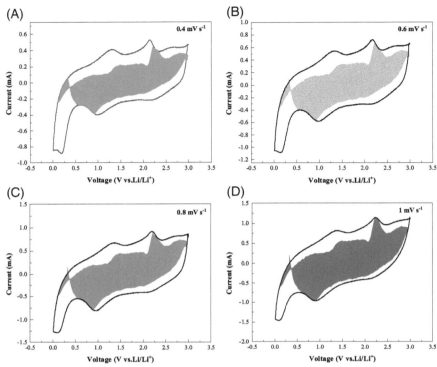

Figure 9.22 CV curve with the pseudocapacitive fraction shown by the shaded area at different scan rate: 0.4 mV s^{-1} (A); 0.6 mV s^{-1} (B); 0.8 mV s^{-1} (C); 1 mV s^{-1} (D).

preparation of novel CMTCFs@MnO$_2$ provided promising anode material for high performance LIBs.

9.5 Tailoring carboxyl tubular carbon nanofibers/MnO$_2$ composites

Meanwhile, in order to enrich the morphology of MnO$_2$ compounded with tubular nanofibers, lamellar MnO$_2$ was grown on the surface of tubular nanofibers by hydrothermal method, and three composite products were obtained through system adjustment. Further, the performances of different samples as cathode of lithium–ion battery were compared by electrochemical test.

9.5.1 Design and characterization of CMTCFs/MnO$_2$

Using carboxyl group-modified tubular carbon nanofibers (CMTCFs) as a carrier, they were composited with δ- MnO$_2$ by hydrothermal method. As shown in Fig. 9.23, CMTCFs coated by δ-MnO$_2$ nanosheet,

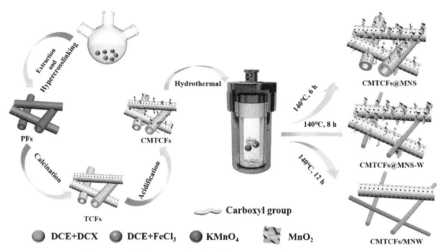

Figure 9.23 Schematic illustration for the fabrication of CMTCFs/MnO$_2$ composite materials.

Figure 9.24 SEM images of CMTCFs/MnO$_2$: CMTCFs@MNS (A); CMTCFs@MNS-W (B); CMTCFs/MNW (C). The ruler was 1 μm. TEM images (D), elemental mapping (E) and high resolution TEM image (F) of the obtained CMTCFs@MNS.

CMTCFs@MNS co-existed with α- MnO$_2$ nanowires and CMTCFs co-existed with α-MnO$_2$ nanowires were obtained by controlling the reaction time, respectively. With the extension of reaction time, MnO$_2$ tended to grow independently. The SEM pictures of the three materials were shown in Fig. 9.24A–C. The δ-MnO$_2$ nanosheets on the surface of CMTCFs@MNS were loosely stacked (Fig. 9.24D and E), the free space between which could not only adapt to the volume change to avoid the rupture of MnO_2 during the discharge/charging process, but also effectively increase the lithium storage sites. The lattice spacing of 0.24 nm in HRTEM (Fig. 9.24F) corresponded to the plane (-111) of birnessite-type MnO_2. The BET specific surface areas of CMTCFs@MNS, CMTCFs@MNS-W, and CMTCFs/MNW were 100.68 m^2/g, 65.38 m^2/g and 88.76 m^2/g, respectively.

9.5.2 Electrochemical performance of CMTCFs/MnO₂

The prepared CMTCFs@MNS, CMTCFs/MNW, and CMTCFs@MNS-W were used as LIBS anode active material to assemble into LIR2016-type half–cells, and their electrochemical performances were compared. The specific capacity of CMTCFs@MNS electrode was up to 1497.1 mAh g^{-1}, which was much higher than that of the electrode CMTCFs@MNS -W (852.1 mAh g^{-1}), CMTCFs/MNW (1240.9 mAh g^{-1}). Fig. 9.25A showed the rate performance of CMTCFs@MNS electrode material. which is superior to the reported composites of MnO_2 and carbon. When the current density finally returned to 100 mA g^{-1} after 12 changes, the average reversible capacity was as high as 1293.93 mAh g^{-1}. Fig. 9.25B shows the test data of the long–term cycling stability of CMTCFs@MNS. After 1000 cycles at a current density of 10,000 mA g^{-1}, the reversible capacity of CMTCFs@MNS reached 400.8 mAh g^{-1}, which was higher than the theoretical capacity of commercial graphite (372 mAh g^{-1}).

9.6 Summary and prospect

In summary, MnO_2 material has a high degree of capacity, but its application is limited because of the strength problem caused by the volume expansion during the charging and discharging process. And one effective solution is to construct free space to offset the volume expansion. The hollow tubular porous fiber itself has excellent electrochemical properties, and the growth of MnO_2 nanomaterials on its surface can combine the advantages of the two to improve the electrochemical properties of the material. In the commercialization process of electrode, high cycle stability, excellent rate performance, and high capacity at high current density are always the prioritized goals. It is not that a certain material meets the requirements alone, but the above properties are perfect on the same material fusion. What is really needed is not a certain material meets the requirements singly, but one independent material in which the above properties are perfectly

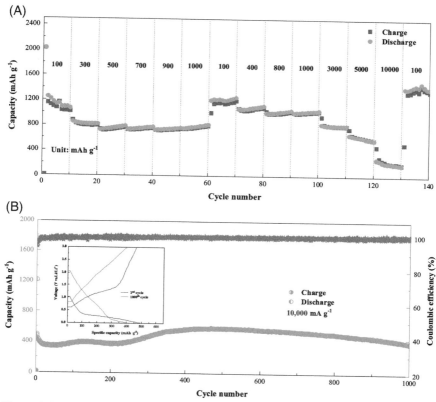

Figure 9.25 Rate performance of the CMTCFs@MNS (A); long-term cycling performance of the CMTCFs@MNS at 10,000 mA g^{-1} (B).

combined. According to the analysis, the preparation of high-performance one-dimensional anode materials by MnO$_2$ and carbon composites would focus on the following directions in the future.

(1) Develop a rational preparation strategy to prepare novel anode materials, taking into account the material design concept of nanometerization, selecting suitable composite carriers, and reserving free space.

(2) Construct ordered pores on the MnO$_2$ component in the negative electrode material, optimize the complete ion channel of the material, and improve the electrochemical performance.

(3) For the proven negative electrode materials, develop a preparation method, which is easy to be industrialized and scaled up, and promote its industrial application.

References

[1] Y. Zhao, L.P. Wang, M.T. Sougrati, Z. Feng, Y. Leconte, A. Fisher, M. Srinivasan, Z. Xu, A Review on Design Strategies for Carbon Based Metal Oxides and Sulfides Nanocomposites for High Performance Li and Na Ion Battery Anodes, Adv. Energy Mater. 7 (2017) 1601424.

[2] L. Sheng, H. Jiang, S. Liu, M. Chen, T. Wei, Z. Fan, Nitrogen-doped carbon-coated MnO nanoparticles anchored on interconnected graphene ribbons for high-performance lithium-ion batteries, J. Power Sources 397 (2018) 325–333.

[3] C. Hou, Z. Tai, L. Zhao, Y. Zhai, Y. Hou, Y. Fan, F. Dang, J. Wang, H. Liu, High performance MnO@C microcages with a hierarchical structure and tunable carbon shell for efficient and durable lithium storage, J. Mater. Chem. A 6 (2018) 9723–9736.

[4] Z. Yan, X. Jiang, Y. Dai, W. Xiao, X. Li, N. Du, G. He, Pulverization Control by Confining Fe_3O_4 Nanoparticles Individually into Macropores of Hollow Carbon Spheres for High-Performance Li-Ion Batteries, ACS Appl. Mater. Inter. 10 (2018) 2581–2590.

[5] L. Guo, H. Sun, C. Qin, W. Li, F. Wang, W. Song, J. Du, F. Zhong, Y. Ding, Flexible Fe_3O_4 nanoparticles/N-doped carbon nanofibers hybrid film as binder-free anode materials for lithium-ion batteries, Appl. Surf. Sci. 459 (2018) 263–270.

[6] J. Shao, H. Zhou, M. Zhu, J. Feng, A. Yuan, Facile synthesis of metal-organic framework-derived Co_3O_4 with different morphologies coated graphene foam as integrated anodes for lithium-ion batteries, J. Alloys Compd. 768 (2018) 1049–1057.

[7] M. Zhong, W.W. He, W. Shuang, Y.Y. Liu, T.L. Hu, X.H. Bu, Metal–Organic Framework Derived Core–Shell Co/Co_3O_4@N-C Nanocomposites as High Performance Anode Materials for Lithium Ion Batteries, Inorg. Chem. 57 (2018) 4620–4628.

[8] X. Yin, C. Zhi, W. Sun, L.P. Lv, Y. Wang, Multilayer NiO@Co_3O_4@graphene quantum dots hollow spheres for high-performance lithium-ion batteries and supercapacitors, J. Mater. Chem. A. 7 (2019) 7800–7814.

[9] G. Wang, Y. Sun, D. Li, W. Wei, X. Feng, K. Mullen, Constructing Hierarchically Hollow Core–Shell MnO_2/C Hybrid Spheres for High-Performance Lithium Storage, Small. 12 (2016) 3914–3919.

[10] S. Li, Y. Zhao, Z. Liu, L. Yang, J. Zhang, M. Wang, R. Che, Flexible Graphene-Wrapped Carbon Nanotube/Graphene@MnO_2 3D Multilevel Porous Film for High-Performance Lithium-Ion Batteries, Small. 14 (2018) 1801007.

[11] X. Zhai, Z. Mao, G. Zhao, D. Rooney, N. Zhang, K. Sun, Nanoflake δ-MnO_2 deposited on carbon nanotubes-graphene-Ni foam scaffolds as self-standing three-dimensional porous anodes for high-rate-performance lithium-ion batteries, J. Power Sources. 402 (2018) 373–380.

[12] D. Wang, Y. Wang, Q. Li, W. Guo, F. Zhang, S. Niu, Urchin-like α-Fe_2O_3/MnO_2 hierarchical hollow composite microspheres as lithium-ion battery anodes, J. Power Sources. 393 (2018) 186–192.

[13] Y. Li, D. Ye, W. Liu, B. Shi, R. Guo, H. Pei, J. Xie, A three-dimensional core-shell nanostructured composite of polypyrrole wrapped MnO_2/reduced graphene oxide/carbon nanotube for high performance lithium ion batteries, J Colloid. Interf. Sci. 493 (2017) 241–248.

[14] F. Yang, W. Li, Y. Zhang, B. Tang, Realizing uniform dispersion of MnO_2 with the post-synthetic modification of metal–organic frameworks (MOFs) for advanced lithium ion battery anodes, Dalton. T. 47 (2018) 13657–13667.

[15] S. Zhu, L. Li, J. Liu, H. Wang, T. Wang, Y. Zhang, L. Zhang, R.S. Ruoff, F. Dong, Structural Directed Growth of Ultrathin Parallel Birnessite on β-MnO_2 for High-Performance Asymmetric Supercapacitors, ACS Nano. 12 (2) (2018) 1033–1042.

[16] S. Akbari, M. Mehdi Foroughi, M. Ranjbar, Solvent-free Synthesis and Characterization of MnO_2 Nanostructures and Investigation of Optical Properties, J. Nanomed. Nanotechnol 9 (3) (2018) 1000498.

[17] P. Liu, Y. Zhu, X. Gao, Y. Huang, Y. Wang, S. Qin, Y. Zhang, Rational construction of bowl-like MnO$_2$ nanosheets with excellent electrochemical performance for supercapacitor electrodes, Chem. Engin. J 350 (2018) 79–88.

[18] B. Yin, S. Zhang, H. Jiang, F. Qu, X. Wu, Phase-controlled synthesis of polymorphic MnO$_2$ structures for electrochemical energy storage, J. Mat. Chem. A. 3 (10) (2015) 5722–5729.

[19] M.H. Alfaruqi, J. Gim, S. Kim, J. Song, D.T. Pham, J. Jo, Z. Xiu, V. Mathew, J. Kim, A layered δ-MnO$_2$ nanoflake cathode with high zinc-storage capacities for eco-friendly battery applications, Electrochem. Commun 60 (2015) 121–125.

[20] T. Barudžija, V. Kusigerski, N. Cvjetićanin, S. Šorgić, M. Perović, M. Mitrić, Structural and magnetic properties of hydrothermally synthesized β-MnO$_2$ and α-K$_x$MnO$_2$ nanorods, J. Alloys Compd. 665 (2016) 261–270.

[21] C.M. Julien, A. Mauger, Nanostructured MnO$_2$ as Electrode Materials for Energy Storage, Nanomaterials 7 (11) (2017) 396.

[22] V. Nulu, A. Nulu, M.G. Kim, K.Y. Sohn, Template-free Facile Synthesis of α-MnO$_2$ Nanorods for Lithium Storage Application, Int. J. Electrochem. Sci 13 (2018) 5565–5574.

[23] L. Liu, Z. Shen, X. Zhang, S. Ma, Facile controlled synthesis of MnO$_2$ nanostructures for high-performance anodes in lithium-ion batteries, J. Mat. Science: Mat. Electron 30 (2) (2018) 1480–1486.

[24] H. Liu, Z. Hu, H. Ruan, R. Hu, Y. Su, L. Zhang, J. Zhang, Nanostructured MnO$_2$ anode materials for advanced lithium ion batteries, J. Mat. Sci.: Mat. Electron 27 (11) (2016) 11541–11547.

[25] C. Jiang, Y. Li, S. Wang, Z. Zhang, Z. Tang, Hierarchical hydrated WO$_3$·0.33H2O/graphene composites with improved lithium storage, Electrochimica. Acta. 278 (2018) 290–301.

[26] S. Wang, W. Quan, Z. Zhu, Y. Yang, Q. Liu, Y. Ren, X. Zhang, R. Xu, Y. Hong, Z. Zhang, K. Amine, Z. Tang, J. Lu, J. Li, Lithium titanate hydrates with superfast and stable cycling in lithium ion batteries, Nat. Commun 8 (1) (2017) 627.

[27] R. Xu, J. Li, A. Tan, Z. Tang, Z. Zhang, Novel lithium titanate hydrate nanotubes with outstanding rate capabilities and long cycle life, J Power Sources. 196 (4) (2011) 2283–2288.

[28] K.W. Nam, S. Kim, E. Yang, Y. Jung, E. Levi, D. Aurbach, J.W. Choi, Critical Role of Crystal Water for a Layered Cathode Material in Sodium Ion Batteries, Chem. Mat 27 (10) (2015) 3721–3725.

[29] C. Jiang, Z. Tang, Z. Zhang, Facile synthesis of a hierarchical manganese oxide hydrate for superior lithium-ion battery anode, Ionics. 25 (8) (2019) 3577–3586.

[30] L. Zhang, J. Song, Y. Liu, X. Yuan, S. Guo, Tailoring nanostructured MnO$_2$ as anodes for lithium ion batteries with high reversible capacity and initial Coulombic efficiency, J. Power Sources 379 (2018) 68–73.

[31] X. Hu, X. Lou, C. Li, Q. Yang, Q. Chen, B. Hu, Green and Rational Design of 3D Layer-by-Layer MnO$_x$ Hierarchically Mesoporous Microcuboids from MOF Templates for High-Rate and Long-Life Li-Ion Batteries, ACS Appl. Mat. Interfaces 10 (17) (2018) 14684–14697.

[32] W. Zhang, B. Zhang, H. Jin, P. Li, Y. Zhang, S. Ma, J. Zhang, Waste eggshell as biotemplate to synthesize high capacity δ-MnO$_2$ nanoplatelets anode for lithium ion battery, Ceramics Int 44 (16) (2018) 20441–20448.

[33] P.M. Ette, K. Selvakumar, S.M. Senthil Kumar, K. Ramesha, Silica template assisted synthesis of ordered mesoporous β–MnO$_2$ nanostructures and their performance evaluation as negative electrode in Li-ion batteries, Electrochimica Acta. 292 (2018) 532–539.

[34] A.A. Voskanyan, C.-K. Ho, K.Y. Chan, 3D δ-MnO$_2$ nanostructure with ultralarge mesopores as high-performance lithium-ion battery anode fabricated via colloidal solution combustion synthesis, J. Power Sources 421 (2019) 162–168.

[35] B.D. Anderson, J.B. Tracy, Nanoparticle conversion chemistry: Kirkendall effect, galvanic exchange, and anion exchange, Nanoscale 6 (21) (2014) 12195–12216.
[36] R. Kas, K.K. Hummadi, R. Kortlever, P. de Wit, A. Milbrat, M.W. Luiten-Olieman, N.E. Benes, M.T. Koper, G. Mul, Three-dimensional porous hollow fibre copper electrodes for efficient and high-rate electrochemical carbon dioxide reduction, Nat. Commun 7 (2016) 10748.
[37] Y. Wang, P. Ding, C. Wang, Fabrication and lithium storage properties of MnO$_2$ hierarchical hollow cubes, J. Alloys Compd 654 (2016) 273–279.
[38] H. Kim, N. Venugopal, J. Yoon, W.-S. Yoon, A facile and surfactant-free synthesis of porous hollow λ-MnO$_2$ 3D nanoarchitectures for lithium ion batteries with superior performance, J. Alloys Compd 778 (2019) 37–46.
[39] S. Hao, B. Zhang, J. Feng, Y. Liu, S. Ball, J. Pan, M. Srinivasan, Y. Huang, Nanoscale ion intermixing induced activation of Fe$_2$O$_3$/MnO$_2$ composites for application in lithium ion batteries, J. Mat. Chem. A. 5 (18) (2017) 8510–8518.
[40] Y. Wang, D. Wang, Q. Li, W. Guo, F. Zhang, Y. Yu, Y. Yang, General Synthesis and Lithium Storage Properties of Metal Oxides/MnO$_2$ Hierarchical Hollow Hybrid Spheres, Particle & Particle Systems Characterization 35 (2) (2018) 1700336.
[41] Y. Su, J. Zhang, K. Liu, Z. Huang, X. Ren, C.-A. Wang, Simple synthesis of a double-shell hollow structured MnO$_2$@TiO$_2$ composite as an anode material for lithium ion batteries, RSC Adv. 7 (73) (2017) 46263–46270.
[42] Y. Zhang, Q. Li, J. Liu, W. You, F. Fang, M. Wang, R. Che, Hierarchical Fe$_2$O$_3$@C@MnO$_2$@C Multishell Nanocomposites for High Performance Lithium Ion Batteries and Catalysts, Langmuir. 34 (18) (2018) 5225–5233.
[43] D.J. Yan, X.D. Zhu, X.T. Gao, L.L. Gu, Y.J. Feng, K.N. Sun, Smartly Designed Hierarchical MnO$_2$@Fe$_3$O$_4$/CNT Hybrid Films as Binder-free Anodes for Superior Lithium Storage, Chem.-An Asian J 13 (20) (2018) 3027–3031.
[44] Y. Wang, W. Guo, Y. Yang, Y. Yu, Q. Li, D. Wang, F. Zhang, Rational design of SnO$_2$@C@MnO$_2$ hierarchical hollow hybrid nanospheres for a Li-ion battery anode with enhanced performances, Electrochimica Acta. 262 (2018) 1–8.
[45] L. Liu, J. Peng, G. Wang, Y. Ma, F. Yu, B. Dai, X.-H. Guo, C.-P. Wong, Synthesis of mesoporous TiO$_2$@C@MnO$_2$ multi-shelled hollow nanospheres with high rate capability and stability for lithium-ion batteries, RSC Adv. 6 (69) (2016) 65243–65251.
[46] D. Li, Y. Zhang, K. Rui, H. Lin, Y. Yan, X. Wang, C. Zhang, X. Huang, J. Zhu, W. Huang, Coaxial-cable hierarchical tubular MnO$_2$@Fe$_3$O$_4$@C heterostructures as advanced anodes for lithium-ion batteries, Nanotechnology. 30 (9) (2019) 094002.
[47] Y. Jiang, Z.-J. Jiang, B. Chen, Z. Jiang, S. Cheng, H. Rong, J. Huang, M. Liu, Morphology and crystal phase evolution induced performance enhancement of MnO$_2$ grown on reduced graphene oxide for lithium ion batteries, J. Mat. Chem. A. 4 (7) (2016) 2643–2650.
[48] Z. Ma, T. Zhao, Reduced graphene oxide anchored with MnO$_2$ nanorods as anode for high rate and long cycle Lithium ion batteries, Electrochimica Acta. 201 (2016) 165–171.
[49] S.-C. Weng, S. Brahma, C.-C. Chang, J.-L. Huang, Synthesis of MnO$_x$/reduced graphene oxide nanocomposite as an anode electrode for lithium-ion batteries, Ceramics Int 43 (6) (2017) 4873–4879.
[50] H. Liu, Z. Hu, L. Tian, Y. Su, H. Ruan, L. Zhang, R. Hu, Reduced graphene oxide anchored with δ-MnO$_2$ nanoscrolls as anode materials for enhanced Li-ion storage, Ceramics Int 42 (12) (2016) 13519–13524.
[51] H. Liu, Z. Hu, Y. Su, H. Ruan, R. Hu, L. Zhang, MnO$_2$ nanorods/3D-rGO composite as high performance anode materials for Li-ion batteries, Appl. Surf. Sci 392 (2017) 777–784.
[52] C. Jiang, C. Yuan, P. Li, H.-g. Wang, Y. Li, Q. Duan, Nitrogen-doped porous graphene with surface decorated MnO$_2$ nanowires as a high-performance anode material for lithium-ion batteries, J. Mat. Chem. A. 4 (19) (2016) 7251–7256.

[53] T. Yang, T. Qian, M. Wang, J. Liu, J. Zhou, Z. Sun, M. Chen, C. Yan, A new approach towards the synthesis of nitrogen-doped graphene/MnO₂ hybrids for ultralong cycle-life lithium ion batteries, J. Mat. Chem. A. 3 (12) (2015) 6291–6296.

[54] D. Wang, K. Wang, L. Sun, H. Wu, J. Wang, Y. Zhao, L. Yan, Y. Luo, K. Jiang, Q. Li, S. Fan, J. Li, J. Wang, MnO₂ nanoparticles anchored on carbon nanotubes with hybrid supercapacitor-battery behavior for ultrafast lithium storage, Carbon. 139 (2018) 145–155.

[55] X. Zhang, T. Wang, C. Jiang, F. Zhang, W. Li, Y. Tang, Manganese Dioxide/Cabon Nanotubes Composite with Optimized Microstructure via Room Temperature Solution Approach for High Performance Lithium-Ion Battery Anodes, Electrochimica Acta. 187 (2016) 465–472.

[56] L. Shen, Q. Dong, G. Zhu, Z. Dai, Y. Zhang, W. Wang, X. Dong, Versatile MnO₂/CNT Putty-Like Composites for High-Rate Lithium-Ion Batteries, Adv. Mat. Interfaces 5 (14) (2018) 1800362.

[57] E. Qu, T. Chen, Q. Xiao, G. Lei, Z. Li, Coaxial MnO₂ Nanoshell/CNFs Composite Film Anode for High-Performance Lithium-Ion Batteries, J. Electrochem. Soc. 165 (2018) A487–A492.

[58] Y. Wu, X. Li, Q. Xiao, G. Lei, Z. Li, J. Guan, The coaxial MnO₂/CNTs nanocomposite freestanding membrane on SSM substrate as anode materials in high performance lithium ion batteries, J. Electroanal. Chem. 834 (2019) 161–166.

[59] C.J. Poonam Sehrawat, S.S. Islam, Carbon nanotubes in Li-ion batteries: A review, Mat. Sci. Eng. B-Adv. 213 (2016) 12–40.

[60] B. Rajagopalan, B. Kim, S.H. Hur, I.K. Yoo, J.S. Chung, Redox synthesis of poly (p–phenylenediamine)–reduced graphene oxide for the improvement of electrochemical performance of lithium titanate in lithium–ion battery anode, J. Alloys Compd. 709 (2017) 248–259.

[61] R. Jin, Y. Cui, S. Gao, S. Zhang, L. Yang, G. Li, CNTs@NC@CuCo₂S₄ nanocomposites: An advanced electrode for high performance lithium-ion batteries and supercapacitors, Electrochimi. Acta. 273 (2018) 43–52.

[62] M.S. Wang, Z.Q. Wang, Z.L. Yang, Y. Huang, J. Zheng, X. Li, Carbon nanotube-graphene nanosheet conductive framework supported SnO₂ aerogel as a high performance anode for lithium ion battery, Electrochimi. Acta. 240 (2017) 7–15.

[63] S. Shi, T. Deng, M. Zhang, G. Yang, Fast facile synthesis of SnO₂/Graphene composite assisted by microwave as anode material for lithium-ion batteries, Electrochimi. Acta. 246 (2017) 1104–1111.

[64] T. Ramireddy, T. Xing, M.M. Rahman, Y. Chen, Q. Dutercq, D. Gunzelmann, A.M. Glushenkov, Phosphorus–carbon nanocomposite anodes for lithium-ion and sodium-ion batteries, J. Mater. Chem. A 3 (2015) 5572–5584.

[65] P. Zhao, M. Yao, H. Ren, N. Wang, S. Komarneni, Nanocomposites of hierarchical ultrathin MnO₂ nanosheets/hollow carbon nanofibers for high-performance asymmetric supercapacitors, Appl. Surf. Sci. 463 (2019) 931–938.

[66] T.H. Lee, D.T. Pham, R. Sahoo, J. Seok, T.H.T. Luu, Y.H. Lee, High energy density and enhanced stability of asymmetric supercapacitors with mesoporous MnO₂@CNT and nanodot MoO₃@CNT free-standing films. Energy Storage Mater, Energy Storage Mater 12 (2018) 223–231.

[67] Z. Jia, J. Wang, Y. Wang, B. Li, B. Wang, T. Qi, X. Wang, Interfacial Synthesis of δ-MnO2 Nano-sheets with a Large Surface Area and Their Application in Electrochemical Capacitors, J. Mater. Sci. Technol. 32 (2016) 147–152.

[68] F. Wang, C. Li, J. Zhong, Z. Yang, A flexible core-shell carbon layer MnO nanofiber thin film via host-guest interaction: Construction, characterization, and electrochemical performances, Carbon. 128 (2018) 277–286.

[69] W. Zhu, H. Huang, W. Zhang, X. Tao, Y. Gan, Y. Xia, H. Yang, X. Guo, Synthesis of MnO/C composites derived from pollen template for advanced lithium-ion batteries, Electrochim. Acta. 152 (2015) 286–293.

[70] Y. Zhong, H. Huang, K. Wang, Z. He, S. Zhu, L. Chang, H. Shao, J. Wang, C.N. Cao, NiO@MnO$_2$ core–shell composite microtube arrays for high-performance lithium ion batteries, RSC Adv. 7 (2017) 4840–4847.

[71] J. Chen, Y. Wang, X. He, S. Xu, M. Fang, X. Zhao, Y. Shang, Electrochemical properties of MnO$_2$ nanorods as anode materials for lithium ion batteries, Electrochim. Acta 142 (2014) 152–156.

[72] J. Wang, J.G. Wang, H. Liu, Z. You, C. Wei, F. Kang, Electrochemical activation of commercial MnO microsized particles for high-performance aqueous zinc-ion batteries, J. Power Sources. 438 (2019) 226951.

[73] J. Wang, J.G. Wang, H. Liu, C. Wei, F. Kang, Zinc ion stabilized MnO$_2$ nanospheres for high capacity and long lifespan aqueous zinc-ion batteries, J. Mater. Chem. A. 7 (2019) 13727–13735.

[74] X. Jiang, X. Yang, Y. Zhu, Y. Yao, P. Zhao, C. Li, Graphene/carbon-coated Fe$_3$O$_4$ nanoparticle hybrids for enhanced lithium storage, J. Mater. Chem. A. 3 (2015) 2361–2369.

[75] Y. Ma, H. Desta Asfaw, C. Liu, B. Wei, K. Edström, Encasing Si particles within a versatile TiO2−xFx layer as an extremely reversible anode for high energy-density lithium-ion battery, Nano Energy. 30 (2016) 745–755.

[76] X.H. Ma, Q.Y. Wan, X. Huang, C.X. Ding, Y. Jin, Y.B. Guan, C.H. Chen, Synthesis of three-dimensionally porous MnO thin films for lithium-ion batteries by improved Electrostatic Spray Deposition technique, Electrochim. Acta. 121 (2014) 15–20.

[77] S. Zhang, L. Zhu, H. Song, X. Chen, J. Zhou, Enhanced electrochemical performance of MnO nanowire/graphene composite during cycling as the anode material for lithium-ion batteries, Nano Energy. 10 (2014) 172–180.

[78] H. Sun, J.G. Wang, X. Zhang, C. Li, F. Liu, W. Zhu, W. Hua, Y. Li, M. Shao, Nanoconfined Construction of MoS$_2$@C/MoS$_2$ Core–Sheath Nanowires for Superior Rate and Durable Li-Ion Energy Storage, ACS Sustain. Chem. Eng. 7 (2019) 5346–5354.

[79] X. Zhang, J.G. Wang, W. Hua, H. Liu, B. Wei, Hierarchical nanocomposite of hollow carbon spheres encapsulating nano-MoO$_2$ for high-rate and durable Li-ion storage, J. Alloys Compd. 787 (2019) 301–308.

[80] J.G. Wang, H. Liu, R. Zhou, X. Liu, B. Wei, Onion-like nanospheres organized by carbon encapsulated few-layer MoS$_2$ nanosheets with enhanced lithium storage performance, J. Power Sources. 413 (2019) 327–333.

CHAPTER 10

Preparation and microwave absorption properties of tubular carbon nanofibers and magnetic nanofibers

Jiqi Wang[a], Fei Wu[a] and Baoliang Zhang[a,b]
[a]School of Chemistry and Chemical Engineering, Northwestern Polytechnical University, Xi'an, China
[b]Xi'an Key Laboratory of Functional Organic Porous Materials, Northwestern Polytechnical University, Xi'an, China

10.1 Introduction

As an efficient information carrier, the electromagnetic wave has been widely used in many fields such as communication, national defense, and military industry. However, the problems of electromagnetic interference, electromagnetic radiation, and information leakage in the application process of an electromagnetic wave are becoming more and more prominent. If the undesired electromagnetic wave is not treated properly, it will not only affect the normal operation of electronic equipment and communication security, but also pose a threat to human health. Researchers have paid close attention to the technology of directional blocking and controllable elimination of electromagnetic waves [1–4].

Microwave absorbing material is a kind of functional material that can convert electromagnetic waves into thermal energy [5–7]. There are two main application ways. Wrapping radiation sources can absorb the outward radiated electromagnetic wave, playing the role of eliminating electromagnetic pollution and blocking signal leakage. In addition, it can also be coated on the surface of the device or target that needs to be protected, absorbing the electromagnetic wave propagated from outside and act as a shield cover or stealth coating. Compared with the general electromagnetic shielding materials, the electromagnetic wave absorbers will absorb almost all electromagnetic waves in the application process, rather than reflection. So, it will not cause reverse interference and secondary pollution. Electromagnetic wave materials have more significant advantages

Fabrication and Functionalization of Advanced Tubular Nanofibers and their Applications.
DOI: https://doi.org/10.1016/B978-0-323-99039-4.00004-8

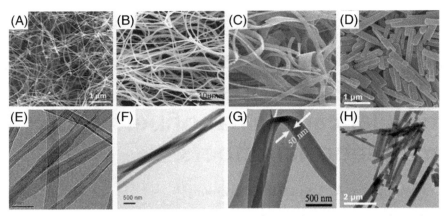

Figure 10.1 SEM (A–D) and TEM (E–H) images of 1D carbon nanomaterials: Carbon nanotubes (A and E), carbon nanofibers (B and F), carbon nanoribbons (C and G), carbon nanorods (D and H). Reprinted (adapted) with permission from Copyright 2022 American Chemical Society [26].

and are considered to be the most efficient solution to electromagnetic radiation [8–12].

Up to now, many materials have been proved to possess excellent microwave absorbing performance including carbon materials, ferrite, magnetic metals, ceramics, semiconductors, and so on, because these materials have a good dielectric loss or magnetic loss ability [13–15]. Among them, carbon materials have the characteristics of high specific surface area, high porosity, low density, good environmental tolerance, and wide material sources. They are more prominent in high efficiency, lightweight, corrosion resistance, and convenience, and have been favored by researchers [16–19].

10.2 One-dimensional carbon nanomaterials

One-dimensional (1D) carbon nanomaterials refer to carbon nanomaterials with large aspect ratios, such as carbon nanotubes, carbon nanofibers, carbon nanoribbons, and carbon nanorods [20–23]. They are widely used in many fields because of their high conductivity, high strength, and high specific surface area. According to the sources, 1D carbon nanomaterials can be divided into carbon nanotubes, polymer-based 1D carbon nanomaterials, electrospinning precursor-derived 1D carbon nanomaterials, and natural polymer-based one-dimensional carbon materials [24,25]. Fig. 10.1 shows the morphologies of some 1D carbon nanomaterials.

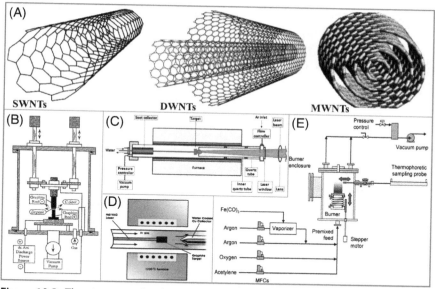

Figure 10.2 The structure of carbon nanotubes (A); schematic diagram of the preparation methods of carbon nanotubes: arc discharge method (B), laser ablation method (C), chemical vapor deposition method (D) and flame method (E). The copyright of the figure obtained from the John Wiley and Sons under the license No: 5271261289675, Mar 17, 2022 [32].

10.2.1 Carbon nanotubes

Since the discovery of nanotubes in 1991, large-scale research has been carried out in the development of synthesis methods and exploration of application properties in the past 30 years [27,28]. According to the difference in tube wall thickness, carbon nanotubes can be divided into single-walled carbon nanotubes (SWNTs), double-walled carbon nanotubes (DWNTs), and multi-walled carbon nanotubes (MWNTs). Their tube diameter can be controlled within 1–50 nm. The structure is shown in Fig. 10.2A. Commonly used preparation methods of carbon nanotubes include arc discharge method, laser ablation method, chemical vapor deposition (CVD), and flame method [29–31]. Schematic diagrams display in Fig. 10.2B–E.

Chrzanowska J, Hoffman J, Małolepszy A, et al. Synthesis of carbon nanotubes by the laser ablation method: Effect of laser wavelength[J]. physica status solidi (b), 2015, 252(8): 1860-1867.
(1) Arc discharge method

In arc discharge method, high purity graphite is used as the electrode. Under inert atmosphere, DC current and 1700°C high temperature, the anode will vaporize and deposit on the surface of cathode and the inside wall of tube furnace, forming carbon nanotubes [33–35]. There are two ways to synthesize CNT by arc discharge deposition: synthesis without catalyst or synthesis with Metal catalysts such as Fe, Co, Ni, Pt, and Pd. Generally, MWCNT synthesis may not use a catalyst, but the synthesis of SWCNT requires a catalyst. The main advantage of arc discharge technology is high production efficiency, but this method is difficult to control the chirality of the product, and the product requires post-treatment to remove impurities such as metal catalysts.

(2) Laser ablation method

Laser ablation is a technology to prepare carbon nanotubes by vaporizing solid graphite with high-energy laser. It needs to be carried out in a tube furnace with inert atmosphere and 1200°C high temperature. Metal powder catalyst is used in the process, and the growth mechanism is similar to arc discharge method. The properties of the product can be controlled by adjusting preparation process parameters, including target material composition, laser properties, inert gas type, and pressure, distance between target and substrate, ambient temperature, and so on [36–38]. For example, the diameter of carbon nanotubes can be controlled by adjusting the laser power, increasing the laser pulse power will reduce the diameter. This method has the potential to produce high purity and high-quality SWCNT on a large scale.

(3) Chemical vapor deposition method

Chemical vapor deposition (CVD) is to introduce carbon-containing gas (hydrocarbon, carbon monoxide, etc.) into a high-temperature tubular furnace containing catalyst, carbon nanotubes will form after catalytic decomposition. In this process, a certain amount of process gas (nitrogen, hydrogen or ammonia, etc.) usually needs to be introduced [39–44]. Compared with arc discharge and laser ablation, CVD method can prepare carbon nanotubes at lower temperature. The growth temperature of MWNTs is between 600 and 900°C, while the growth temperature of SWCNT is between 900 and 1200°C. Higher temperature is conducive to the production of smaller diameter carbon nanotubes. The diameter and wall thickness of carbon nanotubes are not only affected by temperature, but also related to the type and particle size of the metal catalyst, and the type and flow rate of process gas. CVD is an economical and practical large-scale production method. The product has high purity and the reaction process is easy to control.

(4) Flame method

The flame method uses carbon source produced by the combustion of hydrocarbon fuel to synthesize carbon nanotubes under the action of a catalyst [45–47]. Compared with the above-mentioned methods, flame method can simultaneously provide the required carbon source and heat source for the preparation of carbon nanotubes. According to the way of introducing fuel and oxidant, the flame method can be divided into premixed flame method and diffusion flame method. Premixed flame method is to mix the fuel and the oxidant in advance and then introduce it into the burner for combustion. Different from premixed flame, fuel, and oxidant are introduced separately in diffusion flame method, and contact is realized by diffusion. The carbon nanotubes produced by the flame method contain a large number of impurities. In addition to the metal catalyst particles, there is also incompletely reacted carbon, which requires subsequent purification. Carbon nanotube purification usually needs to use hydrogen peroxide and hydrochloric acid solutions to remove amorphous carbon and dissolve the metal catalyst deposited on the surface of carbon nanotubes, respectively. Flame method has the advantages of high efficiency, low cost, and can continuously produce carbon nanotubes in large quantities.

10.2.2 Polymer-derived 1D carbon materials

The polymer precursors with 1D structure are prepared by spinning or chemical synthesis, and the corresponding 1D carbon materials can be transformed by high-temperature carbonization, as shown in Fig. 10.3 [48,49]. This method has been developed to industrial stage, and it can prepare high-performance carbon nanofibers with light weight, high strength, high modulus, corrosion resistance and fatigue resistance in large quantities. The preparation and high-temperature carbonization of polymer nanofiber precursors are the key technologies of this method. The most commonly used spinning technologies include dry-wet spinning, melt spinning and electrostatic spinning. Carbon conversion process generally includes three stages: preoxidation, carbonization and graphitization. In addition, the structural parameters of carbon nanofibers can be controlled by the spinning process. Nanofibers with different shapes and cross sections can be prepared by changing the nozzle structure, and porous carbon nanofibers can be prepared by adding porogen [50,51]. At present, the widely used precursor polymers include polyacrylonitrile (PAN) and asphalt.

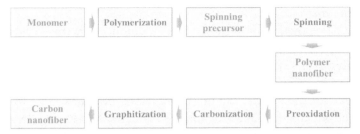

Figure 10.3 Schematic diagram of preparation process of polymer-derived 1D carbon materials.

(1) PAN-derived 1D carbon materials

PAN-derived carbon nanofibers have excellent properties, including high specific strength, heat resistance and corrosion resistance. It has attracted extensive attention in the fields of scientific research and industrial production. It is a kind of carbon nanofiber with the largest production scale and demand, the fastest development and the best performance. The research on PAN-derived carbon nanofibers mainly covers three aspects: precursor polymerization method, preoxidation process and carbonization (graphitization) process, which are the key factors affecting the quality of carbon nanofibers [52–54].

Only by synthesizing eligible PAN precursors can it be possible to prepare high-quality carbon nanofibers. The synthesis of PAN generally follows the radical reaction mechanism. The methods of PAN synthesis include solution polymerization, aqueous precipitation polymerization, mixed solvent precipitation polymerization, aqueous suspension polymerization, emulsion polymerization, atom transfer radical polymerization, and new polymerization methods such as inverse emulsion polymerization, supercritical CO_2 polymerization and radiation polymerization. There are many synthetic methods of PAN, and more new synthetic methods are still being developed and improved. The goal is to synthesize PAN with high molecular weight, narrow molecular weight distribution, and high stereoregularity.

The preoxidation process is time-consuming and cumbersome, which has an important impact on the properties of carbon nanofibers. It is a key step in the whole preparation process. Preoxidation is actually a process in which the plastic PAN linear macromolecular chain transforms into nonplastic heat-resistant trapezoidal structure through a series of polymerization reactions. Preoxidation can improve the mechanical properties of nanofibers and effectively prevent fusing during the carbonization process.

The main factors affecting preoxidation degree include time, temperature and draft. By selecting the optimal preoxidation process, the preoxidized nanofibers with excellent performance can be obtained in a short time and prepare for the later carbonization. However, improper preoxidation process will cause insufficient or excessive preoxidation, easily forming a skin-core structure, affecting the carbonization process, resulting in structure defects and reducing performance. The preoxidation temperature is generally 180–250°C, and the atmosphere is mainly air. When the reaction temperature is lower than 180°C, the reaction rate is slow, consuming more time and low efficiency. When the temperature is higher than 250°C, the reaction is violent and the exothermic concentration, it is easy to cause fusing. Commonly used preoxidation methods include gradient heating preoxidation, constant temperature pre-oxidation, and continuous rising temperature preoxidation.

The industrialization of the carbonization stage is relatively mature. The precursor is placed in an inert atmosphere and gradually heated from 400°C to 1600°C. After low-temperature carbonization (400–1000°C) and high-temperature carbonization (1100–1600 °C), N, O, and H elements are removed, and a two-dimensional layer network graphite structure is gradually formed, so that the nanofibers have strong tensile properties.

(2) Asphalt-derived 1D carbon materials

Asphalt-derived 1D carbon nanofibers refer to the carbon nanofibers prepared by using coal tar pitch, petroleum pitch and other substances rich in polycyclic aromatic hydrocarbons as raw materials through raw material preparation, melt spinning, preoxidation, carbonization, graphitization and carbon nanofiber post-treatment. It has the characteristics of small stress deformation, excellent thermal conductivity, and is not easy to expand when heated. According to performance difference, asphalt-based carbon nanofibers can be divided into general-purpose asphalt carbon nanofibers, mesophase asphalt-based carbon nanofibers with high performance and premesophase asphalt-based carbon nanofibers [55,56].

Asphalt is a substance rich in polycyclic aromatic hydrocarbons and its chemical composition is relatively complex. Therefore, it needs a series of purification and purification treatments to remove quinoline insoluble matter, sulfur element, and other harmful substances, improve its rheological properties and regulate molecular weight, so as to form the raw material for the next spinning. In the spinning process of asphalt, short nanofibers can be produced by spraying method and centrifugal method, and continuous long nanofibers can also be produced by extrusion method. The spraying method is to flow the molten asphalt into the nozzle outlet, and the heated air makes the asphalt and the nanofibers form a certain angle for drafting to form short nanofibers. Centrifugation method is to put the molten asphalt in a centrifuge to rotate at high speed and disperse it into short nanofibers by centrifugal force. Extrusion method uses a pump or pressure to send molten asphalt into the spinning body. Under the action of traction and shear force, the polycyclic aromatic hydrocarbon macromolecular layer is arranged along the nanofibers direction to form continuous filaments. The formed asphalt nanofibers after spinning needs preoxidation treatment. The purpose is to convert the surface layer of asphalt nanofibers from thermoplastic to thermosetting, so as to avoid the adhesion of asphalt nanofibers in the carbonization process and improve the mechanical properties of asphalt nanofibers. The pre-oxidation methods include gas phase method, liquid phase method, and mixed oxidation method. The oxidized asphalt nanofibers need to further carry out a series of dehydrogenation, demethanation, dehydration, polycondensation, and crosslinking reactions in inert atmosphere. Generally, it is carbonized at 1000–2000°C for 30 s–30 min to increase the carbon content in the nanofibers and the tensile strength of monofilament. Graphitization is generally used in the preparation process of mesophase asphalt-derived carbon nanofibers. Under the protection of high-purity argon, the graphitization temperature range is 2500–3000°C, and the residence time is 10–60 s.

10.2.3 Natural polymer-derived 1D carbon materials

Biomass materials come from a wide range of sources and belong to renewable resources. Calcination of biomass materials can be used to prepare carbon materials in large quantities, which has the advantages of low cost and environment-friendly. Many plants in nature have formed one-dimensional tissue structure during the evolution process. These natural polymer nanofibers can be used as precursors for the preparation of carbon nanofibers. At present, the research on the preparation technology and application performance of biomass derived one-dimensional carbon materials has been widely reported. Cotton, wood, sisal, dandelion, and sycamore derived carbon nanofibers are applied in secondary batteries, supercapacitors, water treatment, and microwave absorption [57–60].

Sicheng Li et al. develop a process for preparing carbon nanofibers from wood [61]. Starting from softwood pulp, TEMPO oxidation was followed by mechanical stirring to obtain lignocellulose. After treatment with hydrochloric acid, lignocellulose hydrogel was obtained. Subsequently, wood-derived carbon nanofibers aerogels were prepared through subsequent solvent exchange, freeze-drying, and high-temperature calcination. Its conductivity and specific surface area are 710.9 S/m and 689 m^2/g, respectively. Yan Song et al. synthesize a kind of porous carbon nanofibers with hollow pores using dandelion and investigated its application performance as a lithium-sulfur battery negative electrode. The specific capacity of the material is 950 mAh/g^{-1} at a charge-discharge rate of 0.5 C. Menglin Li et al. extract nanofibers from sisal and obtained carbon nanofibers through high-temperature carbonization. Activated carbon nanofibers were obtained after KOH activation treatment. Its application performance as electrode material of supercapacitor was evaluated. The capacity at 0.5 A/g was 415 F/g. Guoxian Li et al. prepare a flexible carbon nanofibers sponge for water desalination using cotton as a raw material. This material has an ultra-high specific surface area (2680 m^2/g) and a suitable pore size (0.8–4 nm), which is excellent as a capacitive deionization electrode. The electrosorption capacity in 500 mg/L NaCl solution reaches 16.1 mg/g. Our group selects a kind of natural plant nanofibers from sycamore. Hollow carbon nanofibers with helical structure were obtained by vacuum calcination after impregnation with Co ions. A large number of carbon nanotubes were grown on the surface at the same time. Due to its unique structure, this material exhibits good electromagnetic wave attenuation ability, and the minimum reflection loss can reach –61.08 dB under the ultra-low filler content of 5% (Fig. 10.4, Fig. 10.5).

10.3 Microwave absorbers and electromagnetic stealth

Microwave absorbing material is usually composed of matrix and absorbers. Matrix is the film-forming material of microwave absorbing coating, which plays the role of bonding absorbent and other fillers, and determines the physical, mechanical properties and environmental resistance of the absorbing coating. The commonly used matrix materials can be divided into rubber, plastic and resin. Microwave absorber is the functional component that plays a role of attenuating electromagnetic wave. According to the difference of main loss mechanism, microwave absorbers can be divided into three types: resistance loss, dielectric loss and magnetic loss [62–66]. Resistive loss microwave absorbers realize the loss of electromagnetic wave by interacting with electric field. Conductive loss and dielectric loss are the decisive factors for the absorption effect of this type of material [67–70]. Carbon materials and metal powder materials belong to this

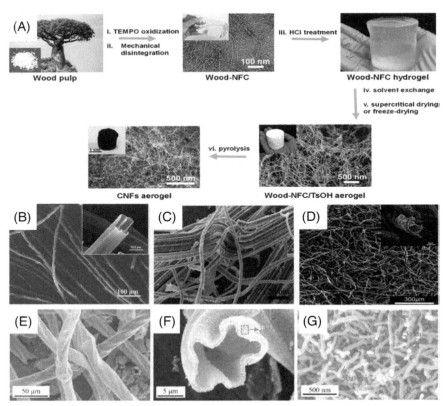

Figure 10.4 Preparation flow chart of wood-derived carbon nanofibers (A), SEM images of carbon nanofibers derived from dandelion (B), sisal (C), cotton (D) and, sycamore (E–G). The copyright of the figure obtained from the John Wiley and Sons under the license No: 5271271094557, Mar 17, 2022 [61].

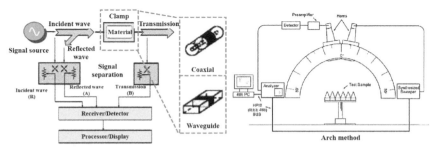

Figure 10.5 Schematic diagram of the device for microwave absorbing performance test.

category. Dielectric absorbing materials such as barium titanate and other ferroelectric ceramics attenuate electromagnetic wave mainly by polarization relaxation of electrons and ions [71–75]. Ferrite is a kind of magnetic absorbing material, and its loss mechanism is the combined action of magnetic loss and electromagnetic resonance [76–78]. Microwave absorbing materials need to meet two characteristics: impedance matching and efficient absorption. One is to make the electromagnetic wave enter the material to the greatest extent, and the other is that the incident electromagnetic wave should be attenuated rapidly when passing through the absorbing layer [79–83].

(1) Impedance matching

If the impedance between the absorbing material and the free space does not match, most of the electromagnetic waves will be strongly reflected, and only a small part can enter the material. Therefore, it cannot produce strong electromagnetic wave absorption effect. If the absorbing material and the free space meet the impedance matching condition, most of the electromagnetic waves will enter the inside of the material rather than being reflected, resulting in a more obvious electromagnetic wave absorption effect [84–89].

The input impedance Z_{in} of a single-layer absorbing material can be expressed as follows:

$$Z_{in} = \sqrt{\frac{\mu}{\varepsilon}} = \sqrt{\frac{\mu_r\mu_0}{\varepsilon_r\varepsilon_0}} = \sqrt{\frac{\mu_0}{\varepsilon_0}}\sqrt{\frac{\mu_r}{\varepsilon_r}} = Z_0\sqrt{\frac{\mu_r}{\varepsilon_r}}$$

Where ε and μ are the dielectric constant and permeability, respectively. Ideally, if $\varepsilon_r = \mu_r$ when electromagnetic wave is incident on the surface of absorbing material, the input impedance Z_{in} of material is the same as impedance Z_0 of free-space, that is, impedance matching is achieved. No reflection occurs on the surface of materials at this time.

(2) Efficient absorption

When the electromagnetic wave enters the interior of the material, it needs to be effectively absorbed and attenuated by electromagnetic loss. If the material has electromagnetic loss capability, the permittivity ($\varepsilon_r=\varepsilon'-j\varepsilon''$) and permeability ($\mu_r = \mu'-j\mu''$) are both complex numbers [90–92]. The real part (ε') and imaginary part (ε'') of permittivity represent the ability to store and attenuate electric fields, respectively. The real part (μ') and imaginary part (μ'') of permeability represent the ability to store and attenuate magnetic fields, respectively. Loss tangent is usually used to stand for the electromagnetic loss. The electric loss tangent value ($\tan\delta\varepsilon=\varepsilon''/\varepsilon'$)

and magnetic loss tangent value ($\tan\delta\mu = \mu''/\mu'$) are used to express the electrical and magnetic loss capabilities of the material [93–96]. The larger the ε'', μ'', and loss tangent, the better the electromagnetic wave absorption effect. Therefore, the electromagnetic loss characteristics of the material to electromagnetic waves can be enhanced by increasing ε'' and μ'', then improving absorbing performance [97–100]. However, the absorbing materials can only attenuate the electromagnetic wave that enter the material. The electromagnetic wave that are reflected on the surface of the material and fail to enter the material are not affected. Better impedance matching can make electromagnetic wave enter the material to the greatest extent. Therefore, when designing an electromagnetic wave absorbing material with excellent microwave absorption performance, it is not only necessary to have larger imaginary parts of complex permittivity and complex permeability for higher electromagnetic loss, but also to make better impedance matching as much as possible. Only by considering the attenuation characteristic and impedance matching characteristic at the same time, can the microwave absorbing effect be enhanced.

Electromagnetic wave attenuation mechanisms include dielectric loss, magnetic loss, and conduction loss. Dielectric loss mainly includes interface polarization, dipole polarization, and Debye relaxation. The interface polarization comes from the charge accumulation of heterogeneous interface under the action of external electric field. Larger the effective interface resulting in stronger the interface polarization. Dipole polarization is a phenomenon caused by the shift of double electric centers in the absorber molecules when the electric field changes. It is related to the characteristics of the molecule structure. Debye relaxation process refers to the hysteresis phenomenon in the process of electric dipole turning with the electric field, which can be described by Debye equation: [101–105]

$$\left(\varepsilon' - \frac{\varepsilon_s + \varepsilon_\omega}{2}\right)^2 + \left(\varepsilon''\right)^2 = \left(\frac{\varepsilon_s + \varepsilon_\omega}{2}\right)^2$$

where ε_s is the static dielectric constant, and ε_∞ is the limit value of the dielectric constant changing with frequency. If there is a Debye relaxation process, the ε''-ε' curve appears as an upper semicircle. Magnetic loss mechanisms include hysteresis loss, eddy current loss, magnetic aftereffect loss, natural resonance, ferromagnetic resonance, size resonance, domain wall resonance, etc. However, for weak electromagnetic waves in GHz band, there are only two ways: eddy current loss and natural resonance. Conduction loss means that after a conductive network is formed inside the absorber, free

Table 10.1 The corresponding relationship between RL and absorptivity.

RL (dB)	0	-5	-10	-15	-20	-30	-40
Absorptivity (%)	0	68.38	90	96.84	99	99.9	99.99

electrons will form a local current under the action of an electric field, and then attenuate electromagnetic wave in the form of heat.

(3) Evaluation index of microwave absorption performance

Electromagnetic wave loss performance of microwave absorbing materials is evaluated by reflection loss (RL), effective bandwidth, matching frequency, matching thickness and other indicators. According to the transmission line theory, RL can be calculated according to the following formula: [106–109]

$$RL(dB) = 20 \log \left| \frac{Z_{in} - Z_0}{Z_{in} + Z_0} \right|$$

$$Z_{in} = Z_0 \sqrt{\frac{\mu_r}{\varepsilon_r}} \tanh \left(j \frac{2\pi f d}{c} \sqrt{\mu_r \varepsilon_r} \right)$$

where f is the frequency of incident electromagnetic wave, d is the thickness of the absorbing layer, and c is the velocity of light in vacuum. Greater absorption intensity resulting in smaller reflection loss value (negative number). At this time, RL is defined as the minimum reflection loss (RL_{min}). The corresponding relationship between RL and absorptivity is shown in Table 10.1. When the RL reaches –10 dB, the absorptivity is 90%, which is considered as effective absorption. Effective absorption bandwidth (EAB) refers to the frequency range covered by RL less than –10 dB, the matching frequency is the frequency corresponding to the maximum absorption intensity, and the matching thickness is the coating thickness when the absorption is the strongest.

(4) Test methods for absorption performance

There are three methods to test the microwave absorbing properties of microwave absorbing materials: coaxial method, waveguide method and arch method. They all need to be tested by a vector network analyzer. The difference lies in the shape and size of the required test fixture and the test samples. The coaxial method takes the annular sample with 3.04 mm inner diameter and 7.00 mm outer diameter as the test piece, which is placed in the air line for test. After the S parameters are measured by the vector network analyzer, and the electromagnetic parameters are obtained by inversion. Finally, RL can be calculated through electromagnetic parameters.

In the waveguide method, a square piece of a certain size is used as a test sample and placed in a waveguide cavity for testing. The process is similar to coaxial method, and the required samples and waveguide fixtures for different wavebands are different. Arch method needs to be tested in a dark room with an arch frame. The test process more truly restores the actual application scenario. RL is the direct test result [110–114].

10.4 1D carbon-based microwave absorbers

1D carbon-based microwave absorber is a very important class of carbon material. Its large aspect ratio makes it easy to disperse in the matrix and lap into a three-dimensional network, which enhances the charge conduction ability between absorbing elements, improves the conductive loss, and makes it possible to obtain higher absorbing performance. In the application process, in addition to 1D pure carbon absorbers, many kinds of composite absorbers have been derived, including 1D carbon/magnetic particle composite absorbers, 1D carbon/dielectric component composite absorbers, and 1D carbon/conductive polymer composite absorbers. Combining dielectric or magnetic functional components are designed to optimize electromagnetic parameters and obtain better performance.

10.4.1 1D pure carbon microwave absorbers

From the perspective of absorbing performance, the current research of 1D pure carbon absorbing agent focuses on the regulation of micro pore structure, introducing micropores or constructing coaxial through holes, increasing specific surface area to obtain stronger interface polarization ability, reducing material density, and reducing absorbing coating. In addition, the dipole polarization center can be introduced by heteroatom doping in order to improve the reflection loss ability [115–117].

(1) Porous carbon nanofibers

Porous carbon nanofibers are prepared by adding porogen into electrospinning precursors. The porogen can be a solid or liquid that is easy to decompose or volatilize. After spinning, it will be separated from the nanofiber's matrix during high-temperature calcination, leaving the occupied volume to form corresponding pores. Hewei Zhen et al. use polyacrylonitrile (PAN) as the precursor, N,N-Dimethylformamide (DMF) as the solvent, and $ZnCl_2$ as the porogen, and prepare porous carbon nanofibers (PCFs) with uniformly distributed micropores through electrospinning and high-temperature calcination processes [118]. Using paraffin as the matrix, and the

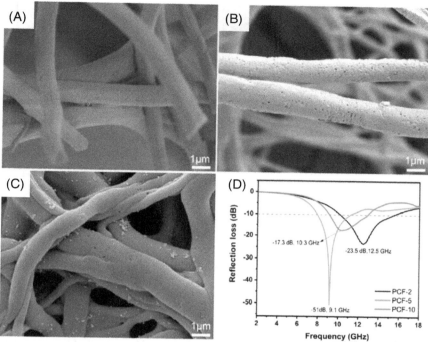

Figure 10.6 SEM images and RLcurve (D) of PCFs obtained at different porogen contents of 2% (A), 5% (B) and 10% (C). The copyright of the figure obtained from Elsevier under the license No: 5271380904064, Mar 17, 2022 [118].

influence of the porogen content on the wave absorption performance of the obtained porous carbon nanofibers was investigated at the filler content of 20%. When the content of $ZnCl_2$ was 5 wt%, PCFs had the most excellent wave absorption performance, and the minimum reflection loss was -51 dB@9.1 GHz, the matching thickness is 3 mm, and the effective absorption frequency range with reflection loss less than -10 dB is 8–11 GHz. Fig. 10.6.

Guang Li et al. also prepare porous carbon nanofibers by spinning blended PAN/PMMA, and investigate the influence of PAN/PMMA ratio on pore properties and microwave absorbing properties [119]. The ratios of PAN/PMMA are 70/30 and 30/70, respectively. Two kinds of porous carbon nanofibers are obtained by preoxidation in air atmosphere at 180–280°C and calcination in nitrogen atmosphere at 1200°C. The microwave absorbing properties of the two kinds of carbon nanofibers have been characterized. The minimum reflection loss of 70/30 porous carbon nanofibers is -31 dB@9.7 GHz, the matching thickness is 2.3 mm, and the minimum

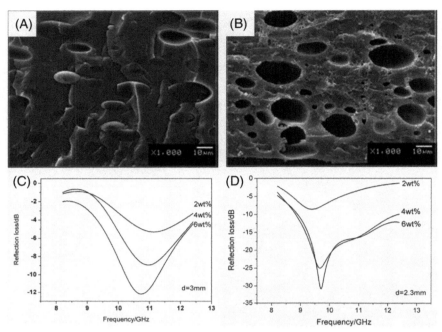

Figure 10.7 SEM images (A and B) and reflection loss curves (C and D) of two nanofibers with PAN/PMMA ratios of 30/70 (A and C) and 70/3 (B and D). Reprinted (adapted) with permission from Copyright 2022 American Chemical Society [119].

reflection loss of 30/70 carbon nanofibers is − 12.2 dB@10.7 GHz, the matching thickness is 3 mm, and the optimal filler amount is 6%. Fig. 10.7.

(2) Hollow carbon nanofibers

The hollow carbon nanofiber precursors are mainly prepared by coaxial electrospinning. It is a new method developed based on the template method, which can prepare continuous core-shell and hollow structure nanofibers in one step. In coaxial electrospinning, the solutions of core and shell materials are respectively installed in two different syringes. The spinneret system is composed of two coaxial capillary tubes with different inner diameters. Under the action of high voltage electric field, the outer liquid flows out and merges with the core liquid. The two liquids will not mix together before solidification. When the shell liquid is stretched and sprayed at high frequency and high speed, strong shear stress will be generated at the interface between the inner and outer solution. Under the action of shear stress, the core solution moves coaxially along the shell, bends, shakes, deforms, and solidifies into ultra-fine coaxial composite nanofibers.

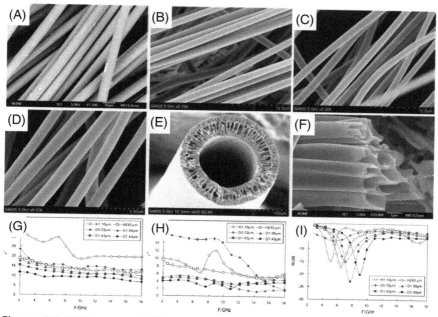

Figure 10.8 SEM images of hollow carbon nanofibers with different diameters: 7.4 μm (A), 1.43 μm (B), 1.08 μm (C), 0.72 μm (D), 1.10 μm (E), 210 μm (F); the real part of permittivity (G); imaginary part of permittivity (H); reflection loss (I). The copyright of the figure obtained from Elsevier under the license No: 5271391092182, Mar 17, 2022 [120].

If the core material is removed by heating or dissolution, leaving the shell material, the hollow nanofibers are obtained.

Zengyong Chu et al. prepare a series of PAN nanofibers with different diameters by means of coaxial electrospinning of dry/wet method [120]. The precursors are pre-oxidized in 523 K air for 60 min and 1073 K nitrogen atmosphere for 120 min and then converted into corresponding hollow carbon nanofibers. The diameters of carbon nanofibers are 7.4, 1.43, 1.08, 0.72, 1.10, and 210 μm, respectively. The microwave absorbing properties of hollow carbon nanofibers with different diameters are characterized, analyzed and compared. It is found that as the diameter decreases, the real and imaginary parts of the dielectric constant increase, and the imaginary part increases more moderately. The reflection loss shows a trend of first increasing and then decreasing. The carbon nanofibers with a diameter of 1.43 μm has the most excellent absorption performance. The reflection loss reaches –23.0 dB and effective absorption bandwidth is 2.3 GHz. Fig. 10.8.

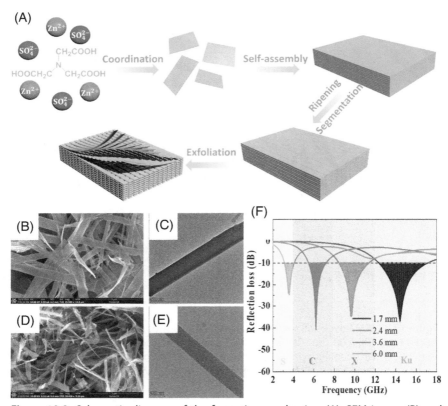

Figure 10.9 Schematic diagram of the formation mechanism (A), SEM image (B) and TEM image (C) of nanoribbon precursor; SEM image (D), TEM image (E) and reflection loss curve (F) of nitrogen-doped carbon nanoribbons. The copyright of the figure obtained from Elsevier under the license No: 5271901056788, Mar 18, 2022 [121].

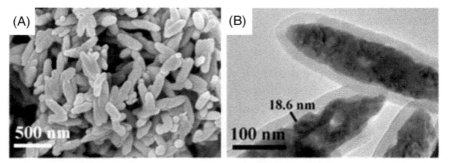

Figure 10.10 SEM (A) and TEM (B) images of Fe_3O_4/C core-shell structured nanocolumns. The copyright of the figure obtained from Elsevier under the license No: 5271841166711, Mar 18, 2022 [122].

(3) 1D carbon nanomaterials with novel structure

With the continuous development of material preparation technology, many 1D carbon materials with novel structures have been found in recent years, including carbon nanoribbons, carbon ball-chains, carbon nanorods and so on. Our group prepared a ribbon-shaped organic/inorganic hybrid precursor in the solvothermal system of N, N-dimethylacetamide with nitrotriacetic acid and zinc sulfate heptahydrate as raw materials [121]. After vacuum calcination, inorganic components were lost and organic components were carbonized, forming a nitrogen doped carbon nanoribbons with rich pores and high specific surface area. The specific surface area was 1242.5 m^2/g. The obtained nitrogen doped porous carbon nanoribbons exhibited significant lightweight advantages in microwave absorption properties, and the optimal filler content is only 4%. The minimum reflection loss was −46.1dB@10.7 GHz, the matching thickness was 2.2 mm, and the effective absorption bandwidth was 3.3 GHz (9.3–12.6 GHz).

10.4.2 1D carbon/magnetic particles composite microwave absorbers

The loss mechanism of carbon material is relatively single, so the absorption performance is limited to some extent. Combining magnetic material is an effective means to improve the microwave absorption performance of carbon materials. It can endow carbon materials with magnetic loss characteristics, reduce coating thickness, increase absorption strength and broaden absorption bandwidth. The current used magnetic components include magnetic metal powder and ferrite. Magnetic metal powders, such as Fe, Co, and Ni, have high snoek limit and saturation magnetization. Therefore, they have strong magnetic loss ability. The electromagnetic parameters can be adjusted by particle size and surface properties. It can enhance magnetic loss while taking into account impedance matching, and become a good magnetic loss microwave absorbing material. Ferrite is a complex material with both dielectric loss and magnetic loss. It produces dielectric loss through self-polarization effect. The main loss mechanism is natural resonance. This kind of material has high microwave absorption efficiency and wide absorption band. It possesses excellent absorption performance in low frequency range. Divided from the composite form, 1D magnetic carbon microwave absorbers can be divided into two kinds: internal loaded magnetic particles and surface coated magnetic components.

(1) 1D magnetic carbon material with magnetic particles inside

The preparation of 1D magnetic carbon material with magnetic particles inside can be achieved by coating magnetic particles with polymer and subsequent calcination, or adding magnetic metal ions into organic components and then being reduced to magnetic particles during calcination process. The magnetic particles in products can be ferrite, single metal or metal carbide.

Xiaofang Liu et al. synthesize phenolic coated Fe_3O_4 nanocolumns through liquid phase reaction, and a series of Fe_3O_4/C core-shell structure nanocolumns are prepared after high-temperature carbonization [122]. The carbon layer of prepared nanocolumns have uniform morphology and thickness. The particle size and carbon layer thickness are flexible and controllable. Using polyvinylidene fluoride as matrix and Fe_3O_4/C as microwave absorber, the reflection loss can reach -38.8 dB at the thickness of 2.1 mm. It is found that the electromagnetic parameters are closely related to the thickness of carbon layer, and there is significant synergistic effect between Fe_3O_4 and carbon in impedance matching.

Huihui Liu et al. combine electrospinning and high-temperature calcination processes to prepare N-doped carbon nanofibers (Co/N-CNFs) loaded with cobalt nanoparticles [123]. If silica microspheres were added to the spinning precursor as porogen, porous magnetic carbon nanofibers with bubble structure could also be obtained. The length of Co/N-CNFs is tens of microns and the average diameter is about 10 μm. Abundant cobalt nanoparticles with the size of about 20 nm are loaded with on the surface and inside the Co/N-CNFs. Co/N-CNFs has a hierarchical pore structure, containing both macropores (500 nm) and mesopores (2–50 nm). The saturation magnetization and coercivity of solid Co/N-CNFs are 28.4 emu g^{-1} and 661 Oe. The saturation magnetization and coercivity of porous Co/N-CNFs are 23.3 emu g^{-1} and 580 Oe, respectively. Solid Co/N-CNFs exhibit excellent electromagnetic wave absorption ability. When the filler content is 5 wt% and the matching thickness is 2 mm, the minimum reflection loss is -25.7 dB@4.3 GHz, and the effective bandwidth is 4.3 GHz. In addition, by adjusting the thickness of coating layer and the content of absorbers, effective absorption can be achieved in the frequency range of 1–18 GHz.

Yajing Wang et al. prepare $Co^{2+}/Fe^{3+}/PVP$ nanofibers by electrospinning by using polyvinylpyrrolidone as the precursor and adding $Fe(NO_3)_3•9H_2O$ and $Co(CH_3COO)_2•4H_2O$ at the same time [124]. The schematic diagram of preparation process is exhibited in Fig. 10.12A. Two kinds of magnetic carbon nanofibers, combining with $CoFe_2O_4$ and

Figure 10.11 Schematic diagram of the synthesis process of Co/N-CNFs. Reprinted (adapted) with permission from Copyright 2022 American Chemical Society [123].

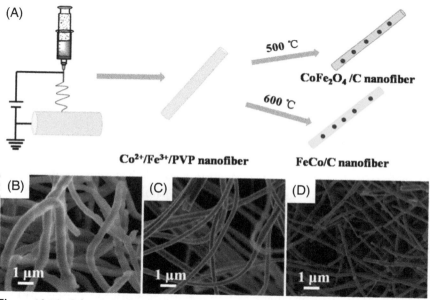

Figure 10.12 Schematic diagram of the preparation process of $CoFe_2O_4/C$ and CoFe/C nanofibers (A); SEM images of $Co^{2+}/Fe^{3+}/PVP$ nanofibers, $CoFe_2O_4/C$ nanofibers (B) and CoFe/C nanofibers (C). The copyright of the figure obtained from Elsevier under the license No: 5271850883362, Mar 18, 2022 [124].

bimetallic CoFe, are obtained by calcining at 500°C and 600°C for 4 h under argon atmosphere. The SEM images of nanofibers are provided in Fig. 10.12B–D. Using paraffin as matrix, the microwave absorbing performance of $CoFe_2O_4/C$ and CoFe/C nanofibers was characterized under 40% filler content. CoFe/C nanofibers show stronger absorbing ability. The minimum reflection loss of $CoFe_2O_4/C$ is only –6.6 dB@4.7 mm, while the minimum reflection loss of CoFe/C is –59.9 dB@2.6 mm. The

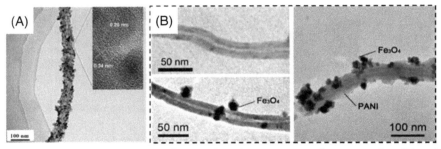

Figure 10.13 The TEM images of MWCNT/Fe$_3$O$_4$ magnetic carbon nanotubes (A) and PANI/Fe$_3$O$_4$/MWCNTs composite nanotubes (B). Reprinted (adapted) with permission from Copyright 2022 American Chemical Society [126].

effective bandwidth is 6.0 GHz (10.1–16.1 GHz). Attenuation mechanism study shows that the electromagnetic wave attenuation mechanisms include natural resonance, exchange resonance, interface polarization, dipole polarization and micro-current loss.

(2) 1D magnetic carbon materials with magnetic components on the surface

Coating magnetic components on the surface can endow carbon with magnetic properties, this process often carries out in the liquid phase system by hydrothermal method, coprecipitation method or pyrolysis method. The magnetic components in the product are mainly ferrite. The magnetic carbon nanofibers can also be prepared by calcining the nanofibers after being immersed in a metal salt solution. The magnetic nanofibers obtained by this approach usually has a lower magnetic content.

Zhijiang Wang et al. realize the growth of Fe$_3$O$_4$ on the surface of multi-walled carbon nanotubes (MWCNT) by pyrolysis [125]. The TEM image of MWCNT/Fe$_3$O$_4$ magnetic carbon nanotubes is displayed in Fig. 10.13A. Structural characterization shows that the diameter is about 80 nm, and the force between surface Fe$_3$O$_4$ particles and carbon nanotubes is strong interaction. In MWCNT/Fe$_3$O$_4$ magnetic carbon nanotubes, the charge transitions from the conduction band of Fe$_3$O$_4$ to the 2p orbital of carbon. The dipole interaction between the magnetic nanoparticles is significantly increased. The synergistic effect of these two materials significantly improves the microwave absorbing performance. The reflection loss reaches -41.61 dB, the matching frequency is 5.5 GHz, and the effective absorption bandwidth is 8.4 GHz.

Maosheng Cao et al. first treat MWCNTs with sulfuric acid, and then prepare Fe$_3$O$_4$/MWCNTs by coprecipitation of NH$_4$Fe(SO$_4$)$_2$•12H$_2$O and

$(NH_4)_2FeSO_4 \cdot 6H_2O$. Finally, $PANI/Fe_3O_4/MWCNTs$ composite nanotubes are obtained by surface polymerization of aniline [126]. The TEM images are shown in Fig. 10.13B. The microwave absorption performance of $Fe_3O_4/MWCNTs$ is better than that of $PANI/Fe_3O_4/MWCNTs$. The analysis shows that Fe_3O_4 increases the magnetic loss performance, but the introduction of polyaniline reduces the dielectric loss. By using paraffin as matrix and at the filler content of 20 wt%, the minimum reflection loss of $Fe_3O_4/MWCNTs$ is -75 dB, and the effective absorption bandwidth reaches 10 GHz.

Yang Liu et al. take biomass sugarcane fibers as raw material [127]. First, sugarcane fibers are heated with 4% KOH solution at 150°C for 10 h for surface treatment. Then, the treated fibers are soaked in $FeCl_3$ solution for 2 h to complete the adsorption of Fe^{3+} ions. Finally, Fe_3O_4/CF magnetic carbon fibers are prepared by calcination at 700°C in argon atmosphere. The preparation process is displayed in Fig. 10.14. At the matching thickness of 1.9 mm, the minimum reflection loss of Fe_3O_4/CF reaches -48.2 dB@15.6 GHz, and the effective absorption bandwidth is 5.1 GHz, covering the frequency range of 12.9–18.0 GHz.

10.4.3 1D carbon/inorganic dielectric nano particles composite microwave absorbers

Combing semiconductors or ceramics can enhance dielectric loss and optimize impedance matching. It is one of the effective means to improve the microwave absorption properties of 1D carbon materials. The commonly used semiconductor particles include transition metal sulfides such as CuS, ZnS, CrS, MoS_2, CoS_x, and transition metal oxides such as TiO_2, ZnO, and MnO_2. Commonly used ceramic microwave absorbers include SiC, Si_3N_4, barium titanate, barium ferrite and so on.

(1) 1D carbon/semiconductors microwave absorbers

Honglin Luo et al. use carboxylated carbon nanofibers as cathode and zinc plate as anode for electrochemical deposition in 0.1 mol/L $Zn(NO_3)_2$ electrolyte [128]. Three kinds of carbon composite nanofibers loaded with different morphologies of ZnO are prepared by adjusting electrolyte temperature and deposition reaction time: flowcus-like ZnO, plate-like ZnO and particle-like ZnO. The preparation process and SEM images of ZnO/carbon composite nanofibers are showed in Fig. 10.15. The effects of surface loaded ZnO morphology on the microwave absorption properties are observed. The minimum reflection losses of floccus-like and particle-like ZnO coated

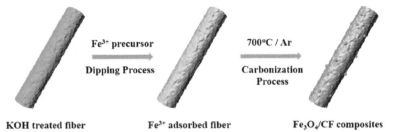

KOH treated fiber   Fe³⁺ adsorbed fiber   Fe₃O₄/CF composites

Figure 10.14 Schematic diagram of the preparation process of Fe_3O_4/CF magnetic carbon fibers. Reprinted (adapted) with permission from Copyright 2022 American Chemical Society [127].

carbon composite nanofibers are -46.6 dB and -46.4 dB, respectively. The corresponding matching thicknesses are 1.21 mm and 1.52 mm.

Weidong Zhang et al. report a preparation method of MoS_2-coated carbon fiber [129]. MoS_2 is in-situ deposited on the surface of carbon fiber with $Na_2MoO_4 \cdot 2H_2O$ and C_2H_5NS as raw materials in hydrothermal system to realize the effective combination of carbon fiber and MoS_2. The SEM and TEM images are exhibited in Fig. 10.16A and B. The conductive loss, interfacial polarization and impedance matching of the hybrid fiber are enhanced synchronously after coating by MoS_2, see Fig. 10.16C. These two components exert significant synergistic effect. The reflection loss is -21.4 dB@3.8 mm, and it has an ultra-wide effective bandwidth of 10.85 GHz (7.15–18.0 GHz).

(2) 1D carbon/ceramics microwave absorbers

Yijing Zhao et al. synthesize a PAN fiber with embedded SiC nanoparticles through electrospinning by mixing SiC nanoparticles into the DMF solution of PAN [6]. SiC nanoparticles loaded carbon nanofibers are prepared by preoxidation in air at 260°C and calcination in argon atmosphere at 800°C. Finally, C-SiC hybrid nanofibers with different SiC loadings are obtained by adjusting the amount of SiC nanoparticles in electrospinning solution. The preparation process is showed in Fig. 10.17. The microwave absorbing properties of pure carbon nanofiber and C-SiC hybrid nanofiber are evaluated. The minimum reflection loss of C-SiC₅ is -53.7 dB@2.15 mm. The microwave absorption performance is greatly improved comparing with pure carbon fiber. The excellent microwave absorption performance is closely related to the stronger interfacial polarization and dipole polarization brought by SiC nanoparticles.

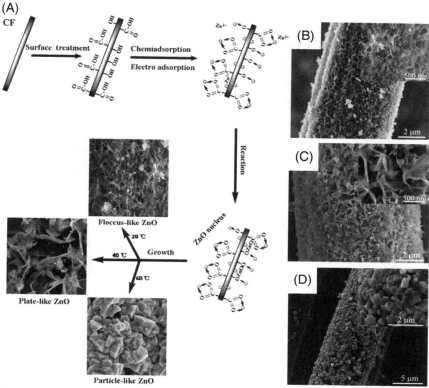

Figure 10.15 Schematic diagram of the preparation process (A) and SEM images (B-D) of ZnO/carbon composite nanofibers with different morphologies: Floccus-like (B), Plate-like (C) and Particle-like (D). The copyright of the figure obtained from Elsevier under the license No: 5271870548041, Mar 18, 2022 [128].

Yaofeng Zhu et al. design and prepare a $BaTiO_3$@MWCNT core-shell heterostructure composite [130]. First, surface-acidified MWCNTs are dispersed in $BaTiO_3$ sol. Then, $BaTiO_3$@MWCNT can be obtained after drying and calcining at 700°C under argon atmosphere for 2 h. The microwave absorption properties of single phase MWCNTs, ingle phase $BaTiO_3$ and $BaTiO_3$@MWCNT composite are studied with paraffin as matrix at the filler content of 20 wt%. $BaTiO_3$@MWCNT composite possesses the strongest absorption capacity with a reflection loss of –11.2 dB. Furthermore, the microwave absorbing performance can be further improved by increasing the filler content of $BaTiO_3$@MWCNT in the matrix. The reflection loss increases to –25.7 dB at a filler content of 40%.

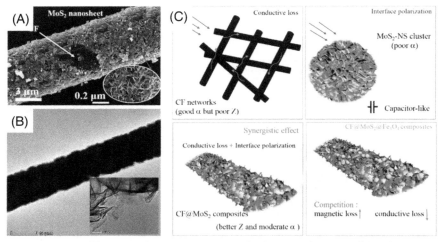

Figure 10.16 SEM image (A), TEM image (B) and schematic diagram of the absorption mechanism (C) of MoS$_2$-coated carbon fiber. The copyright of the figure obtained from Elsevier under the license No: 5271871436084, Mar 18, 2022 [129].

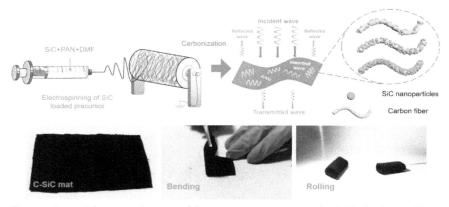

Figure 10.17 Schematic diagram of the preparation process of C-SiC hybrid nanofibers. The copyright of the figure obtained from Elsevier under the license No: 5271881257322, Mar 18, 2022 [6].

10.4.4 1D carbon/conductive polymer microwave absorbers

Conductive polymers mainly refer to polymer materials that have conductive properties or have conductive properties after appropriate doping, such as polyaniline (PANI), polypyrrole (PPy), polythiophene (PTh) and their derivatives. Conductive polymers have a special chemical structure. The molecular main chain contains a conjugated large π bond structure. In an ideal state, electrons delocalize on the entire main chain, the molecular

Figure 10.18 SEM and TEM images of HCNTs (A and D) and PANi@HCNT hybrids doped with HCl (B and E) and D-CSA (C and F). Reprinted (adapted) with permission from Copyright 2022 American Chemical Society [131].

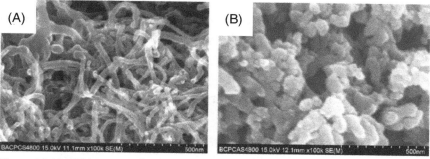

Figure 10.19 SEM images of MWCNTs (A) and PPy/MWCNTs composite fiber with the MWCNT:PPy mass ratio of 0.1 (B). The copyright of the figure obtained from Elsevier under the license No: 5271900225104, Mar 18, 2022 [132].

orbitals of the monomer interact, the highest occupied orbital forms a valence band, and the lowest empty orbital forms a conduction band. There is an energy gap between them. After proper doping, conductive polymer has good conductivity. Combining conductive polymer with carbon materials can enhance the conductivity and improve the conductivity loss. It has been applied in the field of microwave absorption.

Xin Tian et al. coat polyaniline shell on the surface of helical carbon nanotubes (HCNTs, Fig. 10.18 A and D) by in–situ polymerization, and its surface showed concave convex shape [131]. In this process, chiral D-CSA ((1S)-(+)-10-camphorsulfonic acid) and achiral HCl are used as dopants for the preparation of PANI-CSA@HCNTs (Fig. 10.18 B and E) and PANI-HCl@HCNTs (Fig. 10.18 C and F). The influence of the chirality of the dopant molecule on microwave absorbing performance has been studied. Compared with HCNTs without polyaniline and PANI-HCl@HCNTs doped with achiral HCl, PANI-CSA@HCNTs doped with chiral D-CSA exhibits stronger microwave absorption performance with a minimum reflection loss of –32.5 dB at 8.9 GHz. Further study shows that molecular and nanoscale chirality in PANI-CSA@HCNTs play a synergistic role in improving the absorption performance.

Qi Gao et al. realize the organic combination of PPy on the surface of MWCNTs by means of in-situ chemical oxidation polymerization [132]. Six PPy/MWCNTs composite fibers with different proportions have been obtained. Fig. 10.19 shows the SEM images of MWCNTs and PPy/MWCNTs composite fiber with the MWCNT:PPy mass ratio of 0.1. The conductivity, infrared radiation ability and microwave attenuation ability with different PPy loadings are evaluated. With the increase of PPy loading, the conductivity of the composite fiber increases gradually from 0.079 S/cm to 9.14 S/cm, and the infrared emissivity decreases gradually. However, the microwave absorption performance will decrease significantly when the PPy load rate is too high, and the minimum reflection loss is 19 dB@3 mm.

10.5 Tubular carbon nanofibers

Tubular carbon nanofibers (TCNFs) are prepared by a two-step method [133]. Firstly, hypercrosslinked tubular polymer nanofibers are synthesized through our early developed two oil-phase self-polycondensation system. Then, TCNFs with secondary pores can be obtained by using tubular polymer nanofibers as precursors under high-temperature carbonization. The preparation process is shown in Fig. 10.20. Compared with granular or layered carbon materials, TCNFs can be dispersed in the matrix forming an effective three-dimensional conductive network due to their one-dimensional structure. It will produce local current and bring about conductive loss. Compared with other one-dimensional carbon materials, TCNFs not only have axial through pores, but also have abundant small pores on the tube wall. The existence of these pores can not only reflect and scatter electromagnetic wave multiple times, but also serve as a transmission channel for electromagnetic wave and improve impedance matching characteristics. At the same

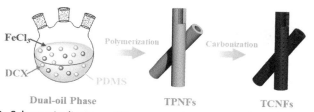

Figure 10.20 Schematic diagram of the preparation process of TCNFs. The copyright of the figure obtained from Elsevier under the license No: 5271910482758, Mar 18, 2022 [133].

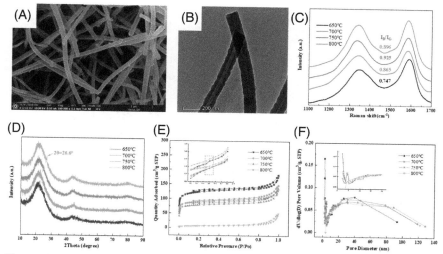

Figure 10.21 SEM and TEM images of TCNFs obtained at 650°C, Raman spectra, XRD curves, nitrogen adsorption-desorption isotherms and pore size distribution curves of TCNFs prepared at different temperature. The copyright of the figure obtained from Elsevier under the license No: 5271910482758, Mar 18, 2022 [133].

time, abundant pores also provide high specific surface area, which makes it have strong interface polarization ability. The above characteristics make it possible for TCNFs to exhibit excellent microwave absorption properties.

Fig. 10.21A and B are the SEM and TEM images of TCNFs obtained at 650°C, respectively. TCNFs still maintain the original tubular morphology of polymer nanofiber precursor without obvious structural change under high temperature calcination, such as collapse. With the increase of calcination temperature, the I_D/I_G value of Raman spectra first increases and then decreases (Fig. 10.21C). This is because the nanofibers transform from amorphous carbon to nanocrystalline graphite and finally to graphite in this stage, and the carbonization degree deepens accordingly. Only wide

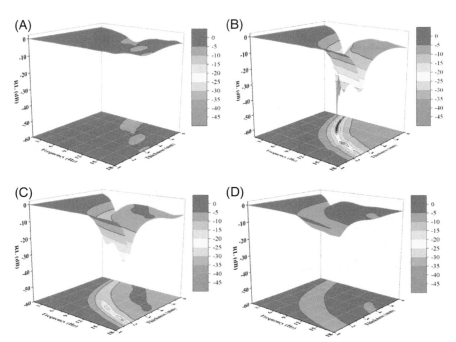

Figure 10.22 3D reflection loss images of TCNFs obtained at different temperatures: 650°C (A), 700°C (B), 750°C (C) and 800°C (D). The copyright of the figure obtained from Elsevier under the license No: 5271910482758, Mar 18, 2022 [133].

peaks of amorphous carbon can be observed in the XRD curves of the samples obtained at 650°C, 700°C and 750°C, but the diffraction peak of crystalline graphite appears at 26.6° in the samples obtained at 800°C, which is consistent with the conclusion obtained by Raman spectroscopy (Fig. 10.21D). Fig. 10.21E and F display the nitrogen adsorption-desorption isotherms and pore size distribution curves, respectively. The adsorption-desorption isotherms belong to type IV with H2 hysteresis loops. The low-temperature sample has higher adsorption capacity and longer low-pressure section, which means larger specific surface area and more abundant. This can be also confirmed by the pore size distribution curves. It indicates that the micropores in tube wall will gradually disappear as calcination temperature increases and graphitization degree will gradually be enhanced, resulting in a decline in pore performance. The specific surface area reduces from 423.06 m^2/g to 24.99 m^2/g.

Fig. 10.22 depicts 3D reflection loss images of TCNFs obtained at different temperatures with paraffin as the matrix at the filler amount of 10 wt% within the thickness of 1–6 mm. It is found that with the increase

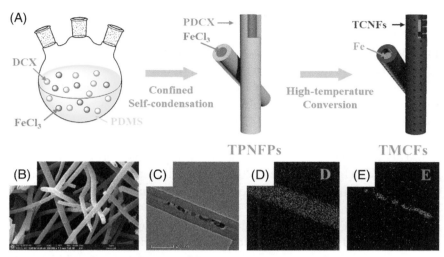

Figure 10.23 Schematic diagram of the preparation process of TMCFs (A), SEM image (B), TEM image (C) and element distributions (D and E). Reprinted (adapted) with permission from Copyright 2022 American Chemical Society [134].

of carbonization temperature, the minimum reflection loss (RL_{min}) shows a trend of first increase and then decrease. The RL_{min} of the products carbonized at 650°C, 700°C, 750°C and 800°C are –7.3 dB, –61.5dB, –26.3dB and –9.6 dB at 8.82 GHz, 7.46 GHz, 16.47 GHz and 11.20 GHz. The matching thicknesses of 6 mm, 4.5 mm, 2.0 mm, and 2.0 mm, respectively. The absorption intensity reaches the maximum at 700°C.

10.6 Tubular magnetic carbon nanofibers with hierarchical pore structure

In order to improve the absorbing performance of TCNFs, magnetic Fe nanoparticles are selected to combine with them. The comprehensive absorbing performance can be enhanced by introducing magnetic components to endow magnetic loss capability. The tubular polymer nanofiber precursors with $FeCl_3$ embedded inside are synthesized by the two–oil phase self-condensation method [134]. Then, tubular magnetic carbon nanofibers (TMCFs) with Fe nanoparticles loaded inside are prepared through carbonization of polymer and in–situ conversion of $FeCl_3$ at high temperature. The preparation process is shown in Fig. 10.23A. The SEM and TEM images of TMCFs are given in Figs. 10.23B and C. The surface morphology and internal structure of TMCFs are clearly visible. The morphology has good uniformity and the diameter is about 150 nm. Coaxial channels and tubular structure can be observed inside. The axial channels are loaded with inorganic nanoparticles with high quality and thickness contrast. It is preliminarily judged as Fe nanoparticles by energy spectrum.

In order to further determine the composition of TMCFs, the samples obtained at different temperatures have been characterized by XRD, and the spectra are provided in Fig. 10.24A. Three sharp diffraction peaks appear in the four spectral lines at 2θ of 44.76°, 65.12°, and 82.46°, corresponding to the (110), (200) and (211) crystal planes of α-Fe, respectively. XRD results confirms that the particles in the axial through holes of TMCFs are Fe nanoparticles. The broad peak at 20.76° is the diffraction peak of amorphous carbon, indicating that the carbon component in TMCFs exists in an amorphous form. Raman spectra show that the I_D/I_G ratio increases from 0.800 to 0.923 (Fig. 10.24B), which means that the increase of carbonization temperature will increase the nanocrystalline graphite. The analysis results of the content of magnetic component Fe nanoparticles in TMCFs and their magnetic response performance are depicted in Fig. 10.24C and D, respectively. With the rise of carbonization temperature, the initial weight loss temperature and the residue mass increase, implying that the TMCFs prepared at high carbonization temperature have better thermal stability and higher magnetic content. This is because more carbon is lost from TMCFs at higher temperature, and the lower carbon content makes the magnetic content rise. The specific saturation magnetization has the same trend as the magnetic content, and the product obtained at high temperature exhibits a higher specific saturation magnetization, up to 33.60 emu/g. The nitrogen adsorption-desorption isotherm is a type IV curve, the low-pressure section of the isotherm is long, and the N_2 adsorption capacity increases rapidly, which are the typical features of rich microporous structure. The presence of H2 hysteresis loops in the isotherm of the medium pressure zone indicates that after the micropores are filled, there are still larger pores that can interact strongly with N_2 molecules and produce pore condensation effect. This phenomenon indicates the existence of secondary pore structure in TMCFs (Fig. 10.24E), corresponding to the double peaks around 30 nm and 5 nm in the pore size distribution curves (Fig. 10.24F).

The calcination temperature and filler content of TMCFs are optimized. The final results show that the optimal carbonization temperature is 700°C and the optimal filler content is 15 wt%, seen Fig. 10.25A and B. At this time, the minimum reflection loss is -47.33 dB@2.2 mm, the effective absorption bandwidth is 6.5 GHz, covering the whole Ku band and part of X band. If the absorbing layer thickness is adjusted to 3 mm, the effective

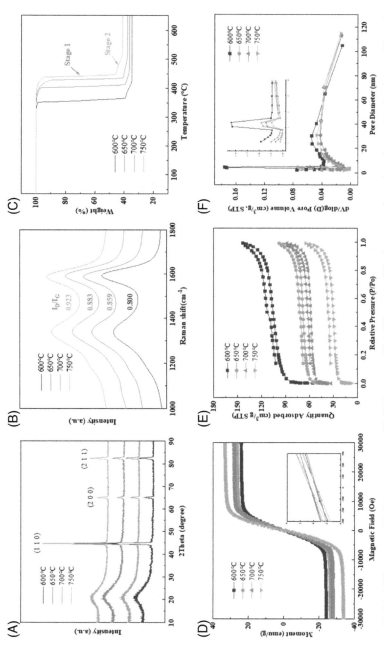

Figure 10.24 XRD spectra (A), Raman spectra (B), TGA curves (C), VSM curves (D), nitrogen adsorption-desorption isotherms (E) and, pore size distribution curves (F) of TMCFs obtained at different temperatures. Reprinted (adapted) with permission from Copyright 2022 American Chemical Society [134].

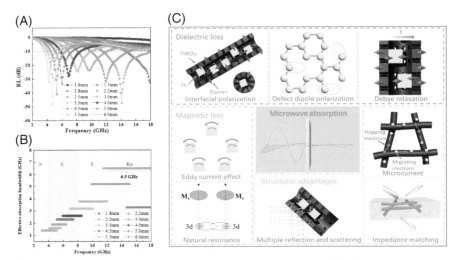

Figure 10.25 The reflection loss curves (A) and effective bandwidth image (B) of TMCFs at the optimal calcination temperature and filler content, the schematic diagram of the absorbing mechanism (C). Reprinted (adapted) with permission from Copyright 2022 American Chemical Society [134].

absorption frequency range covers nearly the entire X-band, and the minimum reflection loss is −23.95 dB. The research of absorbing mechanism shows that the excellent electromagnetic wave loss capability comes from three aspects: dielectric loss, magnetic loss and special structural advantages (Fig. 10.25C). Dielectric loss includes the interface polarization at the adjacent interface, dipole polarization of internal defects in carbon components, and accompanying Debye relaxation process. Magnetic loss comes from eddy current loss and natural resonance effects. In addition, TMCFs have a multi-level pore structure. The abundant pores can cause multiple reflections and scattering of electromagnetic wave inside the absorber, and the abundant internal cavities can improve the impedance matching ability of the material. High specific surface area can produce more interfaces, enhancing interface polarization and Debye relaxation. Due to its one-dimensional structure, TMCFs can be dispersed in matrix form an effective conductive network, which converts electromagnetic wave into heat by generating local current.

10.7 Helical/chiral biomass-derived 3D magnetic porous carbon fibers

Materials with helical morphology have special chiral structure and adjustable chiral parameters in addition to dielectric constant and permeability.

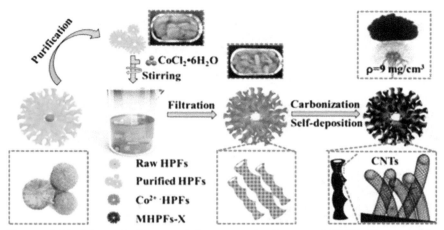

Figure 10.26 Schematic diagram of the preparation process of MHPFs. The copyright of the figure obtained from Elsevier under the license No: 5271920116925, Mar 18, 2022 [135].

They are regarded as advanced microwave absorbing materials. Different from achiral materials, chiral microwave absorbing materials can cause cross-polarization of electromagnetic waves and give microwave absorbers excellent microwave absorption properties. Besides, biomass–derived carbon materials have the characteristics of green, pollution-free, renewable, and easy to obtain, and are considered to be ideal high-yield microwave absorbing materials. Based on the above two considerations, biomass–derived helical hollow fibers (HPFs) are used as the organic framework for the preparation of 3D magnetic porous carbon fibers (MHPFs) with complex helical/chiral structure through a simple soak and in-situ reduction process [135]. The preparation process is shown in Fig. 10.26.

The morphology of the raw HPFs presents a one-dimensional bamboo-like shape with a diameter of about 40 μm. The surface is smooth and the cross-sectional observation shows a solid structure (Fig. 10.27A–C). After high temperature calcination, the HPFs soaked with Co^{2+} show significant volume shrinkage, accompanying by spiralization (Fig. 10.27D, G, and J). The original solid structure also changes into hollow structure and the tube wall thickness decreases (Fig. 10.27E, H, and K), which are caused by the loss of axial organic matter. The cobalt ions loaded on organic framework are in–situ reduced to Co particles under vacuum calcination. Carbon fixation and carbon nanotubes growth are realized by typical CVD method. The growth of carbon nanotubes (CNTs) on MHPFs can be effectively

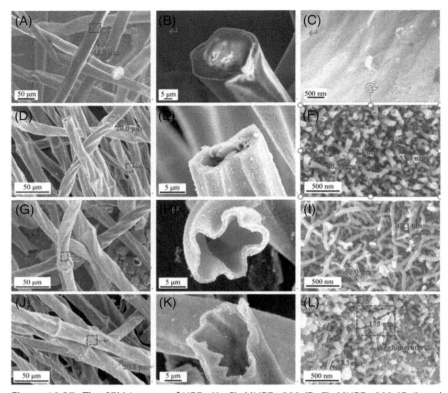

Figure 10.27 The SEM images of HPFs (A–C), MHPFs-800 (D–F), MHPFs-900 (G–I) and MHPFs-1000 (J–L). The copyright of the figure obtained from Elsevier under the license No: 5271920116925, Mar 18, 2022 [135].

controlled by adjusting the carbonization temperature. The morphology of carbon nanotubes on the surface of MHPFs is the most regular at 900°C (Fig. 10.27F, I, and L). It is considered that carbon loss rate and catalytic deposition efficiency jointly affect the growth of CNTs. Carbon loss rate is the escape strength of carbon source at different temperatures. Higher carbonization temperature makes it more difficult for carbon capture, resulting in greater carbon loss rate. Catalytic deposition efficiency refers to the overall effect of coparticles catalyzing the growth of CNTs or fixing the carbon source. The stronger the fixing ability and more regular CNTs growth imply higher the efficiency.

Figure 10.28 displays the 3D plots and 2D curves of RL for MHPFs. Due to the effective synergy among complex 3D helical/chiral structure, hierarchical pore structure and CNTs loaded with Co nanoparticles,

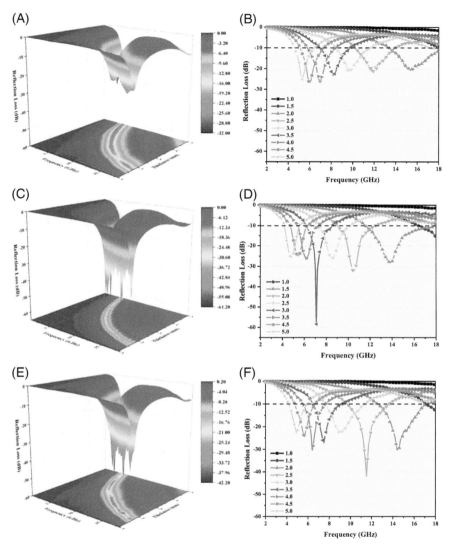

Figure 10.28 3D plots and 2D curves of RL for MHPFs-800 (A and B), MHPFs-900 (C and D) and MHPFs-1000 (E and F) with a loading of 5%. The copyright of the figure obtained from Elsevier under the license No: 5271920116925, Mar 18, 2022 [135].

MHPFs-900 exhibits the most excellent microwave absorption performance in a wide frequency range (4–18 GHz). At the filler content of only 5%, MHPFs-900 can obtain a reflection loss of –61.08 dB loss at low frequency (6.3 GHz). The EAB is 2.1 GHz (5.4–7.5 GHz). At high frequency 13.6 GHz, the EAB reaches 6.2 GHz (11.6–17.8 GHz), covering almost the

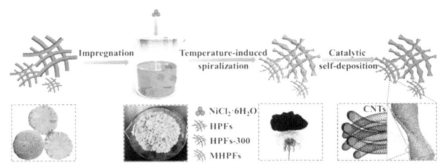

Figure 10.29 Synthesis diagram of MHPFs. The copyright of the figure obtained from Elsevier under the license No: 5271921247505, Mar 18, 2022 [136].

entire Ku band. MHPFs-800 also has a large RL (–30.06 dB) at 6.3 GHz, and EAB is 2.3 GHz (5.3–7.6 GHz). The RL of MHPFs-1000 is as high as –42.09 dB at 11.5 GHz, and the EAB is 3.9 GHz (9.7–13.6 GHz). MHPFs-900 achieves outstanding absorption intensity at low frequency and ultrawide EAB at high frequency. It exhibits dual-band glorious microwave absorption performance.

10.8 Ultralight helical porous carbon fibers with CNTs-confined Ni nanoparticles

Similarly, platanus seed fibers are also used as the matrix. A catalytic self-deposition (CSD) technology is developed to realize the confined growth of Ni nanoparticles in CNTs, and the magnetic spiral porous carbon fibers (MHPFs) are easily prepared [136]. The synthesis diagram of MHPFs is exhibited Fig. 10.29. It does not need to use expensive, flammable and explosive exogenous gas, which effectively avoids the high energy consumption and potential safety problems faced by the traditional energy intensive catalytic chemical vapor deposition (CCVD). The combined use of in-situ soak and CSD technology realizes the in-situ growth and morphology control of CNTs encapsulated with Ni nanocatalysts. By adjusting the loading amount of Ni ions, the growth morphology and electromagnetic parameters of magnetic CNTs are effectively controlled. At the same time, the microwave absorbing mechanism of MHPFs is revealed, and the structure-activity relationship is clarified.

The HPFs loaded with different Ni^{2+} contents are carbonized at 900°C (Fig. 10.30A), and MHPFs with different morphologies are obtained, shown in Fig. 10.30B–M. CNTs are grown on the surface of MHPFs, and the

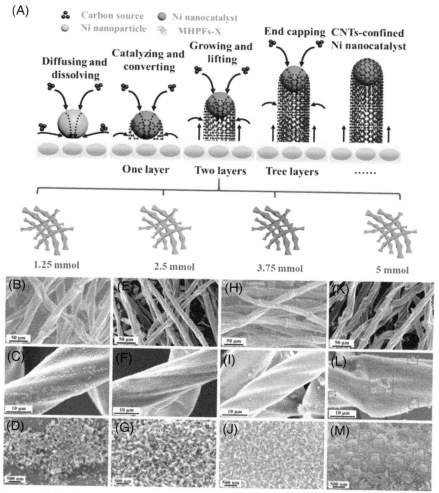

Figure 10.30 Tip growth mechanism of CNTs (A), SEM images of MHPFs-1.25 (B–D), MHPFs-2.5 (E–G), MHPFs-3.75 (H–J) and MHPFs-5 (K–M) under different magnifications. The copyright of the figure obtained from Elsevier under the license No: 5271921247505, Mar 18, 2022 [136].

density of CNTs changed significantly with the increase of Ni^{2+} concentration. This is due to the different content of Ni nanocatalysts triggering the typical tip growth of CNTs. Different from the traditional CCVD method, the CSD process does not require the assistance of exogenous reducing gases and carbon sources such as hydrogen and methane, and the reduction of Ni^{2+} and the catalytic growth of CNTs can be accomplished using organic carbon sources. Therefore, the inherent carbon source and metal catalyst content of the raw material itself play a decisive role in the CSD process. The growth state and quantity of CNTs on MHPFs can be controlled by adjusting the loading amount of Ni nanocatalyst. Specifically, the organic components of the matrix are gradually transformed into carbon components and Ni^{2+} is reduced in situ to elemental nickel particles during the carbonization process, as shown in Fig. 10.30A. Subsequently, the carbon atoms generated by the further pyrolysis of the matrix diffuse to the surface of the molten metal with the assistance of vacuum and are dissolved. After the dissolution reaches saturation, Ni nanocrystals begin to catalyze the growth of CNTs from the bottom. Finally, Ni-coated CNTs are formed. Explicitly, CNTs deposited on the surface of MHPFs-1.25, and the CNTs are relatively sparse (Fig. 10.30B–D). A magnified observation on the surface shows that the growth of CNTs is poor, which can be attributed to the low catalyst concentration. As the loading of Ni^{2+} increases, the surfaces of MPHPFs-2.5 and MPHPFs-3.75 are deposited with well-grown and densely interwoven CNTs (Fig. 10.30E–J). When the Ni^{2+} ion concentration increases to 5 mmol, it can be clearly seen that the CNTs deposited on the surface of MHPFs-5 decrease, and a large number of agglomerated deposits with a particle size of more than 100 nm appears at the mark (Fig. 10.30K–M). The deposits are agglomerated Ni nanoparticles. This result indicates that excessive catalyst concentration is not conducive for the catalytic growth of CNTs.

The microwave absorbing properties of MHPFs-1.25, MHPFs-2.5, MHPFs-3.75 and MHPFs-5 are characterized with paraffin as matrix and at the filler content of 5%. Among them, MHPFs-2.5 exhibited the largest loss intensity, with RL_{min} exceeding -55.39 dB, EAB at 3.8 GHz (7.2–11 GHz), and the corresponding frequency and thickness of 8.4 GHz and 3.4 mm, respectively (Fig. 10.31A and B). The excellent microwave absorption performance of MHPFs is due to the multiple loss mechanism, illustrated in Fig. 10.31C: (1) defects and dipolar polarization caused by heteroatoms; (2) interfacial polarization caused by heterogeneous interfaces such as spiral carbon fiber-CNTs, CNTs-Ni, spiral carbon fiber-Ni, MHPFs-parafin, etc.; (3) 3D conductive network formed between stacked layers; (4) the introduction of nano-Ni effectively optimizes the impedance matching performance, so that the material has a certain magnetic loss capability; (5) the special spiral/chiral structure produces the cross polarization of the incident microwave; (6) optimized impedance matching and multiple reflections and scattering of incident microwave by hollow and tertiary porous structures.

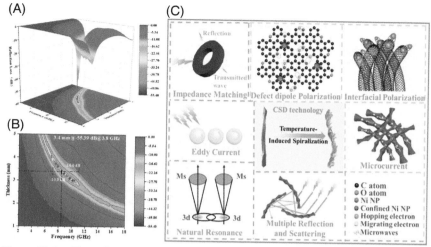

Figure 10.31 The reflection loss images (A and B) and schematic diagram of the absorption mechanism (C) of MHPFs-2.5. The copyright of the figure obtained from Elsevier under the license No: 5271921247505, Mar 18, 2022 [136].

10.9 Summary and outlook

Microwave absorbing materials show significant advantages in electromagnetic radiation protection and information leakage treatment. They can not only block the propagation of electromagnetic waves, but also will not lead to secondary pollution caused by reflection. They are considered to be the most direct and effective means. With the continuous expansion of electromagnetic wave applications and the increase of electronic equipment year by year, the application demand for high-performance absorbing materials is also increasing, and the development of new microwave absorbing agents is expected to be further developed in the future. In addition to the traditional "thinner", "lighter", "wider", and "stronger", the main development directions may include the following:

(1) Lower frequency

The absorbing materials developed at this stage can show good performance in the X-band and Ku-band, but it is difficult to obtain satisfactory results in the lower-frequency, including C-band, S-band and L-band. There is a certain gap in the application of low-frequency absorbing agents in the application scenarios with high loss in the propagation process.

(2) Integration of structure and function

At present, microwave absorbing materials are mostly used in the form of coating, which needs to be coated on the equipment. This process will change the surface morphology and properties of the original materials. The problem can be solved by preparing the absorbing material into the structural parts (shell, skin, etc.) of instruments and equipment. The control of the thickness of the absorbing layer can be effectively relaxed.

(3) Multifunction

Microwave stealth, infrared stealth, visible light stealth, and laser stealth have a wide range of needs, but there is no satisfactory means to integrate them into a single coating. It is difficult to achieve multifunctional stealth. By developing new materials, designing new structures and realizing functional integration, it is expected to further improve the survivability of weapons and equipment.

(4) Integration of technical indicators

Besides the single index of absorbing performance, the comprehensive engineering indexes such as heat resistance, corrosion resistance and aging resistance of the absorbing materials should also be taken into consideration in the design and synthesis process of microwave absorbing agent. It will help to enhance the environmental adaptability and prolong the service life of materials.

References

[1] Q. Liu, Q. Cao, H. Bi, C. Liang, K. Yuan, W. She, Y. Yang, R. Che, CoNi@SiO$_2$@TiO$_2$ and CoNi@Air@TiO$_2$ microspheres with strong wideband microwave absorption, Adv. Mater 28 (3) (2016) 486–490.

[2] H. Sun, R. Che, X. You, Y. Jiang, Z. Yang, J. Deng, L. Qiu, H. Peng, Cross-stacking aligned carbon-nanotube films to tune microwave absorption frequencies and increase absorption intensities, Adv. Mater. 26 (48) (2014) 8120–8125.

[3] M. Qin, L. Zhang, X. Zhao, H. Wu, Defect induced polarization loss in multi-shelled spinel hollow spheres for electromagnetic wave absorption application, Adv. Sci. 8 (8) (2021) 2004640.

[4] K. Li, R. Zhao, J. Xia, G.-L. Zhao, Reinforcing microwave absorption multiwalled carbon nanotube-epoxy composites using glass fibers for multifunctional applications, Adv. Eng. Mater. 22 (3) (2020) 1900780.

[5] M. Han, X. Yin, Z. Hou, C. Song, X. Li, L. Zhang, L. Cheng, Flexible and thermostable graphene/SiC nanowire foam composites with tunable electromagnetic wave absorption properties, ACS Appl. Mater. Interfaces. 9 (13) (2017) 11803–11810.

[6] Y. Zhao, Y. Zhang, C. Yang, L. Cheng, Ultralight and flexible SiC nanoparticle-decorated carbon nanofiber mats for broad-band microwave absorption, Carbon. 171 (2021) 474–483.

[7] X.X. Wang, W.Q. Cao, M.S. Cao, J. Yuan, Assembling nano-microarchitecture for electromagnetic absorbers and smart devices, Adv. Mater. 32 (36) (2020) 2002112.

[8] J. Pan, R. Zhang, Z. Zhen, M. Wang, Z. Li, D. Yin, L. Meng, L. He, H. Zhu, Enhanced microwave absorption of shape anisotropic Fe_3O_4 nanoflakes and their composites, Adv. Eng. Mater. 24 (2) (2021) 2100790.

[9] M. Qin, L. Zhang, X. Zhao, H. Wu, Lightweight Ni foam-based ultra-broadband electromagnetic wave absorber, Adv. Funct. Mater. 31 (30) (2021) 2103436.

[10] Y. Hou, L. Cheng, Y. Zhang, Y. Yang, C. Deng, Z. Yang, Q. Chen, P. Wang, L. Zheng, Electrospinning of Fe/SiC hybrid fibers for highly efficient microwave absorption, ACS Appl. Mater. Interfaces. 9 (8) (2017) 7265–7271.

[11] Y. Zhao, H. Zhang, X. Yang, H. Huang, G. Zhao, T. Cong, X. Zuo, Z. fan, S. Yang, L. Pan, In situ construction of hierarchical core-shell $Fe_3O_4@C$ nanoparticles–helical carbon nanocoil hybrid composites for highly efficient electromagnetic wave absorption, Carbon. 171 (2021) 395–408.

[12] T. Xia, C. Zhang, N.A. Oyler, X. Chen, Hydrogenated TiO_2 nanocrystals: A novel microwave absorbing material, Adv. Mater. 25 (47) (2013) 6905–6910.

[13] Z. Huang, H. Chen, Y. Huang, Z. Ge, Y. Zhou, Y. Yang, P. Xiao, J. Liang, T. Zhang, Q. Shi, G. Li, Y. Chen, Ultra-broadband wide-angle terahertz absorption properties of 3D graphene foam, Adv. Funct. Mater. 28 (2) (2018) 1704363.

[14] Y. Zhang, Y. Huang, T. Zhang, H. Chang, P. Xiao, H. Chen, Z. Huang, Y. Chen, Broadband and tunable high-performance microwave absorption of an ultralight and highly compressible graphene foam, Adv. Mater. 27 (12) (2015) 2049–2053.

[15] X. Xu, S. Shi, Y. Tang, G. Wang, M. Zhou, G. Zhao, X. Zhou, S. Lin, F. Meng, Growth of NiAl-layered double hydroxide on graphene toward excellent anticorrosive microwave absorption application, Adv. Sci. 8 (5) (2021) 2002658.

[16] X. Wang, J. Shu, X. He, X.Wang Minzhang, C. Gao, J. Yuan, M. Cao, Green approach to conductive PEDOT:PSS Decorating magnetic-graphene to recover conductivity for highly efficient absorption, ACS Sustainable Chem. Eng. 6 (2018) 14017–14025.

[17] H. Xu, X. Yin, M. Li, X. Li, X. Li, X. Dang, L. Zhang, L. Cheng, Ultralight cellular foam from cellulose nanofiber/carbon nanotube self-assemblies for ultrabroad-band microwave absorption, ACS Appl. Mater. Interfaces. 11 (25) (2019) 22628–22636.

[18] H. Chen, Z. Huang, Y. Huang, Y. Zhang, Z. Ge, B. Qin, Z. Liu, Q. Shi, P. Xiao, Y. Yang, T. Zhang, Y. Chen, Synergistically assembled MWCNT/graphene foam with highly efficient microwave absorption in both C and X bands, Carbon. 124 (2017) 506–514.

[19] Y. Cheng, Z. Li, Y. Li, S. Dai, G. Ji, H. Zhao, J. Cao, Y. Du, Rationally regulating complex dielectric parameters of mesoporous carbon hollow spheres to carry out efficient microwave absorption, Carbon. 127 (2018) 643–652.

[20] X. Wang, X. Huang, Z. Chen, X. Liao, C. Liu, B. Shi, Ferromagnetic hierarchical carbon nanofiber bundles derived from natural collagen fibers: Truly lightweight and high-performance microwave absorption materials, J. Mater. Chem. C 3 (39) (2015) 10146–10153.

[21] K. Li, H. Sun, X. Zhang, S. Zhang, H. Dong, C. Zhu, Y. Chen, Micro-nanospheres assembled with helically coiled nitrogen-doped carbon nanotubes: Fabrication and microwave absorption properties, Mater. Design. 186 (2020) 108290.

[22] G. Wang, Z. Gao, S. Tang, C. Chen, F. Duan, S. Zhao, S. Lin, Y. Feng, L. Zhou, Y. Qin, Microwave absorption properties of carbon nanocoils coated with highly controlled magnetic materials by atomic layer deposition, ACS Nano. 6 (2012) 11009–11017.

[23] A. Hua, Y. Li, D. Pan, J. Luan, Y. Wang, J. He, S. Tang, D. Geng, S. Ma, W. Liu, Z. Zhang, Enhanced wideband microwave absorption of hollow carbon nanowires derived from a template of $Al_4C_3@C$ nanowires, Carbon 161 (2020) 252–258.

[24] X. Xiao, W. zhu, Z. Tan, W. Tian, Y. Guo, H. Wang, J. Fu, X. Jian, Ultra-small Co/CNTs nanohybrid from metal organic framework with highly efficient microwave absorption, Compos. Part B: Eng. 152 (2018) 316–323.

[25] I.M.D. Rosa, A. Dinescu, F. Sarasini, M.S. Sarto, A. Tamburrano, Effect of short carbon fibers and MWCNTs on microwave absorbing properties of polyester composites containing nickel-coated carbon fibers, Compos. Sci. Technol. 70 (1) (2010) 102–109.

[26] J. Yang, J. Zhang, C. Liang, M. Wang, P. Zhao, M. Liu, J. Liu, R. Che, Ultrathin $BaTiO_3$ nanowires with high aspect ratio: A simple one-step hydrothermal synthesis and their strong microwave absorption, ACS Appl. Mater. Interfaces. 5 (15) (2013) 7146–7151.

[27] M.F.L.D. Volder, S.H. Tawfick, R.H. Baughman, A.J. Hart, Carbon nanotubes: Present and future commercial applications, Science. 339 (6119) (2013) 535–539.

[28] Y. Liu, Y. Zhang, X. Wang, Z. Wang, W. Lai, X. Zhang, X. Liu, Excellent microwave absorbing property of multiwalled carbon nanotubes with skin–core heterostructure formed by outer dominated fluorination, J. Phy. Chem. C. 122 (11) (2018) 6357–6367.

[29] K.S. Ibrahim, Carbon nanotubes-properties and applications: A review, Carbon Lett 14 (3) (2013) 131–144.

[30] I.H. Nwigboji, J.I. Ejembi, Z. Wang, D. Bagayoko, G.-L. Zhao, Microwave absorption properties of multi-walled carbon nanotube (outer diameter 20-30nm)-epoxy composites from 1 to 26.5GHz, Diam. Relat. Mater. 52 (2015) 66–71.

[31] H. Pan, J. Li, Y. Feng, Carbon nanotubes for supercapacitor, Nanoscale Res. Lett. 5 (3) (2010) 654–668.

[32] J. Chrzanowska, J. Hoffman, A. Małolepszy, M. Mazurkiewicz, T.A. Kowalewski, Z. Szymanski, L. Stobinski, Synthesis of carbon nanotubes by the laser ablation method: Effect of laser wavelength, Phys. Status. Solidi. B. 252 (8) (2015) 1860–1867.

[33] N. Arora, N.N. Sharma, Arc discharge synthesis of carbon nanotubes: Comprehensive review, Diam. Relat. Mater. 50 (2014) 135–150.

[34] J. Zhao, L. Wei, Z. Yang, Y. Zhang, Continuous and low-cost synthesis of high-quality multi-walled carbon nanotubes by arc discharge in air, Physica E. 44 (7-8) (2012) 1639–1643.

[35] Z. Shi, Y. Lian, X. Zhou, Z. Gu, Y. Zhang, S. Iijima, L. Zhou, K.T. Yue, S. Zhang, Mass-production of single-wall carbon nanotubes by arc discharge method, Carbon. 37 (1999) 1449–1453.

[36] Y. Zhang, S. Iijima, Formation of single-wall carbon nanotubes by laser ablation of fullerenes at low temperature, Appl. Phys. Lett. 75 (20) (1999) 3087–3089.

[37] W.K. Maser, E. Munoz, A.M. Benito, M.T. Martınez, G.F.d.l. Fuente, Y. Maniette, E. Anglaret, J.-L. Sauvajol, Production of high-density single-walled nanotube material by a simple laser-ablation method, Chem. Phys. Lett. 292 (1998) 587–593.

[38] Y. Sakakibara, S. Tatsuura, H. Kataura, M. Tokumoto, Y. Achiba, Near-infrared saturable absorption of single-wall carbon nanotubes prepared by laser ablation method, Jpn. J. Appl. Phys. 42 (2003) 494–496.

[39] Ç. Öncel, Y. Yürüm, Carbon nanotube synthesis via the catalytic CVD Method: A review on the effect of reaction parameters, Fuller. Nanotube. Car. N. 14 (1) (2006) 17–37.

[40] R. Brukh, S. Mitra, Mechanism of carbon nanotube growth by CVD, Chemical Phys. Lett. 424 (2006) 126–132.

[41] M.L. Terranova, V. Sessa, M. Rossi, The world of carbon nanotubes: An overview of CVD growth methodologies, Chem. Vapor Depos. 12 (6) (2006) 315–325.

[42] C.E. Baddour, F. Fadlallah, D. Nasuhoglu, R. Mitra, L. Vandsburger, J.-L. Meunier, A simple thermal CVD method for carbon nanotube synthesis on stainless steel 304 without the addition of an external catalyst, Carbon. 47 (1) (2009) 313–318.

[43] M.J. Bronikowski, CVD growth of carbon nanotube bundle arrays, Carbon. 44 (13) (2006) 2822–2832.

[44] R.F. Wood, S. Pannala, J.C. Wells, A.A. Puretzky, D.B. Geohegan, Simple model of the interrelation between single- and multiwall carbon nanotube growth rates for the CVD process, Phys. Rev. B. 75 (23) (2007) 235446.

[45] N. Hamzah, M.F.M. Yasin, M.Z.M. Yusop, A. Saat, N.A.M. Subha, Rapid production of carbon nanotubes: a review on advancement in growth control and morphology manipulations of flame synthesis, J. Mater. Chem. A. 5 (48) (2017) 25144–25170.

[46] T.-C. Liu, Y.-Y. Li, Synthesis of carbon nanocapsules and carbon nanotubes by an acetylene flame method, Carbon. 44 (10) (2006) 2045–2050.

[47] L. Yuan, T. Li, K. Saito, Growth mechanism of carbon nanotubes in methane diffusion flames, Carbon. 41 (10) (2003) 1889–1896.

[48] S. Kang, S. Qiao, Y. Cao, Z. Hu, J. Yu, Y. Wang, J. Zhu, Hyper-cross-linked polymers-derived porous tubular carbon nanofibers@TiO$_2$ toward a wide-band and lightweight microwave absorbent at a low loading content, ACS Appl. Mater. Interfaces. 12 (41) (2020) 46455–46465.

[49] Y. Jia, M.A.R. Chowdhury, D. Zhang, C. Xu, Wide-band tunable microwave-absorbing ceramic composites made of polymer-derived SiOC ceramic and in situ partially surface-oxidized ultra-high-temperature ceramics, ACS Appl. Mater. Interfaces. 11 (49) (2019) 45862–45874.

[50] X. Wang, P. Mu, C. Zhang, Y. Chen, J. Zeng, F. Wang, J.X. Jiang, Control synthesis of tubular hyper-cross-linked polymers for highly porous carbon nanotubes, ACS Appl. Mater. Interfaces. 9 (24) (2017) 20779–20786.

[51] N. Li, L. Liu, K. Wang, J. Niu, Z. Zhang, M. Dou, F. Wang, Gelatin-Derived 1D carbon nanofiber architecture with simultaneous decoration of Single Fe-Nx sites and Fe/Fe$_3$C nanoparticles for efficient oxygen reduction, Chemistry. 27 (42) (2021) 10987–10997.

[52] E. Zussman, X. Chen, W. Ding, L. Calabri, D.A. Dikin, J.P. Quintana, R.S. Ruoff, Mechanical and structural characterization of electrospun PAN-derived carbon nanofibers, Carbon. 43 (10) (2005) 2175–2185.

[53] X. Zhang, S. Dong, W. Wu, J. Yang, J. Li, K. Shi, H. Liu, Influence of Lignin units on the properties of Lignin/PAN-derived carbon fibers, J. Appl. Polym. Sci. 137 (42) (2020) 49274.

[54] N. Yusof, A.F. Ismail, Post spinning and pyrolysis processes of polyacrylonitrile (PAN)-based carbon fiber and activated carbon fiber: A review, J. Anal. Appl. Pyrol. 93 (2012) 1–13.

[55] Y. Yang, X. Liu, B. Xu, T. Li, Preparation of vapor-grown carbon fibers from deoiled asphalt, Carbon. 44 (9) (2006) 1661–1664.

[56] Y. Qi, W. Fan, G. Nan, Free-standing, binder-free polyacrylonitrile/asphalt derived porous carbon fiber-A high capacity anode material for sodium-ion batteries, Mater. Lett. 189 (2017) 206–209.

[57] M. Li, H. Xiao, T. Zhang, Q. Li, Y. Zhao, Activated carbon fiber derived from sisal with large specific surface area for high-performance supercapacitors, ACS Sustainable Chem. Eng. 7 (5) (2019) 4716–4723.

[58] S.K. Singh, H. Prakash, M.J. Akhtar, K.K. Kar, Lightweight and high-performance microwave absorbing heteroatom-doped carbon derived from chicken feather fibers, ACS Sustainable Chem. Eng. 6 (4) (2018) 5381–5393.

[59] Z. Wu, K. Tian, T. Huang, W. Hu, F. Xie, J. Wang, M. Su, L. Li, Hierarchically porous carbons derived from biomasses with excellent microwave absorption performance, ACS Appl. Mater. Interfaces. 10 (13) (2018) 11108–11115.

[60] X. Zhou, Z. Jia, A. Feng, X. Wang, J. Liu, M. Zhang, H. Cao, G. Wu, Synthesis of fish skin-derived 3D carbon foams with broadened bandwidth and excellent electromagnetic wave absorption performance, Carbon. 152 (2019) 827–836.

[61] S.C. Li, B.C. Hu, Y.W. Ding, H.W. Liang, C. Li, Z.Y. Yu, Z.Y. Wu, W.S. Chen, S.H. Yu, Wood-derived ultrathin carbon nanofiber aerogels, Angew Chem. Int. Ed. Engl. 57 (24) (2018) 7085–7090.

[62] J. Wang, L. Liu, S. Jiao, K. Ma, J. Lv, J. Yang, Hierarchical carbon Fiber@MXene@MoS$_2$ core-sheath synergistic microstructure for tunable and efficient microwave absorption, Adv. Funct. Mater. 30 (45) (2020) 2002595.

[63] Z. Wu, K. Pei, L. Xing, X. Yu, W. You, R. Che, Enhanced microwave absorption performance from magnetic coupling of magnetic nanoparticles suspended within hierarchically tubular composite, Adv. Funct. Mater. 29 (28) (2019) 1901448.

[64] X. Li, M. Zhang, W. You, K. Pei, Q. Zeng, Q. Han, Y. Li, H. Cao, X. Liu, R. Che, Magnetized MXene microspheres with multiscale magnetic coupling and enhanced polarized interfaces for distinct microwave absorption via a spray-drying method, ACS Appl. Mater. Interfaces. 12 (15) (2020) 18138–18147.

[65] P. Liu, Y. Huang, J. Yan, Y. Yang, Y. Zhao, Construction of CuS nanoflakes vertically aligned on magnetically decorated graphene and their enhanced microwave absorption properties, ACS Appl. Mater. Interfaces. 8 (8) (2016) 5536–5546.

[66] J. Cui, X. Wang, L. Huang, C. Zhang, Y. Yuan, Y. Li, Environmentally friendly bark-derived Co-Doped porous carbon composites for microwave absorption, Carbon. 187 (2022) 115–125.

[67] X. Huang, M. Qiao, X. Lu, Y. Li, Y. Ma, B. Kang, B. Quan, G. Ji, Evolution of dielectric loss-dominated electromagnetic patterns in magnetic absorbers for enhanced microwave absorption performances, Nano Res. 14 (2021) 4006–4013.

[68] Y. Cui, Z. Liu, Y. Zhang, P. Liu, M. Ahmad, Q. Zhang, B. Zhang, Wrinkled three-dimensional porous MXene/Ni composite microspheres for efficient broadband microwave absorption, Carbon. 181 (2021) 58–68.

[69] Y. Cui, K. Yang, J. Wang, T. Shah, Q. Zhang, B. Zhang, Preparation of pleated RGO/MXene/Fe$_3$O$_4$ microsphere and its absorption properties for electromagnetic wave, Carbon. 172 (2021) 1–14.

[70] C. Zhang, X. Li, Y. Shi, H. Wu, Y. Shen, C. Wang, W. Guo, K. Tian, H. Wang, structure engineering of graphene nanocages toward high-performance microwave absorption applications, Adv. Optical Mater. 10 (2) (2021) 2101904.

[71] B. Fan, N. Li, B. Dai, S. Shang, L. Guan, B. Zhao, X. Wang, Z. Bai, R. Zhang, Investigation of adjacent spacing dependent microwave absorption properties of lamellar structural Ti$_3$C$_2$T$_x$ MXenes, Adv. Powder Technol. 31 (2) (2020) 808–815.

[72] L. Guo, Q. An, Z.-Y. Xiao, S.-R. Zhai, W. Cai, H. Wang, Z. Li, Constructing stacked structure of s-doped carbon layer-encapsulated MoO$_2$ NPs with dominated dielectric loss for microwave absorption, ACS Sustainable Chem. Eng. 7 (24) (2019) 19546–19555.

[73] Y. Huo, K. Zhao, P. Miao, J. Kong, Z. Xu, K. Wang, F. Li, Y. Tang, Microwave absorption performance of SiC/ZrC/SiZrOC hybrid nanofibers with enhanced high-temperature oxidation resistance, ACS Sustainable Chem. Eng. 8 (28) (2020) 10490–10501.

[74] Y. Hou, Y. Yang, C. Deng, C. Li, C.F. Wang, Implications from broadband microwave absorption of metal-modified SiC fiber mats, ACS Appl. Mater. Interfaces. 12 (28) (2020) 31823–31829.

[75] M. Ning, Q. Man, G. Tan, Z. Lei, J. Li, R.W. Li, Ultrathin MoS2 nanosheets encapsulated in hollow carbon spheres: A case of a dielectric absorber with optimized impedance for efficient microwave absorption, ACS Appl. Mater. Interfaces. 12 (18) (2020) 20785–20796.

[76] L. Guo, Q.-D. An, Z.-Y. Xiao, S.-R. Zhai, L. Cui, Inherent N-doped honeycomb-like carbon/Fe$_3$O$_4$ composites with versatility for efficient microwave absorption and wastewater treatment, ACS Sustainable Chem. Eng. 7 (10) (2019) 9237–9248.

[77] Z. Li, X. Han, Y. Ma, D. Liu, Y. Wang, P. Xu, C. Li, Y. Du, MOFs-derived hollow Co/C microspheres with enhanced microwave absorption performance, ACS Sustainable Chem. Eng. 6 (7) (2018) 8904–8913.

[78] X. Wang, T. Zhu, S. Chang, Y. Lu, W. Mi, W. Wang, 3D nest-like architecture of core-shell CoFe$_2$O$_4$@1T/2H-MoS$_2$ composites with tunable microwave absorption performance, ACS Appl. Mater. Interfaces. 12 (9) (2020) 11252–11264.

[79] D. Estevez, F.X. Qin, L. Quan, Y. Luo, X.F. Zheng, H. Wang, H.X. Peng, Complementary design of nano-carbon/magnetic microwire hybrid fibers for tunable microwave absorption, Carbon. 132 (2018) 486–494.

[80] W. Feng, Y. Wang, J. Chen, L. Wang, L. Guo, J. Ouyang, D. Jia, Y. Zhou, Reduced graphene oxide decorated with in-situ growing ZnO nanocrystals: Facile synthesis and enhanced microwave absorption properties, Carbon. 108 (2016) 52–60.

[81] L. Li, G. Li, W. Ouyang, Y. Zhang, F. Zeng, C. Liu, Z. Lin, Bimetallic MOFs derived FeM(II)-alloy@C composites with high-performance electromagnetic wave absorption, Chem. Eng. J. 420 (2021) 127609.

[82] X. Li, L. Yu, W. Zhao, Y. Shi, L. Yu, Y. Dong, Y. Zhu, Y. Fu, X. Liu, F. Fu, Prism-shaped hollow carbon decorated with polyaniline for microwave absorption, Chem. Eng. J. 379 (2020) 122393.

[83] Y. Cui, F. Wu, J. Wang, Y. Wang, T. Shah, P. Liu, Q. Zhang, B. Zhang, Three dimensional porous MXene/CNTs microspheres: Preparation, characterization and microwave absorbing properties, Compos. Part A. 145 (2021) 106378.

[84] X. Zhang, J. Zhu, P. Yin, A. Guo, A. Huang, L. Guo, G. Wang, Tunable high-performance microwave absorption of $co_{1-x}s$ hollow spheres constructed by nanosheets within ultralow filler loading, Adv. Funct. Mater. 28 (49) (2018) 1800761.

[85] H. Chen, W. Ma, Z. Huang, Y. Zhang, Y. Huang, Y. Chen, Graphene-based materials toward microwave and terahertz absorbing stealth technologies, Adv. Opt. Mater. 7 (8) (2019) 1801318.

[86] X. Li, W. You, L. Wang, J. Liu, Z. Wu, K. Pei, Y. Li, R. Che, Self-assembly-magnetized mxene avoid dual-agglomeration with enhanced interfaces for strong microwave absorption through a tunable electromagnetic property, ACS Appl. Mater. Interfaces. 11 (47) (2019) 44536–44544.

[87] X. Di, Y. Wang, Y. Fu, X. Wu, P. Wang, Wheat flour-derived nanoporous carbon@$ZnFe_2O_4$ hierarchical composite as an outstanding microwave absorber, Carbon. 173 (2021) 174–184.

[88] L. Gai, G. Song, Y. Li, W. Niu, L. Qin, Q. An, Z. Xiao, S. Zhai, Versatile bimetal sulfides nanoparticles-embedded N-doped hierarchical carbonaceous aerogels (N-Ni_xS_y/Co_xS_y@C) for excellent supercapacitors and microwave absorption, Carbon. 179 (2021) 111–124.

[89] F. Chen, H. Luo, Y. Cheng, R. Guo, W. Yang, X. Wang, R. Gong, Nickel/Nickel phosphide composite embedded in N-doped carbon with tunable electromagnetic properties toward high-efficiency microwave absorption, Compos Part A 140 (2021) 106141.

[90] L. Liang, R. Yang, G. Han, Y. Feng, B. Zhao, R. Zhang, Y. Wang, C. Liu, Enhanced electromagnetic wave-absorbing performance of magnetic nanoparticles-anchored 2D Ti_3C_2Tx MXene, ACS Appl. Mater. Interfaces. 12 (2) (2020) 2644–2654.

[91] D. Ding, Y. Wang, X. Li, R. Qiang, P. Xu, W. Chu, X. Han, Y. Du, Rational design of core-shell Co@C microspheres for high-performance microwave absorption, Carbon. 111 (2017) 722–732.

[92] L. Huang, J. Li, Z. Wang, Y. Li, X. He, Y. Yuan, Microwave absorption enhancement of porous C@$CoFe_2O_4$ nanocomposites derived from eggshell membrane, Carbon. 143 (2019) 507–516.

[93] H. Lv, X. Liang, Y. Cheng, H. Zhang, D. Tang, B. Zhang, G. Ji, Y. Du, Coin-like alpha-Fe_2O_3@$CoFe_2O_4$ core-shell composites with excellent electromagnetic absorption performance, ACS Appl. Mater. Interfaces. 7 (8) (2015) 4744–4750.

[94] Z. Xu, Y. Du, D. Liu, Y. Wang, W. Ma, Y. Wang, P. Xu, X. Han, Pea-like Fe/Fe3C nanoparticles embedded in nitrogen-doped carbon nanotubes with tunable dielectric/magnetic loss and efficient electromagnetic absorption, ACS Appl. Mater. Interfaces. 11 (4) (2019) 4268–4277.

[95] N. Yang, Z.X. Luo, S.C. Chen, G. Wu, Y.Z. Wang, Fe_3O_4 nanoparticle/N-doped carbon hierarchically hollow microspheres for broadband and high-performance microwave absorption at an ultralow filler loading, ACS Appl. Mater. Interfaces. 12 (16) (2020) 18952–18963.

[96] B. Du, J. Qian, P. Hu, C. He, M. Cai, X. Wang, A. Shui, Fabrication of C-doped SiC nanocomposites with tailoring dielectric properties for the enhanced electromagnetic wave absorption, Carbon. 157 (2020) 788–795.

[97] K. Su, Y. Wang, K. Hu, X. Fang, J. Yao, Q. Li, J. Yang, Ultralight and high-strength SiCnw@SiC foam with highly efficient microwave absorption and heat insulation Properties, ACS Appl. Mater. Interfaces. 13 (18) (2021) 22017–22030.

[98] T. Wu, Y. Liu, X. Zeng, T. Cui, Y. Zhao, Y. Li, G. Tong, Facile hydrothermal synthesis of Fe_3O_4/C core-shell nanorings for efficient low-frequency microwave absorption, ACS Appl. Mater. Interfaces. 8 (11) (2016) 7370–7380.

[99] S. Dong, P. Hu, X. Li, C. Hong, X. Zhang, J. Han, $NiCo_2S_4$ nanosheets on 3D wood-derived carbon for microwave absorption, Chem. Eng. J. 398 (2020) 125588.

[100] S. Dong, X. Zhang, X. Li, J. Chen, P. Hu, J. Han, SiC whiskers-reduced graphene oxide composites decorated with MnO nanoparticles for tunable microwave absorption, Chem. Eng. J. 392 (2020) 123817.

[101] K. Manna, S.K. Srivastava, Fe_3O_4@Carbon@polyaniline trilaminar core–shell composites as superior microwave absorber in shielding of electromagnetic pollution, ACS Sustainable Chem. Eng. 5 (11) (2017) 10710–10721.

[102] H. Lv, X. Liang, G. Ji, H. Zhang, Y. Du, Porous Three-dimensional flower-like Co/CoO and its excellent electromagnetic absorption properties, ACS Appl. Mater. Interfaces. 7 (18) (2015) 9776–9783.

[103] X. Yuan, R. Wang, W. Huang, L. Kong, S. Guo, L. Cheng, Morphology design of co-electrospinning MnO-VN/C nanofibers for enhancing the microwave absorption performances, ACS Appl. Mater. Interfaces. 12 (11) (2020) 13208–13216.

[104] S. Gao, S.-H. Yang, H.-Y. Wang, G.-S. Wang, P.-G. Yin, Excellent electromagnetic wave absorbing properties of two-dimensional carbon-based nanocomposite supported by transition metal carbides Fe_3C, Carbon. 162 (2020) 438–444.

[105] C. Han, M. Zhang, W.-Q. Cao, M.-S. Cao, Electrospinning and in-situ hierarchical thermal treatment to tailor $C-NiCo_2O_4$ nanofibers for tunable microwave absorption, Carbon. 171 (2021) 953–962.

[106] Y. Sun, J. Zhang, Y. Zong, X. Deng, H. Zhao, J. Feng, M. He, X. Li, Y. Peng, X. Zheng, Crystalline-amorphous permalloy@iron oxide core-shell nanoparticles decorated on graphene as high-efficiency, lightweight, and hydrophobic microwave absorbents, ACS Appl. Mater. Interfaces. 11 (6) (2019) 6374–6383.

[107] W. Zhang, J. Zhang, P. Wu, G. Chai, R. Huang, F. Ma, F. Xu, H. Cheng, Y. Chen, X. Ni, L. Qiao, J. Duan, Parallel aligned nickel nanocone arrays for multiband microwave absorption, ACS Appl. Mater. Interfaces. 12 (20) (2020) 23340–23346.

[108] N. He, X. Yang, L. Shi, X. Yang, Y. Lu, G. Tong, W. Wu, Chemical conversion of Cu_2O/PPy core-shell nanowires (CSNWs): A surface/interface adjustment method for high-quality Cu/Fe/C and Cu/Fe_3O_4/C CSNWs with superior microwave absorption capabilities, Carbon. 166 (2020) 205–217.

[109] N. He, Z. He, L. Liu, Y. Lu, F. Wang, W. Wu, G. Tong, Ni^{2+} guided phase/structure evolution and ultra-wide bandwidth microwave absorption of Co Ni1- alloy hollow microspheres, Chem. Eng. J. 381 (2020) 122743.

[110] R. Wang, M. He, Y. Zhou, S. Nie, Y. Wang, W. Liu, Q. He, W. Wu, X. Bu, X. Yang, Self-assembled 3D flower-like composites of heterobimetallic phosphides and carbon for temperature-tailored electromagnetic wave absorption, ACS Appl. Mater. Interfaces. 11 (41) (2019) 38361–38371.

[111] J. Pan, H. Guo, M. Wang, H. Yang, H. Hu, P. Liu, H. Zhu, Shape anisotropic Fe_3O_4 nanotubes for efficient microwave absorption, Nano Res. 13 (3) (2020) 621–629.

[112] K. Hu, H. Wang, X. Zhang, H. Huang, T. Qiu, Y. Wang, C. Zhang, L. Pan, J. Yang, Ultralight $Ti_3C_2T_x$ MXene foam with superior microwave absorption performance, Chem. Eng. J. 408 (2021) 127283.

[113] J. Luo, K. Zhang, M. Cheng, M. Gu, X. Sun, MoS_2 spheres decorated on hollow porous ZnO microspheres with strong wideband microwave absorption, Chem. Eng. J. 380 (2020) 122625.

[114] F. Wang, X. Li, Z. Chen, W. Yu, K.P. Loh, B. Zhong, Y. Shi, Q.-H. Xu, Efficient low-frequency microwave absorption and solar evaporation properties of γ-Fe_2O_3 nanocubes/graphene composites, Chem. Eng. J. 405 (2021) 126676.

[115] Z. Fang, C. Li, J. Sun, H. Zhang, J. Zhang, The electromagnetic characteristics of carbon foams, Carbon. 45 (15) (2007) 2873–2879.

[116] C.P. Neo, V.K. Varadan, Optimization of carbon fiber composite for microwave absorber, IEEE T. Electromagn. C. 46 (1) (2004) 102–106.

[117] S.K. Singh, M.J. Akhtar, K.K. Kar, Hierarchical carbon nanotube-coated carbon fiber: Ultra lightweight, thin, and highly efficient microwave absorber, ACS Appl. Mater. Interfaces 10 (29) (2018) 24816–24828.

[118] H. Zhen, H. Wang, X. Xu, Preparation of porous carbon nanofibers with remarkable microwave absorption performance through electrospinning, Mater. Lett. 249 (2019) 210–213.

[119] G. Li, T. Xie, S. Yang, J. Jin, J. Jiang, Microwave absorption enhancement of porous carbon fibers compared with carbon nanofibers, J. Phys. Chem. C 116 (16) (2012) 9196–9201.

[120] Z. Chu, H. Cheng, W. Xie, L. Sun, Effects of diameter and hollow structure on the microwave absorption properties of short carbon fibers, Ceram. Int. 38 (6) (2012) 4867–4873.

[121] J. Wang, F. Wu, Y. Cui, A. Zhang, Q. Zhang, B. Zhang, Efficient synthesis of N-doped porous carbon nanoribbon composites with selective microwave absorption performance in common wavebands, Carbon. 175 (2021) 164–175.

[122] X. Liu, X. Cui, Y. Chen, X.-J. Zhang, R. Yu, G.-S. Wang, H. Ma, Modulation of electromagnetic wave absorption by carbon shell thickness in carbon encapsulated magnetite nanospindles-poly(vinylidene fluoride) composites, Carbon. 95 (2015) 870–878.

[123] H. Liu, Y. Li, M. Yuan, G. Sun, H. Li, S. Ma, Q. Liao, Y. Zhang, Situ preparation of cobalt nanoparticles decorated in N-doped carbon nanofibers as excellent electromagnetic wave absorbers, ACS Appl. Mater. Interfaces. 10 (26) (2018) 22591–22601.

[124] Y. Wang, Y. Sun, Y. Zong, T. Zhu, L. Zhang, X. Li, H. Xing, X. Zheng, Carbon nanofibers supported by FeCo nanocrystals as difunctional magnetic/dielectric composites with broadband microwave absorption performance, J. Alloy. Compd. 824 (2020) 153980.

[125] Z. Wang, L. Wu, J. Zhou, W. Cai, B. Shen, Z. Jiang, Magnetite nanocrystals on multiwalled carbon nanotubes as a synergistic microwave absorber, J. Phys. Chem. C 117 (10) (2013) 5446–5452.

[126] M.S. Cao, J. Yang, W.L. Song, D.Q. Zhang, B. Wen, H.B. Jin, Z.L. Hou, J. Yuan, Ferroferric oxide/multiwalled carbon nanotube vs polyaniline/ferroferric oxide/multiwalled carbon nanotube multiheterostructures for highly effective microwave absorption, ACS Appl. Mater. Interfaces 4 (12) (2012) 6949–6956.

[127] Y. Liu, Z. Chen, W. Xie, S. Song, Y. Zhang, L. Dong, In-Situ growth and graphitization synthesis of porous Fe_3O_4/carbon fiber composites derived from biomass as lightweight microwave absorber, ACS Sustainable Chem. Eng. 7 (5) (2019) 5318–5328.

[128] H. Luo, G. Xiong, X. Chen, Q. Li, C. Ma, D. Li, X. Wu, Y. Wan, ZnO nanostructures grown on carbon fibers: Morphology control and microwave absorption properties, J. Alloy. Compd. 593 (2014) 7–15.

[129] W. Zhang, X. Zhang, Q. Zhu, Y. Zheng, L.F. Liotta, H. Wu, High-efficiency and wide-bandwidth microwave absorbers based on MoS2-coated carbon fiber, J. Colloid Interface Sci. 586 (2021) 457–468.

[130] Y.-F. Zhu, Q.-Q. Ni, Y.-Q. Fu, One-dimensional barium titanate coated multi-walled carbon nanotube heterostructures: Synthesis and electromagnetic absorption properties, RSC Adv. 5 (5) (2015) 3748–3756.

[131] X. Tian, F. Meng, F. Meng, X. Chen, Y. Guo, Y. Wang, W. Zhu, Z. Zhou, Synergistic enhancement of microwave absorption using hybridized polyaniline@helical CNTs with dual chirality, ACS Appl. Mater. Interfaces. 9 (18) (2017) 15711–15718.

[132] Q. Gao, Y. Wang, D. He, L. Gao, Y. Zhou, M. Fu, Infrared and microwave properties of polypyrrole/multi-walled carbon nanotube composites, J. Lumin. 152 (2014) 117–120.

[133] J. Wang, Y. Huyan, Z. Yang, A. Zhang, Q. Zhang, B. Zhang, Tubular carbon nanofibers: Synthesis, characterization and applications in microwave absorption, Carbon. 152 (2019) 255–266.

[134] J. Wang, F. Wu, Y. Cui, J. Chen, A. Zhang, Q. Zhang, B. Zhang, Facile synthesis of tubular magnetic carbon nanofibers by hypercrosslinked polymer design for microwave adsorption, J. Am. Ceram. Soc. 103 (10) (2020) 5706–5720.

[135] F. Wu, K. Yang, Q. Li, T. Shah, M. Ahmad, Q. Zhang, B. Zhang, Biomass-derived 3D magnetic porous carbon fibers with a helical/chiral structure toward superior microwave absorption, Carbon. 173 (2021) 918–931.

[136] F. Wu, Z. Liu, T. Xiu, B. Zhu, I. Khan, P. Liu, Q. Zhang, B. Zhang, Fabrication of ultralight helical porous carbon fibers with CNTs-confined Ni nanoparticles for enhanced microwave absorption, Compos. Part B. 215 (2021) 108814.

CHAPTER 11

Multiple composite tubular carbon nanofibers: Synthesis, characterization, and applications in microwave absorption

Fei Wu[a], Jiqi Wang[a] and Baoliang Zhang[a,b]
[a]School of Chemistry and Chemical Engineering, Northwestern Polytechnical University, Xi'an, China
[b]Xi'an Key Laboratory of Functional Organic Porous Materials, Northwestern Polytechnical University, Xi'an, China

11.1 Introduction

With the realization of the 5G application, the emerging electronic equipment not only greatly facilitates human life, but also brings serious electromagnetic pollution [1]. To ensure information security and people's health, the development of new microwave absorption materials is imminent. According to the loss mechanisms of electromagnetic waves, absorbers can be divided into dielectric loss materials and magnetic loss materials [2]. Common dielectric loss materials mainly include carbon materials: graphene [3], carbon nanotubes [4], carbon fibers [5], and so on [6]. Generally, they have the advantage of low density. Magnetic loss materials (Fe, Co, Ni, their alloys, and oxides) need a high amount of filler to achieve the ideal microwave absorption performance [7–11]. However, the traditional absorbers always exhibit limited reflection loss (RL) and narrow effective absorption bandwidth (EAB, RL<-10 dB). In practical applications, materials with different loss mechanisms are integrated to effectively improve microwave absorption abilities. The novel multishell composite absorbers possess unique configurations and excellent properties [2,12]. It is a kind of material that can solve the above problems and has great development prospects.

In general, the design and fabrication of multishell composite absorbers mainly focus on two aspects: the selection of appropriate components and

Fabrication and Functionalization of Advanced Tubular Nanofibers and their Applications.
DOI: https://doi.org/10.1016/B978-0-323-99039-4.00007-3
299

the construction of specific core-shell structures [13]. According to the previous works [4–6], Benefiting from anisotropy and high aspect ratio, one-dimensional (1D) microwave absorption materials can easily form an axial carrier transmission path under electromagnetic fields. It is conducive to conductive loss to dissipate the energy of electromagnetic waves. In the past few decades, various 1D materials, such as nanofibers [14], nanowires [15], nanotubes [16], and nanoribbons [17], have been widely applied in the field of microwave absorption due to their unique physicochemical properties. However, these absorbers usually possess a mismatching impedance, exhibiting narrow absorption bandwidth. The common strategy is to integrate it with other materials, such as magnetic particles (Fe, Co, Ni, and their oxides) and functional shells (C, SiO_2, and polypyrrole (PPY)) [18–20]. They are assembled into 1D multishell composite absorbers to realize the complementarity of loss mechanisms and the adjustment of impedance matching, so as to realize synergistic effects between various materials and complement advantages.

In brief, according to the impedance matching and loss mechanisms of absorbers, it is difficult to improve microwave absorption abilities only by improving dielectric loss or magnetic loss performance. Fortunately, combining various types of microwave absorption materials can achieve desirable comprehensive effects. Besides, the introduction of various shell materials can improve the thermal stability and corrosion resistance of materials. The selection of core and core-shell layers with various functions can effectively optimize microwave absorption performance.

11.2 1D multishell composite materials

1D multishell composite materials refer to composite materials with an ordered structure formed by chemical bonds or other interactions with specific 1D materials as the core. Generally, it is composed of a core, middle layer, shell layer, and possible void layer. It possesses the characteristics of distinct layers, large specific surface area, and complementary functions, showing great application potential in microwave absorption, catalysis, batteries, biomedicine, and other fields. Since the absorption bands of different material components can be complementary, the 1D multishell composite material can broaden the total absorption band, making up for the narrow absorption bandwidth of traditional absorbers. Thanks to the superiorities of the multishell structure, different materials can realize the integration of functions, resulting in synergistic effects. Therefore, 1D multishell composite absorbers have become the focus of research in the field of microwave absorption.

11.2.1 Fiber-based multishell absorbers

For 1D multishell composite absorbers, the most common core is carbon fiber [21–25]. Carbon fibers can be formed by the transformation of various types of polymer fibers and spinning fibers after carbonization at high temperature and can display solid or tubular structures. However, the loss mechanism of carbon fibers is relatively single, limiting microwave absorption abilities. Fabricating composite materials is an effective way to improve the microwave absorption performance, which can reduce the coating thickness, increase the RL and broaden the EAB.

(1) Carbon paper core

Wang et al. [21] designed a ternary core-shell composite with commercial carbon paper fibers as the core, MXene as the intermediate dielectric layer, and MoS_2 as the modified layer (CF@MXene@MoS_2) to achieve high-performance microwave absorption (Fig. 11.1). Among them, carbon fibers with excellent mechanical properties and an easy-to-modify surface could firmly support MXene to form 1D heterostructures. On the other hand, the outermost MoS_2 could improve impedance matching and protect MXene from oxidation. The results indicated that the EAB between X-band and Ku band was realized by adjusting the load of MoS_2, and CF@MXene@MoS_2 exhibited desirable microwave absorption performance.

(2) Carbon cloth core

Similarly, Han et al. [22] synthesized a novel CF@MXene@ZnO multishell structure by self-assembly and hydrothermal process, exhibiting tunable and efficient microwave absorption performance (Fig. 11.2). Due to the synergistic effect, the introduction of the interlayer MXene and the outer layer ZnO enhanced the conductance loss and the impedance matching of the absorber. The minimum RL (RL_{min}) value was −67.35 dB at 9.0 GHz with a matching thickness of 3.5 mm. The optimal EAB was 5.44 GHz, covering the entire X-band, corresponding to a matching thickness of 4.0 mm. In addition, due to the synergistic effect of the high photothermal conversion efficiency of MXene, the high photocatalytic activity of ZnO, and the high thermal conductivity of CF, it could also be used as an energy converter and molecular heater.

(3) Polymer fiber core

Wang et al. [23] prepared magnetic hybrid nanofibers (MHNFs: $Fe/Fe_3O_4@TCNFs@TiO_2$) with a three-layer core-shell structure by a three-step method (Fig. 11.3). The electromagnetic parameters of MHNFs were tuned and optimized by adjusting the TiO_2 loading, so that they possessed both magnetic loss and dielectric loss capabilities, and exhibited excellent microwave absorption performance. The results showed that the TiO_2 layer could optimize impedance matching characteristics and improved the microwave absorption performance. At a filler loading of 15 wt%, the RL_{min} reached -44.8 dB@13.9 GHz, and the thickness was only 1.6 mm.

11.2.2 Magnetic nanochain-based multishell absorbers

Magnetic absorbers such as Fe_3O_4, CoNi, $ZnCo_2O_4$, and $ZnFe_2O_4$ have the advantages of high magnetic permeability, non-toxicity, and low cost [26–30]. However, due to the limitations of high density and narrow EAB, traditional magnetic particles are difficult to meet the requirements of high-performance microwave absorption materials. Since their remarkable anisotropy and strong dielectric attenuation, 1D magnetic composites have received extensive attention in the field of microwave absorption. Meanwhile, the development of multishell composite structures combining magnetic and dielectric loss is considered an effective approach to improve microwave absorption properties.

(1) Magnetic Fe₃O₄ nanochain core

Qiao et al. [26] developed two new electromagnetic nanocomposites: core-shell $Fe_3O_4@void@SiO_2$ nanochains and $Fe_3O_4@void@SiO_2@PPy$ nanochains (Fig. 11.4). Studies have shown that $Fe_3O_4@void@SiO_2@PPy$ nanochains exhibited stronger absorption capacity and wider EAB than $Fe_3O_4@void@SiO_2$ nanochains. This was due to the introduction of the PPY shell, enhancing conductive loss, dipole polarization, interfacial polarization, and multiple reflections. Its RL_{min} was -54.2 dB@17.70 GHz, and the maximum EAB could reach 5.90 GHz (11.49–17.39 GHz). This study proved that the core-shell structure was beneficial to improve the microwave absorption performance of the chain-like magnetic-dielectric composite, laying a good foundation for the development of the same type of microwave absorption materials.

(2) Magnetic ZnFe₂O₄ nanochain core

Ma et al. [27] fabricated 1D $ZnFe_2O_4@SiO_2$ (ZS) nanochains by the Stöber method with the assistance of an external magnetic field.

Subsequently, 1D $ZnFe_2O_4@SiO_2@C$ (ZSC) nanochains were obtained using dopamine polymerization and pyrolysis techniques (Fig. 11.5). Finally, 1D flower-like $ZnFe_2O_4@SiO_2@C@NiCo_2O_4$ (ZSCNC) nanochains were prepared by a hydrothermal method. The results indicated that the nitrogen-doped carbon layer endowed the ZSC nanochains with strong dielectric loss, resulting in an ultrawide EAB of 6.22 GHz. The 1D ZSCNC nanochains exhibited superior microwave absorption properties due to their larger specific surface area and abundant heterogeneous multi-interfaces. At a filler loading of 30 wt%, the ZSCNC nanochains displayed an RL_{min} value of −54.29 dB@11.14 GHz and an EAB of 5.66 GHz (11.94–17.60 GHz), corresponding to a matching thickness of 2.39 mm.

(3) Magnetic Ni nanochain core

Bi et al. [28] synthesized Ni@Co/C@PPy composite nanochains with integrated composition and microstructural advantages by a modified Stöber method and a simple hydrothermal method (Fig. 11.6). The results implied that the addition of Co/C and PPy could effectively optimize the impedance matching and enhance the electromagnetic attenuation capability. Under the combined effects of multiple reflection and scattering, conductive loss, and interface polarization, the Ni@Co/C@PPy composite nanochains exhibited excellent microwave absorption. Its RL_{min} value could reach − 48.76 dB, and the EAB was 5.10 GHz, corresponding to a matching thickness of 2.0 mm.

11.2.3 Inorganic nanowire-based multishell absorbers

1D inorganic dielectric nanowire absorbers, such as Cu nanowires, SiC nanowires, MnO_2 nanowires, and TiO_2 nanowires, have high aspect ratios similar to carbon fibers, are easy to generate conductive networks, and possess outstanding dielectric loss capability [31–35]. At the same time, they possess excellent physicochemical stability and have attracted much attention in the field of microwave absorption. To further optimize the microwave absorption properties of inorganic nanowires, the novel multishell inorganic nanowire absorbers came into being.

(1) Cu nanowire core

To improve the microwave absorption properties and oxidation resistance of Cu nanowires, He et al. [31] synthesized Cu@Fe@C core-shell nanowires with solid structures and $Cu@Fe_3O_4@C$ core-shell nanowires (CSNWs) with hollow structures by the in-situ carbothermic reduction-CVD process (Fig. 11.7). The research showed that Cu@Fe@C CSNWs possessed good impedance matching, complementary dielectric and magnetic loss capabilities, and exhibited excellent microwave absorption performance. At 13.07 GHz, its RL_{min} was −43.09 dB for a matching thickness of 1.8 mm. And the RL values below −20 dB could be obtained in the 3.3–4.5 GHz and 5.11–15.9 GHz ranges.

(2) TiO$_2$ nanowire core

Ding et al. [32] prepared a TiO_2@Fe_3O_4@PPy composite absorber with a multilayer core-shell structure by a solvothermal method and polymerization strategies (Fig. 11.8). The complex permittivity of the ternary composites could be optimized by adjusting the shell thickness of PPy to improve the loss capability. The results showed that the heterointerfaces constructed by PPy@Fe_3O_4 and Fe_3O_4@TiO_2 contribute to the enhancement of polarization loss. Furthermore, magnetic Fe_3O_4 was conducive to magnetic dielectric synergy. The RL_{min} of TiO_2@Fe_3O_4@PPy could reach −61.8 dB for a thickness of 3.2 mm, and the EAB could cover 6.0 GHz.

(3) SiC nanowire core

To improve the microwave absorption performance of SiC nanowires, Xu et al. [33] designed double interfaces on the surface of SiC nanowires to form SiC@C@PPy heterostructures (Fig. 11.9). The inner carbon layer and outer PPy layer were synthesized by gas etching and oxidative polymerization. The results showed that the SiC@C@PPy composite absorber exhibited ideal electromagnetic attenuation and impedance matching properties, benefiting from the polarization enhancement of the double interfaces and the strong conductive network in the material. For a thickness of 2.78 mm, the optimal EAB of SiC@C@PPy could reach 8.4 GHz, corresponding to a filler loading of 15 wt%. Furthermore, by modulating the thickness of the absorber, the EAB could cover the entire X and Ku bands.

11.3 Magnetic tubular fiber with multilayer heterostructure

We synthesized hypercrosslinked tubular polymer fibers through the dual-oil phase confined self-polycondensation system developed earlier first, and prepared precursors with multilayer heterostructures by pyrolysis of iron acetylacetonate and self-polymerization of dopamine, respectively [36]. Subsequently, the precursors were carbonized at high temperatures to obtain tubular carbon nanofibers with a multishell structure. The preparation process was shown in Fig. 11.10. 1D nanofibers could be stacked to form

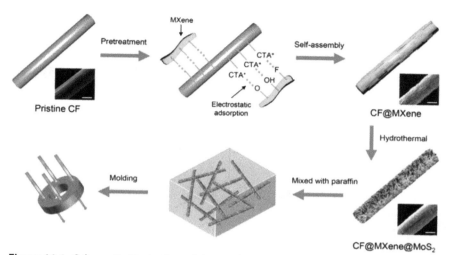

Figure 11.1 Schematic illustration of the synthesis process of CF@MXene@MoS$_2$. [Reproduced with permission from ref. 21, Wiley, 5277410907317].

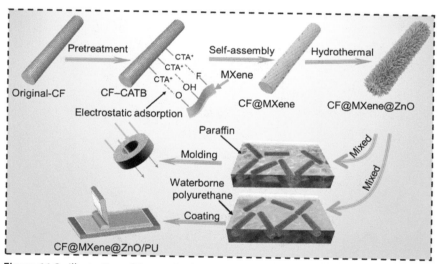

Figure 11.2 Illustration of the synthetic process of CF@MXene@ZnO-based composites. [Reproduced with permission from ref. 22, Elsevier, 5278091254548].

an efficient three-dimensional (3D) conductive network, resulting in a conductive loss. Magnetic nanoparticles endowed materials with a certain magnetic loss capability, enriching the loss mechanisms. The nitrogen-doped carbon layer improved the impedance matching and enriched the heterointerfaces. The existence of hierarchical pores could not only reflect and scatter electromagnetic waves but also bring a large specific surface area.

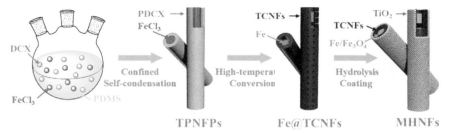

Figure 11.3 Schematic diagram of the preparation process of MHNFs. [Reproduced with permission from ref. 23, Elsevier, 5278101028785].

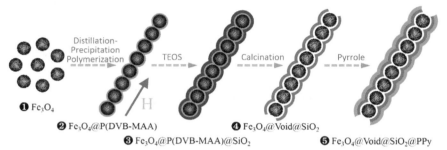

Figure 11.4 Schematic diagram of the preparation of Fe_3O_4@void@SiO_2@PPy nano-chains. [Reproduced with permission from ref. 26, Springer Nature, 5278130056274].

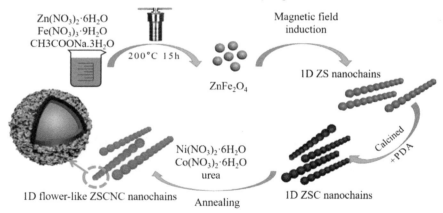

Figure 11.5 Preparation diagram of ZSCNC nanochains. [Reproduced with permission from ref. 27, Elsevier, 5278130856525].

The above properties provided the basis for the multishell tubular carbon fibers to exhibit excellent microwave absorption properties.

The TEM and EDS spectra of the precursor PF@Fe_3O_4@PDA were shown in Fig. 11.11A–C. The significantly increased diameter and the presence of N and O elements indicated that PDA was successfully coated

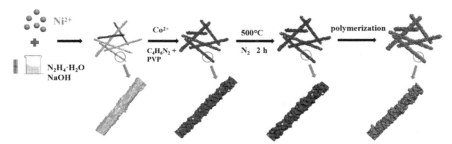

Figure 11.6 Schematic diagram of the preparation of Ni@Co/C@PPy nanochains. [Reproduced with permission from ref. 28, Elsevier, 5278140321987].

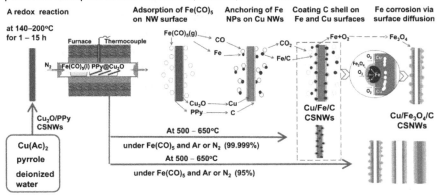

Figure 11.7 Schematic diagram of the preparation of CSNWs. [Reproduced with permission from ref. 31, Elsevier, 5278221118214].

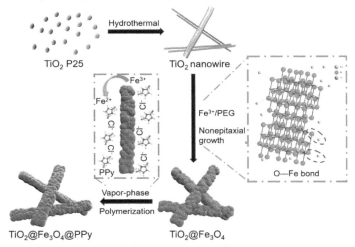

Figure 11.8 Schematic diagram of the preparation of TiO_2@Fe_3O_4@PPy nanowires. [Reproduced with permission from ref. 32, Wiley, 5278210930192].

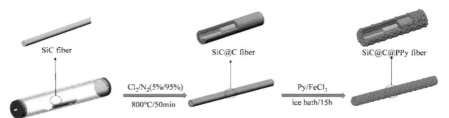

Figure 11.9 Schematic diagram of the preparation of SiC@C@PPy. [Reproduced with permission from ref. 33, ACS Publications].

on the surface of PF@Fe$_3$O$_4$. The surface roughness of PF@Fe$_3$O$_4$@PDA increased significantly, and the independent nanoparticles disappeared (Fig. 11.11D). This was because the exposed Fe$_3$O$_4$ nanoparticles on the surface of PF@Fe$_3$O$_4$ had been coated by the PDA layer. The average diameter of PF@Fe$_3$O$_4$@PDA was about 280 nm, and the calculated PDA layer thickness was 70 nm. After carbonization, the three-layer tubular morphology of the product could be observed (Fig. 11.11E). Inorganic nanoparticles were distributed inside the tube and between the two layers with low mass thickness contrast. The results suggested that the inner layer was the carbon layer left by the thermal decomposition of the hypercrosslinked polymer, and the outer layer was the nitrogen-doped carbon layer generated by the pyrolysis of PDA. This could be corroborated by the EDS mapping of C and N elements (Fig. 11.11F and G). TCF@Fe$_3$O$_4$@NCLs maintained defined fiber morphology (Fig. 11.11H). The average diameter had no significant change compared with PF@Fe$_3$O$_4$@PDA. The thickness of the nitrogen-doped carbon layer was about 65 nm. The diameter of Fe$_3$O$_4$ transformed by FeCl$_3$ in situ was about 20 nm, and the diameter of Fe$_3$O$_4$ introduced by the solvothermal method was about 7 nm.

According to the transmission line theory, microwave absorption performance of the material at 2–18GHz was characterized. To investigate the contribution of each component in the material to microwave absorption properties, the RL of TCF, TCF@Fe$_3$O$_4$, and TCF@Fe$_3$O$_4$@NCLs were measured. When the filler loading was 10%, the 3D and 2D RL values of absorbers under the corresponding thickness were shown in Fig. 11.12. Fig. 11.12A and B illustrated the RL$_{min}$ and EAB of TCF were −13.9 dB (17.3 GHz, 1.9 mm) and 3.9 GHz (14.1–18 GHz). Due to the introduction of Fe$_3$O$_4$ nanoparticles, the impedance matching of TCF@Fe$_3$O$_4$ was optimized, and its microwave absorption performance in the low-frequency band was significantly improved (Fig. 12C and D). Its RL$_{min}$=-52.8 dB (7.2 GHz, 4.6 mm) with an EAB of 4.4 GHz (5.8–10.2 GHz). However,

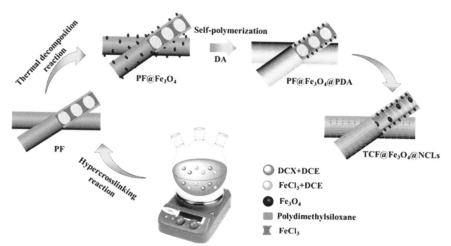

Figure 11.10 Schematic diagram of the preparation process of multi-shell carbon nanofibers. [Reproduced with permission from ref. 36, Elsevier, 5278231463216].

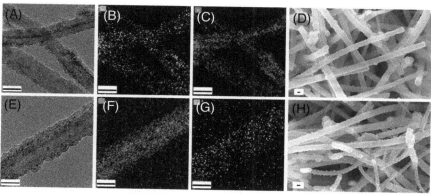

Figure 11.11 TEM image (A), HR-EDS image of N (B) and O (C) elements, SEM image (D) of PF@Fe$_3$O$_4$@PDA; TEM image (E), HR-EDS image of C (F) and N (G) elements, SEM image (H) of TCF@Fe$_3$O$_4$@NCLs. The ruler was 100 nm. [Reproduced with permission from ref. 36, Elsevier, 5278231463216].

the thickness of the material used had also increased. The RL$_{min}$ and EAB of TCF@Fe$_3$O$_4$@NCLs was −43.6 dB (9.9 GHz, 3.3 mm) and 4.6 GHz (8.2–12.8 GHz). As illustrated in Fig. 11.12E and F, the sample exhibited excellent microwave absorption ability at various thicknesses and frequency ranges. In addition, the matching thickness of the sample was effectively reduced. After the nitrogen–doped carbon layer was coated, the introduced defects and polarization interfaces were increased, and the overall impedance matching of the material was greatly improved.

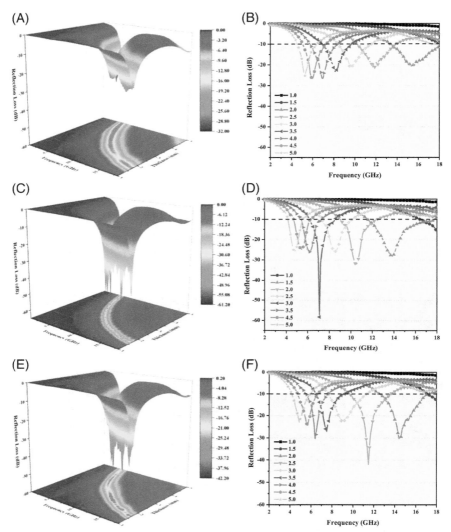

Figure 11.12 Three-dimensional and two-dimensional reflection loss patterns of TCF (A and B), TCF@Fe₃O₄ (C and D), and TCF@Fe₃O₄@NCLs (E and F) with different thicknesses within 2–18 GHz. [Reproduced with permission from ref. 36, Elsevier, 5278231463216].

Fig. 11.13 summarized all potential microwave absorption mechanisms of TCF@Fe₃O₄@NCLs: (1) The multishell structure optimized the impedance matching of the material; (2) The multiple interfaces equipped absorbers with more polarization centers, inducing multiple polarization effects; (3) The 3D conductive network formed by the overlapping of 1D

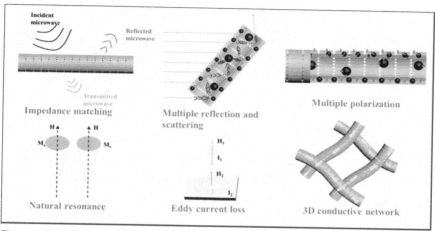

Figure 11.13 Schematic diagram of the absorption mechanisms of TCF@Fe$_3$O$_4$@NCLs. [Reproduced with permission from ref. 36, Elsevier, 5278231463216].

materials could improve the conductive loss of absorbers; (4) The introduction of magnetic Fe$_3$O$_4$ particles endowed the material with magnetic loss capabilities, including natural resonance and eddy current loss; (5) The tubular structure, porous structure, and Fe$_3$O$_4$ nanoparticles could cause multiple scatterings and reflections of incident electromagnetic waves, which was beneficial to the attenuation and dissipation of electromagnetic waves.

11.4 Intertwined one-dimensional heterostructure obtained by MXene and MOF

Transition metal carbides or nitrides (MXenes) have attracted much attention due to their special structures and surface chemistry. However, like other 2D materials, its application properties are severely affected by stacking and agglomeration. Therefore, as shown in Fig. 11.14, we fabricated 1D MXene fibers using the hydroxylation self-assembly technique and induced the 1D directional growth of bimetallic MOFs [37]. Through a simple catalytic deposition process, the construction of double 1D heterostructures was realized in one step. Without the assistance of any template and rigid framework, it could form a nanoscale 3D porous structure. The study indicated that the growth of CNTs could be effectively controlled by the loading of MOFs on 1D MXene fibers, thereby realizing the regulation of electromagnetic parameters. The 3D cross-linked network constructed by the intertwined double 1D hetero components (MXene fibers/CoNi/C and CNTs/CoNi) exhibited a large heterointerface, hierarchical pore structure,

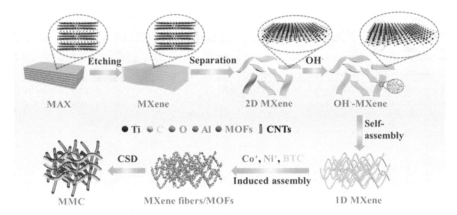

Figure 11.14 Schematic diagram of fabrication strategy. [Reproduced with permission from ref. 37, Elsevier, 5278241150074].

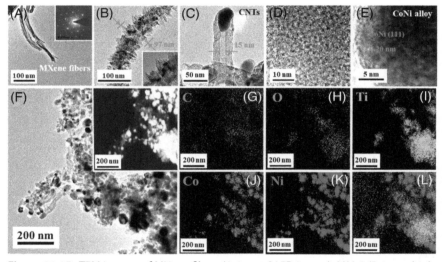

Figure 11.15 TEM images of MXene fibers (A, inset: SAED image), MM-3 (B, inset: high-resolution image), and CNTs (C); HR-TEM images of CNTs (D and E); TEM image of MMC-3 (F, inset: HAADF-STEM photo), and EDS-mappings of C (G), O (H), Ti (I), Co (J), Ni (K) as well as overlaid element graph (L). [Reproduced with permission from ref. 37, Elsevier, 5278241150074].

and good electrical conductivity, showing superior microwave absorption performance.

We used TEM to characterize the internal fine structure and composition of the samples. As shown in Fig. 11.15A, the 1D MXene was a solid fibrous structure with a diameter of less than 20 nm. Selected area diffraction

(SAED) of the MXene fiber portion exhibited a typical MXene single crystal diffraction pattern (inset). After MXene fibers were composited with CoNi-bimetal MOFs, the diameter of the samples increased significantly (marked in Fig. 11.15B). A lamellar MOF array was found outside the MXene fibers (inset). In addition, the tip growth of CNTs was successfully triggered by catalytic self-deposition technology, and inorganic nanoparticles were present at the tips with higher mass-thickness contrast (Fig. 11.15C). The high-resolution photos of the tube wall and tip were shown in Fig. 11.15D and E, where two sets of clear lattice fringes appeared, with spacings of 0.34 nm and 0.20 nm, corresponding to the (002) facet and (111) facet of graphitized carbon and the CoNi alloy, respectively. The TEM image of MMC was shown in Fig. 15F, the existence of double 1D heterostructures could be observed. Further EDS analysis of this region (Fig. 11.15G–K) indicated that carbon elements were uniformly distributed in the sample, and Ti and oxygen elements were concentrated in stacked MXene fibers. At the same time, there were overlapping Co and Ni elements around the Ti element, corresponding to the tip portion of the CoNi nanoparticles formed by the transformed MOF and the tip-grown CNTs. As illustrated in Fig. 11.15L, the coexistence of C, O, Ti, Co, and Ni proved that there were abundant heterointerfaces in the double 1D heterostructure, which was propitious to the generation of the multi-polarization effects of the absorber.

After calculation, the 3D RL curves (Fig. 11.16A–C) and 2D RL curves (Fig. 11.16D–F) of MMC-1, MMC-2, and MMC-3 were obtained. MMC-2 and MMC-3 displayed better microwave absorption properties. Specifically, it could be intuitively seen that at the low frequency of 5.9 GHz, the RL_{min} of MMC-1 exceeded −33.8 dB, and the EAB reached 3.3 GHz (4.6–6.9 GHz). MMC-2 exhibited the RL_{min} (-53.7 dB) at 8.6 GHz with an EAB of 3.0 GHz (7.3–10.3 GHz). The RL_{min} of MMC-3 at high frequency 15.1 GHz was as high as −51.6 dB, and EAB included 4.5 GHz (13.2–17.7 GHz). Compared with MMC-1 and MMC-2 (Fig. 11.16G and H), MMC-3 had absolute advantages in terms of matching thickness and EAB while ensuring excellent RL. In addition, the three groups of MMC materials could achieve excellent microwave absorption performance at low frequency, medium frequency, and high frequency, respectively, and displayed desirable microwave absorption ability in three frequency bands (C, X, and Ku) (Fig. 11.16I).

Finally, as shown in Fig. 11.17, the outstanding microwave absorption performance of MMC could be attributed to the following points: (1) The

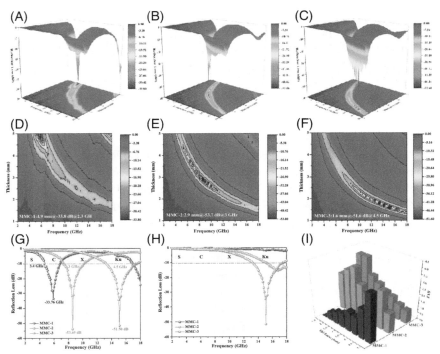

Figure 11.16 3D plots of RL (A–C), 2D curves of RL (D–F), comparison of the optimized RL (G), the RL curves for the thickness of 1.6 mm (H), and EAB for various MMC (I). [Reproduced with permission from ref. 37, Elsevier, 5278241150074].

synergistic effect of double 1D heterostructures (multishell structures) endowed MMC with excellent impedance matching and attenuation characteristics; (2) Tertiary porous structures could induce multiple reflections and scatterings of incident electromagnetic waves and adjust impedance matching; (3) Multishell structures with a large number of heterointerfaces, such as CoNi alloy/CNTs, MXene fiber/CoNi Alloys, MXene fibers/CNTs, etc. provided strong interfacial polarization; (4) Pores, defects and CoNi nanoparticles introduced by carbonization could induce dipolar polarization; (5) Magnetic CoNi nanoparticles could cause natural resonances and eddy currents (6) Double 1D heterostructures were intertwined to form a 3D conductive network, providing a strong conductive loss.

11.5 Core-shell MnO_2@NC@MoS_2 nanowires

MnO_2 nanowires are a typical class of dielectric materials, but when they are used as absorbers, they usually exhibit narrow EAB related to their

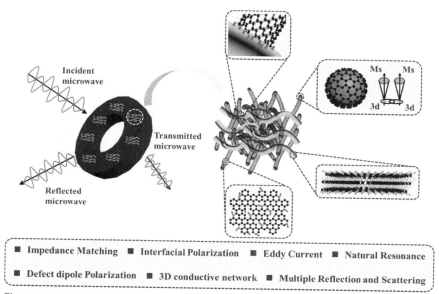

Figure 11.17 Schematic illustration of the microwave absorption mechanism for MMC. [Reproduced with permission from ref. 37, Elsevier, 5278241150074].

weak impedance matching. A common strategy is to dope it with other materials to optimize performance. MoS_2 nanosheets, as layered transition metal dichalcogenide (TMD) nanostructures, have been widely used in microwave absorption due to their excellent physical and chemical properties as well as tunable sheet structures. In this work, as shown in Fig. 11.18, we used MnO_2 nanowires as a template to construct a 3D hierarchical structure of $MnO_2@NC@MoS_2$ absorber through the self-polymerization of dopamine and the surface growth of MoS_2 nanosheets, assisted by vacuum carbonization [38]. At the same time, the MnO_2 nanowire template was etched with an acid solution, and the $NC@MoS_2$ nanotube absorber was obtained. Through the layer-by-layer structure design, microwave absorption performance of the material was effectively improved and the EAB was broadened.

The SEM images of $MnO_2@NC@MoS_2$ and $NC@MoS_2$ nanotubes were shown in Fig. 11.19A–C, $MnO_2@PDA@MoS_2$ was successfully prepared by annealing treatment at 700°C. It could be seen that the carbonization process could not change the morphology of the material, indicating that the synthesized $MnO_2@PDA@MoS_2$ possessed good thermal stability. The MnO_2 template in the $MnO_2@NC@MoS_2$ precursor was etched and removed with an oxalic acid solution to obtain $NC@MoS_2$ nanotubes.

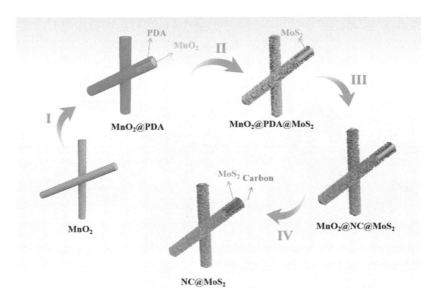

Figure 11.18 Schematic illustration of NC@MoS$_2$ nanotubes preparation. [Reproduced with permission from ref. 38, Elsevier, 5278680447996].

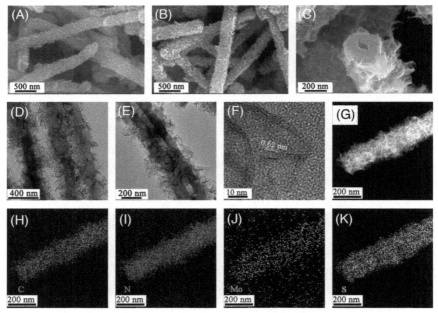

Figure 11.19 SEM images of MnO$_2$@NC@MoS$_2$ (A) and NC@MoS$_2$ nanotubes (B and C). TEM images (D–E) of NC@MoS$_2$ nanotubes and HR-TEM images of MoS$_2$ nanosheets (F). HAADF image (G) and corresponding element mappings (H–K) of NC@MoS$_2$ nanotubes. [Reproduced with permission from ref. 38, Elsevier, 5278680447996].

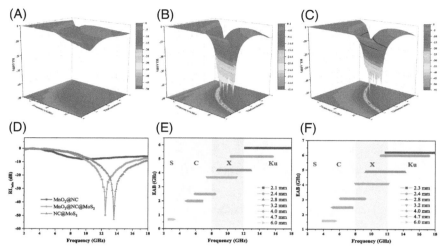

Figure 11.20 The 3D (A–C) and 2D (D) reflection loss patterns of MnO_2@NC, MnO_2@NC@MoS_2, and NC@MoS_2. The effective absorption band at a different thicknesses of MnO_2@NC@MoS_2 (E) and NC@MoS_2 (F). [Reproduced with permission from ref. 38, Elsevier, 5278680447996].

The SEM image was shown in Fig. 11.19B, and the morphology of the material remained unchanged before and after the etching. At this time, the diameter of NC@MoS_2 nanotubes was about 250 nm, which was comparable to that before etching. The 85 nm-diameter pores left after the MnO_2 nanowires were etched could be seen in Fig. 11.19C, and the hollow structure could also be seen in the TEM images (Fig. 11.19D and E), proving the effective removal of the template. HR-TEM gave an edge view of interconnected MoS_2 nanosheets with spatial lattice fringes. The interplanar spacing was measured to be 0.62 nm corresponding to the (002) crystal plane of hexagonal MoS_2. The elemental mapping analysis of the NC@MoS_2 nanotubes in Fig. 11.19G–K exhibited the presence of C, N, Mo, and S elements, proving the uniform composite of the components in the material.

Fig. 11.20 presented the 3D and 2D RL models and corresponding bandwidth plots for MnO_2@NC, MnO_2@NC@MoS_2, and NC@MoS_2 at 25% filler content. At this time, the MnO_2@NC nanowires had almost no microwave absorption capability (Fig. 11.20A). It was due to weak impedance matching. It could be seen from Fig. 11.20B and C that both MnO_2@NC@MoS_2 and NC@MoS_2 possessed excellent microwave absorption properties. The RL_{min} reached −49.50 dB@12.5 GHz and −52.56 dB@13.6 GHz, respectively (Fig. 11.20D). The corresponding

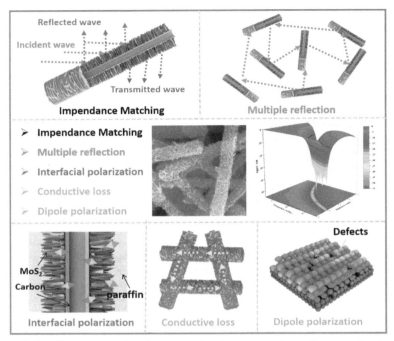

Figure 11.21 The schematic diagram of microwave absorption mechanism of NC@MoS$_2$. [Reproduced with permission from ref. 38, Elsevier, 5278680447996].

matching thicknesses were all 2.4 mm. It could be seen from Fig. 11.20E and F that at 2.4 mm, the EABs of MnO$_2$@NC@MoS$_2$ and NC@MoS$_2$ reached 5.2 GHz (10.4–15.6 GHz), and 6.0 GHz (11.2–17.2 GHz), respectively. After the MnO$_2$ was etched, a hollow tubular structure appeared. It was beneficial to the multiple reflection loss of electromagnetic waves while reducing the material density. Therefore, under the same thickness, the RL$_{min}$ value became smaller, and the EAB became wider. It was worth mentioning that NC@MoS$_2$ exhibited the widest bandwidth (6.2 GHz) at 2.3 mm, covering the entire Ku-band.

Fig. 11.21 summarized the attenuation mechanisms of the multishell composite to microwaves. Excellent impedance matching enabled electromagnetic waves to penetrate the material and be absorbed; The effective construction of the hollow tubular structure not only reduced the material density but also prolonged the transmission path of microwaves. It was conducive to the occurrence of multiple reflection loss. At the same time, depending on the special surface morphology of MoS$_2$, the multiple reflections between the lamellae also effectively attenuated electromagnetic waves.

The existence of heterointerfaces and the accumulation of a large number of nonuniformly distributed space charges between the interfaces made the interfacial polarization and dipole polarization occur continuously. A large number of crystal defects and atomic doping further enhanced the dipole polarization. Finally, the overlapping layered structure with a large aspect ratio provided long enough channels for the persistent flow of electrons, enhancing the conductive loss and making the material possess excellent microwave absorption properties.

11.6 Summary and outlook

To meet the technical requirements of electromagnetic materials in daily life and military applications, it is of great significance to develop microwave absorption materials with excellent comprehensive properties. 1D multishell composite materials, as a class of microwave absorption materials, possess unique hierarchical structures and synergistic properties, exhibiting great research potential. In the future, the main development directions of 1D multishell composite absorbers might focus on the following aspects:

(1) Designing novel compositions and structures of absorbers, optimizing the comprehensive properties, and achieving goals including "thinner, lighter, wider, and stronger" are the focus of subsequent development;

(2) On the premise of ensuring comprehensive performance, simplifying the assembly steps of the multishell structures to achieve engineering and practicality is still a difficult and far-reaching task;

(3) Researchers have summarized several microscopic mechanisms for the excellent properties of multishell structures, such as impedance matching, polarizations, voids, and multiple reflections and scatterings. How to characterize and perfect the above-mentioned mechanisms is the difficulty of future research.

Reference

[1] B. Wang, Q. Wu, Y. Fu, T. Liu, A review on carbon/magnetic metal composites for microwave absorption, J. Mater. Sci. Technol. 86 (2021) 91–109.
[2] Y. Bhattacharjee, S. Bose, Core–shell nanomaterials for microwave absorption and electromagnetic interference shielding: A review, ACS Appl. Mater. Interfaces 4 (2021) 949–972.
[3] X. Wang, Y. Lu, T. Zhu, S. Chang, W. Wang, $CoFe_2O_4$/N-doped reduced graphene oxide aerogels for high-performance microwave absorption, Chem. Eng. J. 388 (2020) 124217.
[4] F. Wu, Z. Liu, T. Xiu, B.L. Zhu, I. Khan, P. Liu, Q. Zhang, B. Zhang, Fabrication of ultralight helical porous carbon fibers with CNTs-confined Ni nanoparticles for enhanced microwave absorption, Compos. B Eng. 215 (2021) 108814.

[5] F. Wu, K. Yang, Q. Li, T. Shah, M. Ahmad, Q. Zhang, B. Zhang, Biomass-derived 3D magnetic porous carbon fibers with a helical/chiral structure toward superior microwave absorption, Carbon 173 (2021) 918–931.

[6] W. Cao, X. Wang, J. Yuan, W. Wang, M. Cao, Temperature dependent microwave absorption of ultrathin graphene composites, J. Mater. Chem. C 3 (38) (2015) 10017–10022.

[7] X. Xie, B. Wang, Y. Wang, C. Ni, X. Sun, W. Du, Spinel structured MFe_2O_4 (M= Fe Co, Ni, Mn, Zn) and their composites for microwave absorption: A review, Chem. Eng. J. 428 (2022) 131160.

[8] Z. Wu, K. Pei, L. Xing, X. Yu, W. You, R. Che, Enhanced microwave absorption performance from magnetic coupling of magnetic nanoparticles suspended within hierarchically tubular composite, Adv. Funct. Mater. 29 (2019) 1901448.

[9] Z. Xiang, Y. Song, J. Xiong, Z. Pan, X. Wang, L. Liu, R. Liu, H. Yang, W. Lu, Enhanced electromagnetic wave absorption of nanoporous Fe_3O_4@carbon composites derived from metal-organic frameworks, Carbon 142 (2019) 20–31.

[10] N. Gao, W. Li, W. Wang, D. Liu, Y. Cui, L. Guo, G. Wang, Balancing dielectric loss and magnetic loss in $Fe-NiS_2$/NiS/PVDF composites toward strong microwave reflection loss, ACS Appl. Mater. Interfaces 12 (2020) 14416–14424.

[11] B. Zhao, X. Guo, W. Zhao, J. Deng, G. Shao, B. Fan, Z. Bai, R. Zhang, Yolk-shell Ni@SnO_2 composites with a designable interspace to improve the electromagnetic wave absorption properties, ACS Appl. Mater. Interfaces 8 (2016) 28917–28925.

[12] K. Yuan, R. Che, Q. Cao, Z. Sun, Q. Yue, Y. Deng, Designed fabrication and characterization of three-dimensionally ordered arrays of core-shell magnetic mesoporous carbon microspheres, ACS Appl. Mater. Interfaces 7 (2015) 5312 –2319.

[13] T. Wu, Y. Liu, X. Zeng, T. Cui, Y. Zhao, Y. Li, G. Tong, Facile hydrothermal synthesis of Fe_3O_4/C core-shell nanorings for efficient low-frequency microwave absorption, ACS Appl. Mater. Interfaces 8 (2016) 7370–7380.

[14] J. Wang, Y. Huyan, Z. Yang, A. Zhang, Q. Zhang, B. Zhang, Tubular carbon nanofibers: Synthesis, characterization and applications in microwave absorption, Carbon 152 (2019) 255–266.

[15] L. Yana, M. Zhang, S. Zhao, T. Sun, B. Zhang, M. Cao, Y. Qin, Wire-in-tube ZnO@carbon by molecular layer deposition: Accurately tunable electromagnetic parameters and remarkable microwave absorption, Chem. Eng. J. 382 (2020) 122860.

[16] N. Yang, Z. Luo, G. Zhu, S. Chen, X. Wang, G. Wu, Y. Wang, Ultralight three-dimensional hierarchical cobalt nanocrystals/N-doped CNTs/carbon sponge composites with a hollow skeleton toward superior microwave absorption, ACS Appl. Mater. Interfaces 11 (2019) 35987–35998.

[17] J. Wang, F. Wu, Y. Cui, A. Zhang, Q. Zhang, B. Zhang, Efficient synthesis of N-doped porous carbon nanoribbon composites with selective microwave absorption performance in common wavebands, Carbon 175 (2021) 164–175.

[18] A. Hazarika, B. Deka, D. Kim, Y. Park, H. Park, Microwave-induced hierarchical iron-carbon nanotubes nanostructures anchored on polypyrrole/graphene oxide-grafted woven Kevlar® fiber, Compos. Sci. Technol. 129 (2016) 137–145.

[19] J. Ma, J. Shu, W. Cao, M. Zhang, X. Wang, J. Yuan, M. Cao, A green fabrication and variable temperature electromagnetic properties for thermal stable microwave absorption towards flower-like Co_3O_4@rGO/SiO_2 composites, Compos. Pt. B-Eng. 166 (2019) 187–195.

[20] Y. Li, F. Meng, Y. Mei, H. Wang, Y. Guo, Y. Wang, F. Peng, F. Huang, Z. Zhou, Electrospun generation of Ti_3C_2Tx MXene@graphene oxide hybrid aerogel microspheres for tunable high-performance microwave absorption, Chem. Eng. J. 391 (2020) 123512.

[21] J. Wang, L. Liu, S. Jiao, K. Ma, J. Lv, J. Yang, Hierarchical carbon fiber@MXene@MoS$_2$ core-sheath synergistic microstructure for tunable and efficient microwave absorption, Adv. Funct. Mater. 30 (2020) 2002595.

[22] X. Han, Y. Huang, S. Gao, G. Zhang, T. Li, P. Liu, A hierarchical carbon Fiber@MXene@ZnO core-sheath synergistic microstructure for efficient microwave absorption and photothermal conversion, Carbon 183 (2021) 872–883.

[23] J. Wang, Y. Cui, F. Wu, T. Shah, M. Ahmad, A. Zhang, Q. Zhang, B. Zhang, Core-shell structured Fe/Fe$_3$O$_4$@TCNFs@TiO$_2$ magnetic hybrid nanofibers: Preparation and electromagnetic parameters regulation for enhanced microwave absorption, Carbon 165 (2020) 275–285.

[24] B. Wang, S. Li, F. Huang, S. Wang, H. Zhang, F. Liu, Q. Liu, Construction of multiple electron transfer paths in 1D core-shell hetetrostructures with MXene as interlayer enabling efficient microwave absorption, Carbon 187 (2022) 56–66.

[25] S. Zhao, Z. Gao, C. Chen, G. Wang, B. Zhang, Y. Chen, J. Zhang, X. Li, Y. Qin, Alternate nonmagnetic and magnetic multilayer nanofilms deposited on carbon nanocoils by atomic layer deposition to tune microwave absorption property, Carbon 98 (2016) 196–203.

[26] M. Qiao, D. Wei, X. He, X. Lei, J. Wei, Q. Zhang, Novel yolk–shell Fe$_3$O$_4$@void@SiO$_2$@PPy nanochains toward microwave absorption application, J. Mater. Sci. 56 (2021) 1312–1327.

[27] M. Ma, W. Li, Z. Tong, Y. Ma, Y. Bi, Z. Liao, J. Zhou, G. Wu, M. Li, J. Yue, X. Song, X. Zhang, NiCo$_2$O$_4$ nanosheets decorated on one-dimensional ZnFe$_2$O$_4$@SiO$_2$@C nanochains with high-performance microwave absorption, J. Colloid Interface Sci. 578 (2020) 58–68.

[28] Y. Bi, M. Ma, Z. Liao, Z. Tong, Y. Chen, R. Wang, Y. Ma, G. Wu, One-dimensional Ni@Co/C@PPy composites for superior electromagnetic wave absorption, J. Colloid Interface Sci. 605 (2022) 483–492.

[29] Y. Fu, H. Liao, B. Wang, Q. Wu, T. Liu, Constructing yolk-shell Co@void@PPy nanocomposites with tunable dielectric properties toward efficient microwave absorption, J. Alloys Compd. 890 (2021) 161715.

[30] M. Ma, W. Li, Z. Tong, Y. Yang, Y. Ma, Z. Cui, R. Wang, P. Lyu, W. Huang, 1D flower-like Fe$_3$O$_4$@SiO$_2$@MnO$_2$ nanochains inducing RGO self-assembly into aerogels for high-efficiency microwave absorption, Mater. Des. 188 (2020) 108462.

[31] N. He, X. Yang, L. Shi, X. Yang, Y. Lu, G. Tong, W. Wu, Chemical conversion of Cu$_2$O/PPy core-shell nanowires (CSNWs): A surface/interface adjustment method for high-quality Cu/Fe/C and Cu/Fe$_3$O$_4$/C CSNWs with superior microwave absorption capabilities, Carbon 166 (2020) 205–217.

[32] J. Ding, L. Wang, Y. Zhao, L. Xing, X. Yu, G. Chen, J. Zhang, R. Che, Boosted interfacial polarization from multishell TiO$_2$@Fe$_3$O$_4$@PPy heterojunction for enhanced microwave absorption, Small 15 (2019) 1902885.

[33] C. Xu, F. Wu, L. Duan, Z. Xiong, Y. Xia, Z. Yang, M. Sun, A. Xie, Dual interfacial polarization enhancement to design tunable microwave absorption nanofibers of SiC@C@PPy, ACS Appl. Electron. Mater. 2 (2020) 1505–1513.

[34] W. Zhang, X. Zhang, Y. Zheng, C. Guo, M. Yang, Z. Li, H. Wu, H. Qiu, H. Yan, S. Qi, Preparation of polyaniline@MoS$_2$@Fe$_3$O$_4$ nanowires with a wide band and small thickness toward enhancement in microwave absorption, ACS Appl. Nano Mater. 1 (2018) 5865–5875.

[35] L. Yan, L. Li, X. Ru, D. Wen, L. Ding, X. Zhang, H. Diao, Y. Qin, Core-shell, wire-in-tube and nanotube structures: Carbon-based materials by molecular layer deposition for efficient microwave absorption, Carbon 173 (2021) 145–153.

[36] F. Wu, P. Liu, J. Wang, T. Shah, M. Ahmad, Q. Zhang, B. Zhang, Fabrication of magnetic tubular fiber with multi-layer heterostructure and its microwave absorbing properties, J. Colloid Interface Sci. 577 (2020) 242–255.

[37] F. Wu, Z. Liu, J. Wang, T. Shah, P. Liu, Q. Zhang, B. Zhang, Template-free self-assembly of MXene and CoNi-bimetal MOF into intertwined one-dimensional heterostructure and its microwave absorbing properties, Chem. Eng. J. 422 (2021) 130591.

[38] K. Yang, Y. Cui, Z. Liu, P. Liu, Q. Zhang, B. Zhang, Design of core–shell structure $NC@MoS_2$ hierarchical nanotubes as high-performance electromagnetic wave absorber, Chem. Eng. J. 426 (2021) 131308.

CHAPTER 12

Biomedical applications of multifunctional tubular nanofibers

Idrees Khan[a] and Baoliang Zhang[a,b]
[a]School of Chemistry and Chemical Engineering, Northwestern Polytechnical University, Xi'an, China
[b]Xi'an Key Laboratory of Functional Organic Porous Materials, Northwestern Polytechnical University, Xi'an, China

12.1 Introduction

Nanotechnology deals with the fabrication, characterization, and application of materials in nanometer range dimensions. Nanotechnology has a substantial influence on the healthcare industry, specifically in disease diagnosis and treatment [1]. Nanofibers (NFs) have received increased attention in medicine and biotechnology [2] due to their high surface-area-to-volume ratio which results in enhanced drug solubility, superior mechanical strength, high and tunable porosity, surface functionalization, and their similarity to the extracellular matrix, that promotes their utilization for wound dressings. NFs have been potentially explored as ultrasensitive biosensors for the diagnosis of cancer and malaria, tissue-engineered scaffolds, biomedical devices, drug delivery systems, dressings for wound healing, detection of circulating cancer cells, and detection of glucose, urea, cholesterol, bacteria, etc. [3,4]. The small pore size and high porosity of the nanofiber mats could facilitate bacterial isolation and gaseous exchange during the wound-repairing state [5]. Nanofibers for various biomedical applications are commonly produced through the electrospinning technique [6]. Polymeric electrospun nanofibers are used to deliver anticancer and antibiotic agents, proteins, RNA, DNA, and growth factors [7]. Almost all types of human tissues are fibers ranging from micro to nanoscale in bundles like structures, whose function is to conduct nervous impulses, provide elasticity and strength enforcement, and the movement of the whole body. Thus NFs have close similarities and connections with the human body [8].

The electrospinning technique is the most applied approach worldwide for the preparation of NFs with unique properties [9]. Electrospinning

Fabrication and Functionalization of Advanced Tubular Nanofibers and their Applications.
DOI: https://doi.org/10.1016/B978-0-323-99039-4.00008-5
323

324 Fabrication and functionalization of advanced tubular nanofibers and their applications

is a commonly used method to change the microscopic morphology of polymers [10]. Electrospinning can prepare fiber having nanoscale diameters ranging from 40 and 2000 nm, which are the most suitable materials for biomedical applications [11]. The evolution in electrospinning technology and their ability to synthesize tunable NFs have to lead to the development of innovative applications of NFs in the health care fields. Resolutions like organ repair, wound dressing and burns, crucial vital monitoring, purification of blood, and treatment of many diseases were offered by the NFs technology [12]. Electrospinning enables the preparation of nanofiber scaffolds with a high surface-to-volume ratio and porosity, essentially required for the desired diffusibility of the bioengineered kidney tubes. Solution electrospinning is a common method employed for fabricating tubular poly-ε-caprolactone (PCL) scaffolds for neuronal or vascular grafts that have proven sufficient mechanical stability and biocompatibility in dog, mouse, and sheep models [13].

NFs can be synthesized from various natural or synthetic polymer materials, or a combination of both types. Natural polymers have lower immunogenicity and better biocompatibility, and synthetic polymers have greater flexibility in their production as well as modification. Proteins and polysaccharides are the most commonly used natural polymers efficiently used in the electrospinning and NFs production for the delivery of various biological products. Synthetic polymers predominate in the production of nanofibers that contain biological products. Similarly, PEO, PVA, PCL, and their copolymers, poly-lactic acid (PLA), and polyvinylpyrrolidone (PVP) are the most common synthetic polymers for NFs production. Polyethylene oxide (PEO) is a nontoxic, neutral, biocompatible, and hydrophilic polymer, that used widely used in tissue engineering and drug delivery [14].

Collagen is, the main component of ECM in the native blood vessels, due to its extensive availability, simple processing methods, and good biomechanical properties, is extensively utilized as a scaffold biomaterial for the tubular organs. The customized collagen scaffold prepared by the electrospinning technique can attain a similar topology architecture to the ECM. Furthermore, electrospinning can synthesize scaffold materials with some specific components for achieving controlled degradation in the remodeling process [15].

In the delivery system, NFs can offer an adequate matrix for the encapsulation and incorporation of the therapeutic agents and are also able to prevent destruction before reached to their targeting sites with high efficiency and low side effects. Such fibrous structures have high flexibility in fabricating

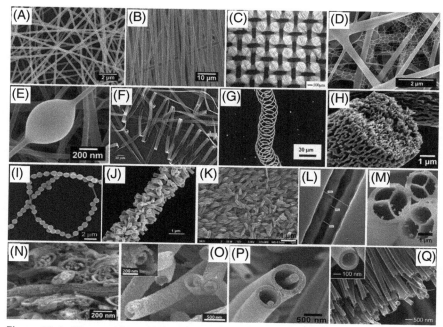

Figure 12.1 Fibers with multifarious morphologies prepared by electrospinning. (A–D) Different NF assembly morphologies: (A) random oriented, (B) aligned as well as (C) patterned and (D) spider-web-like nano-fiber/net structures. (E–Q) Various single NFs with (E) bead-on-string, (F) ribbon-like, (G) helical, (H) porous [30], (I) necklace-like, (J) firecracker-shaped, (K) rice grain-shaped, (L) core—shell, (M) multichannel tubular, (N) multicore cable-like, (O) tube-in-tube, (P) nanowire-in-microtube and (Q) hollow structures [16].

several morphologies as shown in Fig. 12.1. The figure shows fibers with various morphologies that could be fabricated by controlling the processing condition and modifying the standard setup of electrospinning to synthesize nonwoven fibers having straight aligned, randomly aligned, ribbon, porous structures, core–shell, etc [16].

The nonwoven mats of electrospun NFs can be processed further into the scaffolds with different and complex architectures like tubular conduits and stacked arrays. Stacked arrays of the electrospun NFs are composed of multiple layers of the fibers, which are deposited sequentially on top of one another. The key advantages of producing this type of stacked arrays scaffold are to mimic certain natural tissue concerning its structural, mechanical, and biochemical properties, and also enable the creation of multilayered tissues. Tubular conduits composed of electrospun NFs are used usually for

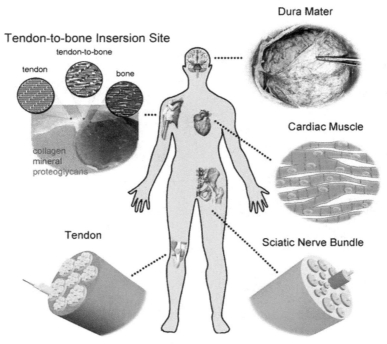

Figure 12.2 Illustration of some typical examples of tissues in the human body whose regeneration would benefit from the use of nanofiber-based scaffolds with anisotropic structures that could be readily fabricated by electrospinning [17].

applications in neural and vascular tissue engineering as these NFs look like the hollow structures of different human tissues as observed in Fig. 12.2 [17].

Nanostructures with hollow interiors have received much attention because of their various potential applications in different emerging fields [18]. Hollow fiber membranes have a unique and important form of membrane configuration owing to their high packing density, high specific surface area, and self-supporting nature [19]. Tubular nanofiber is reported for different applications such as osmotic membrane [20], metals adsorption [21], catalysis, [22], sensing [23] microwave absorption [24], lithium–ion batteries [25], oil/water separation, and dye degradation [19,26], etc. as discussed in the previous chapters. Tubular nanofibers have promising potential in biomedical applications [27]. Tubular nanofibres for biomedical applications are also prepared through electrospinning methods. Electrospinning technology has received increased attention because it can offer a biomimetic environment with the nano–scale to microscale diameter fibers, which may easily form tubular scaffolds with desirable range diameter [28]. For example, bilayer and multilayer tubular PLA nanofibers have been prepared through the electrospinning method for cardiovascular applications. Aligned PLA (polylactic acid) NFs were prepared to utilize 11.5 wt% of the PLA solution, applying a 15 kV voltage, by spinning onto a rotating mandrel with a speed of 5000 rpm. The solution flow rate was 1 mL/h and the tip-target distance was 10 cm. After the electrospinning process completion, the aligned NFs mat was removed from the mandrel and then fused into a tubular NFs form having a single (bilayer) and also to the multiple layers [29].

Tubular nanofibers have potential biomedical applications in different fields such as tissue regeneration, bio-sensing, drug delivery, medical devices, etc. Some of the potential applications of tubular nanofibers are discussed below:

12.2 Tissue regeneration

Tissues as well as organs having tubular structures are omnipresent throughout the whole human body, present in the vasculature (e.g., veins, arteries, capillaries), urinary (e.g., ureter, urethra, bladder), respiratory (e.g., trachea, esophagus) and gastrointestinal systems. Due to this widespread occurrence of these tissues and organs, tissue engineering of the tubular organs is of great interest as a huge number of surgeries are annually operated on these organs [30]. NFs technology getting more and more attention in the tissue engineering field [31]. Electrospun materials have been successfully applied in the tissue engineering field for replacing the damaged tissue and repairing the function of the native tissue such as vascular, nerve tissues, and soft bone [32]. Electrospun matrices of suitable biocompatible and biodegradable polymers can be easily seeded with the cells and can act as a scaffold for tissue regeneration [33]. NFs mimic the porous topography of the natural extracellular matrix which is advantageous for the regeneration of tissue [34]. Controlled fabrication of the 3D tubular biomimetic scaffolds shows potential applications in different biomedical fields such as engineering complex tissue models, 3D cell culture, hemostasis, and tissue regeneration and repairing [35]. Tubular scaffolds fabricated from aligned electrospun fibers can regulate cellular alignment as well as relevant functional expression, with applications in the field of tissue engineering. The tissue-engineered NFs may hold several advantages such as versatility for bio-functionalization and promotion of the desired cell behaviors [36]. The tubular structure is applied for developing artificial tissues such as blood vessels, axons, nerves, or other organs which need the tubular structure [37]. In peripheral nerve graft and vascular graft, a tubular scaffold is usually required [38]. Some examples of tissue regeneration are discussed below:

12.2.1 Blood vessels

Aligned nanofibers for fabricating tubular scaffolds can be employed for blood vessels engineering [39,40]. Designing multilayered tubular scaffolds can effectively mimic the native architecture and can also approach the functional features of the artery [41]. Tubular conduits might be fabricated with several properties which are appropriate for vascular tissue engineering [42]. The electrospinning process was employed widely for fabricating tubular scaffolds having different diameters and lengths of synthetic and natural polymers for the vascular grafts. Such tubular scaffolds designed could be flexible, strong, and looks like blood vessels. Tubular scaffolds synthesized from polyglycolic acid, elastin, and collagen with 1 mm thickness and 12 cm length, have characteristics similar to the native arteries [43]. PCL/collagen tubular vascular tissue engineering scaffold produced through coaxial electrospinning shows good biocompatibility and cell affinity and has a promising application in vascular tissue engineering [44]. In a study four-layer, tubular scaffolds (FLTSs) were fabricated and used for simulating the functions and structures of native blood vessels. The FLTSs exhibit outstanding mechanical properties with higher circumferential as well as longitudinal tensile properties. The FLTSs are much more biocompatible than the random fibers layer due to the high viability of the human umbilical vein endothelial cells (HUVECs) on the FLTSs. The results show that FLTS enhanced both cell viability and attachment as compared to the random fibers layer. The comparison of the live/dead, as well as immunostaining images, indicated that the FLTSs inner layer contained higher numbers of live cells than the other random fibers layer. The random and aligned composite structured FLTSs are favorable for promoting the HUVECs growth. The cell adhesion and also their proliferation on FLTSs were observed to be superior on the random fibers layer. This demonstrates that FLTSs hold the potential application in the regeneration of vascular tissue as well as in replacements of clinical arterial processes [45]. The tubular scaffold of collagen-coated poly(L-lactic acid-polycaprolactone) NFs was engineered to a vascular graft of small diameter for replacing a rabbit vein which shows that the scaffold could sustain suturing and the process of implantation [1]. Gelatin blended with polyglycolic acid (PGA) tubular nanofibers were analyzed for biocompatibility using human umbilical artery smooth muscle cells and human umbilical vein endothelial cells. The study proposed that tubular scaffolds having an inner layer of PGA/10 wt% gelatin and an outer layer of PGA/30 wt% gelatin are hopeful scaffolds for vascular tissue engineering [46]. Seeding of human coronary artery endothelial cells (HCAECs) onto the poly(l-lactic acid)-poly(e-caprolactone) (PLLA–PCL) (70:30) tubular nanofibrous scaffold displayed that the HCAECs were subconfluent in just after one day and had well spread on a scaffold in seven days. The results confirmed that the collagen-coated PLLA–PCL scaffold promotes stable and fast in vitro endothelialization [47].

Tubular HA/collagen nanofibrous vessel scaffolds having hierarchical architecture were fabricated with collagen and HA through sequential electrospinning. The tube has a wall thickness of ~0.62 mm and an inner diameter of 3.0 mm. The tube inner wall is a layer of pure HA NFs having a thickness of ~0.22 mm, while the tube outer wall is layered of pure collagen NFs having a thickness of ~0.4 mm. The morphological analysis of tubular scaffolds shows that the inner wall surface of collagen and HA/collagen NFs tubes have a porous nanofiber network-like nano-topography. The single NFs have a continuous cylindrical shape and the average diameter of an HA/collagen NFs tube is (541.3 ± 19.3) nm, which is smaller slightly than that of pure collagen NFs (580 ± 16) nm. The average pore diameter of the inner wall of the pure collagen and HA/collagen and NFs is (2159 ± 77) and (2139 ± 81) nm, respectively. The characteristics are represented in Fig. 12.3. The HA/collagen NFs have good blood compatibility and stability, which highlights their appropriateness for tissue engineering vascular implants. These HA/collagen NFs can promote significantly the proliferation, elongation, and phenotypic shape expression of PAECs [48].

12.2.2 Nerves cells

An alternative to autographs, nerve guidance conduits may be potentially used for promoting neuronal growth and guiding axonal extension after the nerve injury. A synthetic nerve guidance conduit was synthesized to promote as well as facilitate the regeneration of nerves [49]. Injuries in peripheral nerves are currently repaired with sensory nerve autograft, processed nerve allograft, or tubular conduits, but due to the limitations of these options, multiple synthetic tubular conduits have been developed for segmental peripheral nerve repair. For example, poly-L-lactide-co-caprolactone (PLCL) tubular NFs perform an improved sensory function as compared to the autograft repair in a critical-size defect in the sciatic nerve in a rat model [50].

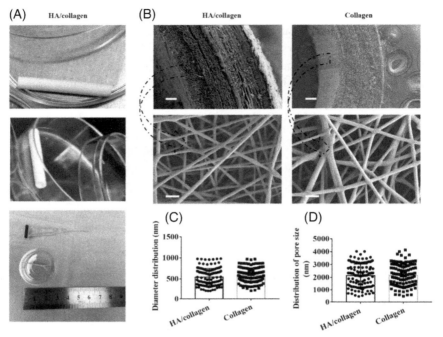

Figure 12.3 Physical characteristics of electrospun nanofibers. (A) The lateral (upper panel), cross-section (middle panel) and strip nanofiber concentric axis film (after cutting along the long axis) view of an HA/collagen nanofiber tube. (B) SEM micrographs of the cross-section (upper panel) views and the inner wall surface (lower panel) views of HA/collagen and collagen nanofibers before cross-linking. Scale bars: 100 m (upper panel), 2 m (lower panel). Statistical data of the diameter (C) and pore size (D) distribution of various nanofibers [48].

The surface geometry of tubular scaffolds affects the arrangement of the fibrin matrix and cellular response within the conduit immediately after the sciatic nerve injury [51]. Multitubular conduit of PCL nanofiber with a honeycomb structure shows potential repairing of large defects in thick nerves. The seeded bone marrow stem cells can proliferate efficiently in all the tubes with even longitudinal and circumferential distributions. These tubular NFs can also examine the basic aspects that are involved in the development of the peripheral nervous system and cells migration [52]. Electrospun poly (L-lactide-co-glycolide) biodegradable tubular NFs are feasible for in vivo nerve regeneration. The tubular nanofiber conduits were permeable, flexible, and display no swelling and thus can be very effective aids for the repair and regeneration of the nerve. These fibers are biodegradable and degraded from their implantation site after their purpose is performed [53]. Chitosan/PCL tubular nanofibrous scaffolds are well-suited nerve guides for the regeneration of peripheral nerves, where severed nerve endings cannot be repaired with sutures [54].

12.2.3 Urethral epithelial cells

A tubular multifaceted bio-interfacing tissue-engineered autologous scaffolds composed of alternating block polyurethane (abbreviated as PU-alt) have been reported for promoting neo-vascularization for the urethral regeneration in a rabbit. The tubular scaffold has NFs hierarchical architecture, flexible mechanical properties, and a hydrophilic PEGylation interface which has the capability of promoting oriented elongation, adhesion, and proliferation of rabbit autologous urethral epithelial cells and smooth muscle cells simultaneously. The tubular NFs can also upregulate the expression of keratin (AE1/AE3) in epithelial cells and contractile protein (α-SMA) in smooth muscle cells, as also the subsequent synthesis of elastin. The structure of the urethral tubular scaffolds remained stable during the urethral reconstruction with the scaffold degradation and can help the hosts regenerate functional new urethral tissues with intact tissue structure which was close to that of autografts [55].

12.2.4 Cardiac tissue

Mimicking the nanofibrous structure like the extracellular matrix and conductivity for electrical propagation of native myocardium might be greatly advantageous for cardiac tissue engineering as well as cardiomyocytes-based bioactuators. Tubular PLA/PANI 3D bioactuators with folding and tubular shapes conductive nanofibrous were developed for cardiac tissue engineering that displays the great potential in CMs-based 3D bioactuators [56].

12.2.5 Bone regeneration

NFs have been studied medically as a scaffold for bone regeneration therapy [57] and as a scaffold to promote the bone healing process [58]. Nanofibrous tubular poly (L-lactic acid) (PLLA) scaffolds were successfully fabricated for bone tissue regeneration. The surface of the nanofibrous tubular scaffold offers a suitable environment for the attachment as well as a proliferation of MC3T3-E1 subclone 14 cells, which prove their significant potential for regeneration of long bone tissue [59]. The polylactic acid nanofibers coated and uncoated 3D-printed tubular scaffold shows brilliant cellular biocompatibility for applications in bone tissue. The response of cell adhesion of

Figure 12.4 SEM micrographs of the 3D tubular scaffold coated with nanofibers show the surface seeded with hFOB cells, whereas the good spread of the cells with a spindle-shaped morphology was observed typical of osteoblast cells with the projection of the cell-material interaction [60].

human fetal osteoblast (hFOB) cells seeded onto the 3D tubular scaffolds coated with and without fibers shows that more cells get adhered to nanofibers coated 3D tubular scaffolds than the uncoated 3D tubular scaffold. The cell attachment increased by 13% after 4 h and 15% after 24 h on a nanofibers-coated 3D tubular scaffold. These results indicate that combining the fiber membrane and rough surface morphologies paid key role in the attachment of human osteoblast cells as observed in the SEM images in the Fig. 12.4 [60]. Similarly, tubular PCL scaffolds combined with human mesenchymal cells and rhBMP-7 have been developed for a humanized bone organ. The tubular composite scaffold caused homing as well as a proliferation of the human prostate cancer cells, and the development of the macro metastases [61].

12.3 Bio-sensing applications

Nanofibers have been applied in creating nanosensors for the detection of specific biomarkers of certain illnesses and diseases. The high surface area to volume ratio of NFs improves the selectivity, sensitivity, and detection limit of the sensor [62]. The characteristics of electrospun NFs such as high surface area, flexibility, porosity, portable nature, and cost-effectiveness lead them ideal materials for sensing applications [63]. Electrospun nanocomposite has emerged as a promising candidate for biosensors due to due to their high specific surface area, high catalytic as well as electron transfer, superior electric conductivity, controllable surface modification and conformation, and unique mat structure [64]. Bao et al. [65] fabricated hollow Nanotubular polyaniline (PANI) nanotubes/Au hybrid nanostructures and utilized them as biosensing substrate materials for H_2O_2 detection and found them very sensitive with a detection limit of 0.25 μM. Similarly, urease immobilized Pt/PANi hybrid NFs hollow tubular structure was utilized as biosensors for detection of urea, which shows wide linear range detection and brilliant anti-interference property against the chloride ion [66]. Tubular structure CuO/Co_3O_4 NFs were designed as a sensor for the highly sensitive detection of cancer cells. The tubular NFs showed high sensitivity and rapid identification of cancer cells through pressure signals irradiating under visible light. This method provided a brilliant dynamic range of $50–10^5$ cells mL^{-1} for the Hela cells with a detection limit of at least 50 cells mL-1 only in 15 mins. The brilliant performance of this technique is attributed to the highly efficient photocatalytic activity of CuO/Co_3O_4 NFs during the AB dehydrogenation as well as the high hydrogen capacity of AB [67]. Similarly, TiO_2 hollow nanofibers had tubular structures prepared and explored for glucose biosensing. The nanotubular and porous TiO_2 offer a well-defined hierarchical nanostructure for glucose oxidase (GOD) loading, and the fine TiO_2 nanocrystals facilitate the direct transfer of an electron from GOD to the electrode, also the strong interaction between hollow TiO_2 and GOD significantly improves the biosensor stability. These biosensors show very well bio-sensing performances both in O_2-containing and O_2-free situations with very good sensitivity, long-term stability, sound reliability, and satisfactory selectivity [68].

12.4 Drug delivery applications

Drug–delivery systems based on nanofibers are extensively applicable for the specific drug release, according to the timing and target location, and to attain the desired therapeutic effects [69]. Drug delivery is an auspicious tool in the field of pharmaceutical science because it can maximize the therapeutic effects of the drug delivered and minimize the undesired side effects. NFs produced with biocompatible and biodegradable polymers received increased attention due to their effectiveness, flexibility, and exceptional physicochemical properties for example, small diameter, high aspect ratio, and large surface area. Nanofibrous scaffolds could minimize the disadvantages of the systemic perfusion with the free drug or other drug delivery systems and can maximize drug action pharmaceutical by a sustained and controlled release directly at the action site. The Nanofibrous scaffolds can decrease the threat of antibiotic-resistant bacteria and multi-drug resistance in cancer therapy by dose-specific, site-specific, and timed release of different drugs [70].

Mesoporous silica nanoparticles doped gelatin/poly(ester-urethane) urea (C-PEEUU) bilayered vascular scaffold were used for the encapsulation and release of salvianolic acid (SAL). The loaded SAL within the tubular scaffold represents good mechanical properties and a sustained release profile. The tubular bilayered vascular scaffold was found efficient to slow down the release rate of SAL. The long-term anticoagulant efficacy of tubular bilayered vascular scaffold suggests that the tubular scaffold might be useful for the prevention of acute thrombosis and intimal hyperplasia [71]. Cilostazol-loaded Poly(ε-Caprolactone) tubular nanofibers were employed in the drug delivery system for cardiovascular applications. The drug release of CIL from tubular electrospun PCL materials followed Peppas-Sahlin and Ritger-Peppas models and indicate a diffusion-controlled release mechanism in combination with the relaxation of polymer. The SEM micrographs demonstrated that the electrospun PCL fibers have retained their integrity and structure after dissolution testing with no deformations, however crystals of CIL at the fiber's surface were almost completely eluted as observed in Fig. 12.5 [72].

12.5 Antimicrobial applications

The synthesis of NFs with some specific properties through electrospinning has received significant attraction for antibacterial applications [73]. Electrospun NFs could be applicable for the development of a biodegradable antimicrobial layer, while through the coaxial electrospinning technique, it is possible to incorporate some antimicrobial agents in a spatially-controlled bilayer format [74]. 3D tubular-shaped triple antibiotic–eluting Polydioxanone nanofibers were investigated against a multispecies biofilm (Actinomyces naeslundii, Streptococcus sanguinis, and Enterococcus faecalis) on human dentin. Upon exposure to 3D tubular-shaped triple antibiotic–eluting nanofibrous construct, viable bacteria are almost eliminated on the surface of dentin and from inside the dentinal tubules, which was almost similar to the bacterial death rate promoted by the triple antibiotic paste. The incorporation of antibiotics into the electrospun NFs leads to the fabrication of a more cell-friendly and localized intracanal medication. The

Before [0 h]   After [48 h]

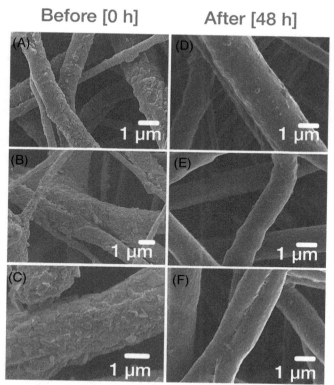

Figure 12.5 SEM micrographs before (A, B, C) and after (D, E, F) release studies of PCL materials loaded with 6.25% (A, D), 12.50% (B, E) and 18.75% (C, F) CIL [72].

observed antimicrobial effect against such multispecies biofilm suggests their substantial clinical potential as a disinfection strategy before regenerative endodontics [75]. Similarly, pure SnO_2 and Ag-doped SnO_2 hollow tubular NFs were successfully fabricated via a combination of an electrospinning technique and a calcination procedure, and their antibacterial activity was evaluated against *E. coli* and *S. aureus* strains in terms of their growth curves and minimum inhibitory concentration. Pure SnO_2 NFs display poor antibacterial activity towards both *E. coli* and *S. aureus* while Ag-doped SnO_2 hollow NFs exhibit as high as about 98% when 100 mg NFs were used against *E. coli* and above 80% for *S. aureus*. The possible antibacterial mechanism of Ag-doped SnO_2 hollow NFs is attributed to the good dispersion of nanoparticles which are required for effective antibacterial activities. As NFs are one-dimensional nanomaterials, hence the materials are dispersed and the agglomeration effect is not important. So the Ag-doped SnO_2 hollow NFs can stably release Ag^+ at the nanolevel, thus

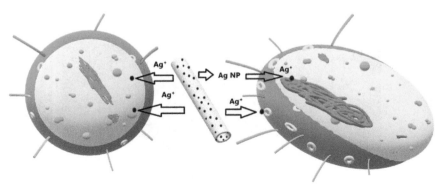

Figure 12.6 Schematic of the antibacterial mechanism of Ag-doped hollow nanofibers. Reprinted by permission from [Springer Nature Customer Service Centre Gmb] [Springer] [Electronic Materials Letters] [76] Copyright (2020) under the license No: 5266920940510, March 13, 2022.

demonstrating brilliant antibacterial activity. The particles size of Ag affects directly the effective surface of Ag and then also affects their dissolution rate and the resulting antibacterial potential. The membrane separation from the cytoplasm of Ag-treated *E. coli* may result from the death of a bacterial cell. The considerable negative charge on the cell membrane gets neutralized by the Ag^+, which causes the cell membrane destruction and results in cytoplasm leakage. When Ag reacts with the bacterial cell membrane, the Ag^+ released enters the bacterial cell and may damage DNA, which leads to mutation or death. Ag NPs can also destroy the mitochondria of the bacterial cell, which cause cell death because of insufficient supply of energy. Ag^+ can interact with the ROS produced in the cells which leads to the destruction of the structure of the internal cells and finally cell death. These four possible antibacterial processes against *E. coli* cells by Ag-doped SnO_2 tubular nanofibers are outlined in Fig. 12.6 [76].

12.6 Applications in blood dialysis

The Nanofibrous membrane has a high specific surface area, high porosity, high interconnected porosity, and exceptional permeability and therefore excellent candidate for filtration applications. Polysulfone tubular nanofibers membrane is reported for the removal of creatinine and urea from the blood serums and urine of dialysis patients. The tubular NFs membrane with a diameter of 3 mm removes 90.4% and 100% urea and creatinine, respectively. The tubular membrane also removed creatinine and urea from patients' blood serums efficiently and the membrane removed water continuously

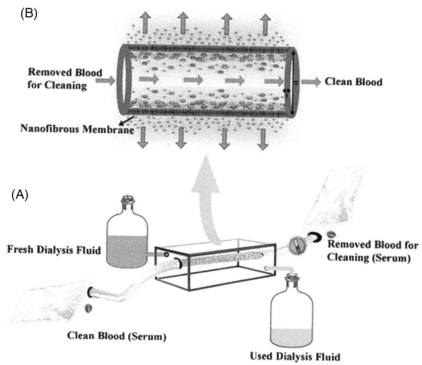

Figure 12.7 A Schematic view of the dialysis device. b Schematic view of the diffusion process through dialysis device. Reprinted by permission from [Springer Nature Customer Service Centre Gmb] [Springer] [Applied Biochemistry and Biotechnology] [77] Copyright (2018) under the license No: 5266910509622, March 13, 2022.

with a low flow rate of blood, and the patients felt less annoyed. The schematic demonstration of the simulated hemodialysis device along with the diffusion process in dialysis is revealed in Fig. 12.7 [77].

12.7 Applications in medical devices

Magnetically functionalized Nickel-polyurethane based tubular nanofibers have been fabricated through electrospinning and their main features in terms of direct elastomagnetic response have been examined. The resultant devices display good elastomagnetic stretchability at room temperature-longitudinal relative strain/exciting field $\sim 4 \cdot 10^{-3}/(1.5 \cdot 10^4$ A/m). These results and their unique attitude suggest their potential use as micro-conduits for innovative biomedical applications such as stone-eliminating urinary stents, vibration therapy, fabrication of cardiac patches, stents, artificial nerve

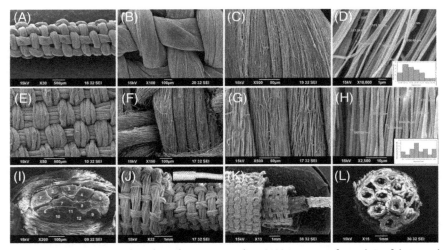

Figure 12.8 Hierarchical multiscale nano textiles. SEM images of conduits fabricated using yarns constituted of well-oriented fibers. (A–D) PCL collagen; (E–H) PLLA; (I) cross-sectional view showing hierarchical bundling of fibers; (J) conduits with a step-change in diameter (the inset shows its optical image); and (K, L) hierarchical concentric and multi-channeled structures. The red stars indicate individual conduits of the same or different diameters. Reprinted (adapted) with permission from [80] Copyright {2022} American Chemical Society.

guides, magneto–active components, antithrombus vascular grafts, sensors and actuators [78]. Tubular magnetic polymer nanocomposite can protect industrial robots and medical apparatus from electromagnetic interference and keep the human body from radiation [79].

12.8 Other medical applications

Tubular woven nano textiles were prepared from low-strength nano yarns which consist of thousands of bundles of nanofibers. The tubular woven nano textiles lead to a super hydrophilic material that contributed to the enhanced adsorption of protein and consequent cell attachment and they are spreading. The in vivo testing verified that the textile conduit was suturable, robust, nonthrombogenic, and kink-proof, and could act as an effective embolizer when installed into an artery. Such properties and applications of tubular woven nanotextile in the medical field are expected to open a new area of nano textiles for advanced applications in several fields. Fig. 12.8 shows the SEM micrographs of hierarchical multiscale nano textiles. The research group further demonstrates the practical applicability of this weaving technology for the development of nano textiles using low-strength mono as well as hybrid materials (PLLA and PCL: Collagen

75:25 respectively) as observed in Fig. 12.8 (A–H). To meet the force requirements during NFs weaving these yarns were successfully packed into bundles as observed in Fig. 12.8I. Similarly, complex multiscale conduits were also prepared as shown in Fig. 12.8J–L, which had to be used in other potential biomedical applications [80].

Tubular NFs are also used in cell encapsulation and cell therapies. Cell encapsulation has extensive potential in the treatment of several hormone-deficient diseases as well as in endocrine disorders. Various highly porous and tubular NFs membranes of premade planar or tubular devices were impregnated with hydrogel precursor solutions followed by crosslinking to obtain various nanofiber-enabled hydrogel devices. The nanofiber-enabled encapsulation devices (NEEDs) hold on the properties of both the NFs and hydrogel. The devices were employed for the encapsulation and culture of some different cell types including both single cells as well as cell aggregates (islets). The therapeutic potential of the devices was also studied through a type 1 diabetic mouse model applying primary rat islets. These devices can overcome potentially some challenges in the field of cell encapsulation and can contribute in the future to the development of cell therapies [81].

12.9 Conclusion

The potential applications of tubular NFs in different biomedical fields like engineering complex tissue models, 3D cell culture, tissue regeneration and repair, and hemostasis are due to their several unique properties. The tubular structure NFs are used for the development of artificial tissues such as blood vessels, axons, nerves, or also in other organs where the tubular structure is needed. Besides tissue regeneration, tubular nanofibers are also utilized as potential materials in drug delivery, antibiotic agent, blood dialysis, and many other fields. Multifunctional tubular nanofibers cab be utilized for innovative biomedical applications such as stone-eliminating urinary stents, vibration therapy, and fabrication of cardiac patches, stents, artificial nerve guides, magneto–active components, antithrombus vascular grafts, sensors, and actuators.

References

[1] R. Sridhar, R. Lakshminarayanan, K. Madhaiyan, V.A. Barathi, K.H.C. Limh, S. Ramakrishna, Electrosprayed nanoparticles and electrospun nanofibers based on natural materials: Applications in tissue regeneration, drug delivery and pharmaceuticals, Chem. Soc. Rev. 44 (2015) 790–814, doi:10.1039/c4cs00226a.

[2] A. Eatemadi, H. Daraee, N. Zarghami, H. Melat Yar, A. Akbarzadeh, Nanofiber: Synthesis and biomedical applications, Artif. Cells, Nanomedicine, Biotechnol 44 (2016) 111–121, doi:10.3109/21691401.2014.922568.

[3] P. Prabhu, Nanofibers for medical diagnosis and therapy, Handb. Nanofibers, Springer International Publishing, 2019, pp. 831–867, doi:10.1007/978-3-319-53655-2_48.

[4] R. Rasouli, A. Barhoum, M. Bechelany, A. Dufresne, Nanofibers for biomedical and healthcare applications, Macromol. Biosci. 19 (2019), doi:10.1002/mabi.201800256.

[5] F. Wang, S. Hu, Q. Jia, L. Zhang, Advances in electrospinning of natural biomaterials for wound dressing, J. Nanomater. 2020 (2020), doi:10.1155/2020/8719859.

[6] S. Chen, J.V. John, A. McCarthy, J. Xie, New forms of electrospun nanofiber materials for biomedical applications, J. Mater. Chem. B. 8 (2020) 3733–3746, doi:10.1039/d0tb00271b.

[7] R.S. Bhattarai, R.D. Bachu, S.H.S. Boddu, S. Bhaduri, Biomedical applications of electrospun nanofibers: Drug and nanoparticle delivery, Pharmaceutics 11 (2019), doi:10.3390/pharmaceutics11010005.

[8] J. Sharma, M. Lizu, M. Stewart, K. Zygula, Y. Lu, R. Chauhan, X. Yan, Z. Guo, E.K. Wujcik, S. Wei, Multifunctional nanofibers towards active biomedical therapeutics, Polymers (Basel) 7 (2015) 186–219, doi:10.3390/polym7020186.

[9] V. Khunová, D. Pavliňák, I. Šafařík, M. Škrátek, F. Ondreáš, Multifunctional electrospun nanofibers based on biopolymer blends and magnetic tubular halloysite for medical applications, Polym 13 (2021) 3870 Page13 (2021) 3870, doi:10.3390/POLYM13223870.

[10] X. Li, Z. Zhuang, D. Qi, C. Zhao, High sensitive and fast response humidity sensor based on polymer composite nanofibers for breath monitoring and non-contact sensing, Sensors Actuators B Chem 330 (2021) 129239, doi:10.1016/J.SNB.2020.129239.

[11] W.K. Essa, S.A. Yasin, I.A. Saeed, G.A.M. Ali, Nanofiber-based face masks and respirators as COVID-19 protection: A review, Membr 11 (2021) 250 Page11 (2021) 250, doi:10.3390/MEMBRANES11040250.

[12] V.S. Reddy, Y. Tian, C. Zhang, Z. Ye, K. Roy, A. Chinnappan, S. Ramakrishna, W. Liu, R. Ghosh, A review on electrospun nanofibers based advanced applications: from health care to energy devices, Polym. 13 (2021) 3746 Page13 (2021) 3746, doi:10.3390/POLYM13213746.

[13] K. Jansen, M. Castilho, S. Aarts, M.M. Kaminski, S.S. Lienkamp, R. Pichler, J. Malda, T. Vermonden, J. Jansen, R. Masereeuw, Fabrication of kidney proximal tubule grafts using biofunctionalized electrospun polymer scaffolds, Macromol. Biosci. 19 (2019) e1800412, doi:10.1002/MABI.201800412.

[14] S. Stojanov, A. Berlec, Electrospun nanofibers as carriers of microorganisms, stem cells, proteins, and nucleic acids in therapeutic and other applications, Front. Bioeng. Biotechnol. 8 (2020) 130, doi:10.3389/FBIOE.2020.00130/BIBTEX.

[15] Y. Niu, M. Galluzzi, M. Fu, J. Hu, H. Xia, In vivo performance of electrospun tubular hyaluronic acid/collagen nanofibrous scaffolds for vascular reconstruction in the rabbit model, J. Nanobiotechnology. 19 (2021) 1–13, doi:10.1186/S12951-021-01091-0/FIGURES/5.

[16] H. Maleki, K. Khoshnevisan, S.M. Sajjadi-Jazi, H. Baharifar, M. Doostan, N. Khoshnevisan, F. Sharifi, Nanofiber-based systems intended for diabetes, J. Nanobiotechnology 19 (2021) 1–34 2021 191, doi:10.1186/S12951-021-01065-2.

[17] W. Liu, S. Thomopoulos, Y. Xia, Electrospun nanofibers for regenerative medicine, Adv. Healthc. Mater. 1 (2012) 10–25, doi:10.1002/ADHM.201100021.

[18] C. Wang, J. Wang, L. Zeng, Z. Qiao, X. Liu, H. Liu, J. Zhang, J. Ding, Fabrication of electrospun polymer nanofibers with diverse morphologies, Molecules 24 (2019), doi:10.3390/molecules24050834.

[19] Y.J. Wu, C.F. Xiao, J. Zhao, Preparation of an electrospun tubular PU/GE nanofiber membrane for high flux oil/water separation, RSC Adv. 9 (2019) 33722–33732, doi:10.1039/c9ra04253a.

[20] S. Arslan, M. Eyvaz, S. Güçlü, A. Yüksekdağ, İ. Koyuncu, E. Yüksel, Investigation of water and salt flux performances of polyamide coated tubular electrospun nanofiber membrane under pressure, J. Environ. Sci. Heal. - Part A Toxic/Hazardous Subst. Environ. Eng. 55 (2020) 606–614, doi:10.1080/10934529.2020.1724011.

[21] M. Ahmad, K. Yang, L. Li, Y. Fan, T. Shah, Q. Zhang, B. Zhang, Modified tubular carbon nanofibers for adsorption of uranium(VI) from water, ACS Appl. Nano Mater. 3 (2020) 6394–6405, doi:10.1021/acsanm.0c00837.

[22] J. Lee, J. Gil Kim, J. Young Chang, Fabrication of a conjugated microporous polymer membrane and its application for membrane catalysis OPEN, (n.d.). 10.1038/s41598-017-13827-w.

[23] E.G. Pineda, M.J.R. Presa, C.A. Gervasi, A.E. Bolzán, Tubular-structured polypyrrole electrodes decorated with gold nanoparticles for electrochemical sensing, J. Electroanal. Chem. 812 (2018) 28–36, doi:10.1016/j.jelechem.2018.01.047.

[24] J. Wang, Y. Huyan, Z. Yang, A. Zhang, Q. Zhang, B. Zhang, Tubular carbon nanofibers: Synthesis, characterization and applications in microwave absorption, Carbon N. Y. 152 (2019) 255–266, doi:10.1016/j.carbon.2019.06.048.

[25] Y. Huyan, J. Wang, J. Chen, Q. Zhang, B. Zhang, Magnetic tubular carbon nanofibers as anode electrodes for high-performance lithium-ion batteries, Int. J. Energy Res. 43 (2019) er.4821, doi:10.1002/er.4821.

[26] X. Wang, C. Xiao, H. Liu, M. Chen, H. Xu, W. Luo, F. Zhang, Robust functionalization of underwater superoleophobic PVDF-HFP tubular nanofiber membranes and applications for continuous dye degradation and oil/water separation, J. Memb. Sci. 596 (2020) 117583, doi:10.1016/j.memsci.2019.117583.

[27] H. Niu, H. Zhou, H. Wang, Electrospinning: An advanced nanofiber production technology, Energy Harvest. Prop. Electrospun Nanofibers, IOP Publishing, 2019, doi:10.1088/978-0-7503-2005-4ch1.

[28] S. Wang, Y. Zhang, H. Wang, G. Yin, Z. Dong, Fabrication and properties of the electrospun polylactide/silk fibroin-gelatin composite tubular scaffold, Biomacromolecules 10 (2009) 2240–2244, doi:10.1021/BM900416B.

[29] K.T. Shalumon, P.R. Sreerekha, D. Sathish, H. Tamura, S.V. Nair, K.P. Chennazhi, R. Jayakumar, Hierarchically designed electrospun tubular scaffolds for cardiovascular applications, J. Biomed. Nanotechnol. 7 (2011) 609–620, doi:10.1166/JBN.2011.1337.

[30] S. Chen, H. Wang, A. McCarthy, Z. Yan, H.J. Kim, M.A. Carlson, Y. Xia, J. Xie, Three-dimensional objects consisting of hierarchically assembled nanofibers with controlled alignments for regenerative medicine, Nano Lett. 19 (2019) 2059–2065, doi:10.1021/ACS.NANOLETT.9B00217/SUPPL_FILE/NL9B00217_SI_011.AVI.

[31] X. Xu, S. Ren, L. Li, Y. Zhou, W. Peng, Y. Xu, Biodegradable engineered fiber scaffolds fabricated by electrospinning for periodontal tissue regeneration, J. Biomater. Appl. (2020) 088532822095225, doi:10.1177/0885328220952250.

[32] H. Liu, C.R. Gough, Q. Deng, Z. Gu, F. Wang, X. Hu, Recent advances in electrospun sustainable composites for biomedical, environmental, energy, and packaging applications, Int. J. Mol. Sci. 21 (2020), doi:10.3390/ijms21114019.

[33] R. Dorati, E. Chiesa, S. Pisani, I. Genta, T. Modena, G. Bruni, C.R.M. Brambilla, M. Benazzo, B. Conti, The effect of process parameters on alignment of tubular electrospun nanofibers for tissue regeneration purposes, J. Drug Deliv. Sci. Technol. 58 (2020), doi:10.1016/j.jddst.2020.101781.

[34] R. Sridhar, J.R. Venugopal, S. Sundarrajan, R. Ravichandran, B. Ramalingam, S. Ramakrishna, Electrospun nanofibers for pharmaceutical and medical applications, J. Drug Deliv. Sci. Technol. 21 (2011) 451–468, doi:10.1016/S1773-2247(11)50075-9.

[35] S. Chen, J.V. John, A. McCarthy, M.A. Carlson, X. Li, J. Xie, Fast transformation of 2D nanofiber membranes into pre-molded 3D scaffolds with biomimetic and oriented porous structure for biomedical applications, Appl. Phys. Rev. 7 (2020) 021406, doi:10.1063/1.5144808.

[36] D. Gugulothu, A. Barhoum, S.M. Afzal, B. Venkateshwarlu, H. Uludag, Structural multifunctional nanofibers and their emerging applications, Handb. Nanofibers, Springer International Publishing, 2019, pp. 693–732, doi:10.1007/978-3-319-53655-2_16.

[37] M. Qamar Khan, D. Kharaghani, I.S. Kim, Z. Khatri, Nanofibers for medical textiles, Handb. Nanofibers, Springer International Publishing, 2019, pp. 1–17, doi:10.1007/978-3-319-42789-8_57-1.

[38] S. Ramakrishna, R. Jose, P.S. Archana, A.S. Nair, R. Balamurugan, J. Venugopal, W.E. Teo, Science and engineering of electrospun nanofibers for advances in clean energy, water filtration, and regenerative medicine, J. Mater. Sci. 45 (2010) 6283–6312 2010 4523, doi:10.1007/S10853-010-4509-1.

[39] R. Vasita, D.S. Katti, Nanofibers and their applications in tissue engineering, Int. J. Nanomedicine. 1 (2006) 15–30, doi:10.2147/nano.2006.1.1.15.

[40] D. Sousa Coelho, B. Veleirinho, T. Alberti, A. Maestri, R. Yunes, P. Fernando Dias, M. Maraschin, Electrospinning technology: Designing nanofibers toward wound healing application, Nanomater. - Toxicity, Hum. Heal. Environ., IntechOpen, 2020, doi:10.5772/intechopen.81530.

[41] A. Alessandrino, A. Chiarini, M. Biagiotti, I. Dal Prà, G.A. Bassani, V. Vincoli, P. Settembrini, P. Pierimarchi, G. Freddi, U. Armato, Three-layered silk fibroin tubular scaffold for the repair and regeneration of small caliber blood vessels: From design to in vivo pilot tests, Front. Bioeng. Biotechnol. 7 (2019) 356, doi:10.3389/fbioe.2019.00356.

[42] S. Agarwal, J.H. Wendorff, A. Greiner, Use of electrospinning technique for biomedical applications, Polymer (Guildf) 49 (2008) 5603–5621, doi:10.1016/j.polymer. 2008.09.014.

[43] A.M. Al-Enizi, M.M. Zagho, A.A. Elzatahry, Polymer-based electrospun nanofibers for biomedical applications, Nanomaterials 8 (2018), doi:10.3390/nano8040259.

[44] N. Duan, X. Geng, L. Ye, A. Zhang, Z. Feng, L. Guo, Y. Gu, A vascular tissue engineering scaffold with core-shell structured nano-fibers formed by coaxial electrospinning and its biocompatibility evaluation, Biomed. Mater. 11 (2016) 035007, doi:10.1088/1748-6041/11/3/035007.

[45] Q. Hu, C. Su, Z. Zeng, H. Zhang, R. Feng, J. Feng, S. Li, Fabrication of multilayer tubular scaffolds with aligned nanofibers to guide the growth of endothelial cells, J. Biomater. Appl. 35 (2020) 553–566, doi:10.1177/0885328220935090.

[46] P. GU, Biomedical applications of natural polymer based nanofibrous scaffolds, Int. J. Med. Nano Res. 2 (2015), doi:10.23937/2378-3664/1410010.

[47] Y.F. Goh, I. Shakir, R. Hussain, Electrospun fibers for tissue engineering, drug delivery, and wound dressing, J. Mater. Sci. 48 (2013) 3027–3054, doi:10.1007/s10853-013-7145-8.

[48] Y. Niu, M. Galluzzi, Hyaluronic acid/collagen nanofiber tubular scaffolds support endothelial cell proliferation, phenotypic shape, and endothelialization, Nanomaterials 11 (2021) 2334, doi:10.3390/NANO11092334/S1.

[49] D.J. Lee, A. Fontaine, X. Meng, D. Park, Biomimetic nerve guidance conduit containing intraluminal microchannels with aligned nanofibers markedly facilitates in nerve regeneration, ACS Biomater. Sci. Eng. 2 (2016) 1403–1410, doi:10.1021/acsbiomaterials.6b00344.

[50] J. Jin, S. Limburg, S.K. Joshi, R. Landman, M. Park, Q. Zhang, H.T. Kim, A.C. Kuo, Peripheral nerve repair in rats using composite hydrogel-filled aligned nanofiber conduits with incorporated nerve growth factor, Tissue Eng. - Part A. 19 (2013) 2138–2146, doi:10.1089/ten.tea.2012.0575.

[51] J. Jin, M. Park, A. Rengarajan, Q. Zhang, S. Limburg, S.K. Joshi, S. Patel, H.T. Kim, A.C. Kuo, Functional motor recovery after peripheral nerve repair with an aligned nanofiber tubular conduit in a rat model, Regen. Med. 7 (2012) 799–806, doi:10.2217/rme.12.87.

[52] J. Xue, H. Li, Y. Xia, Nanofiber-based multi-tubular conduits with a honeycomb structure for potential application in peripheral nerve repair, Macromol. Biosci. 18 (2018) 1800090, doi:10.1002/mabi.201800090.

[53] T.B. Bini, S. Gao, T.C. Tan, S. Wang, A. Lim, L.Ben Hai, S. Ramakrishna, Electrospun poly(L-lactide-co-glycolide) biodegradable polymer nanofibre tubes for peripheral nerve regeneration, Nanotechnology 15 (2004) 1459–1464, doi:10.1088/0957-4484/15/11/014.

[54] C. Mahoney, Nanofibrous structure of chitosan for biomedical applications, (2012). 10.4172/2155-983X.1000102.

[55] Y. Niu, G. Liu, M. Fu, C. Chen, W. Fu, Z. Zhang, H. Xia, F.J. Stadler, Designing a multifaceted bio-interface nanofiber tissue-engineered tubular scaffold graft to promote neo-vascularization for urethral regeneration, J. Mater. Chem. B. 8 (2020) 1748–1758, doi:10.1039/c9tb01915d.

[56] L. Wang, Y. Wu, T. Hu, B. Guo, P.X. Ma, Electrospun conductive nanofibrous scaffolds for engineering cardiac tissue and 3D bioactuators, Acta Biomater. 59 (2017) 68–81, doi:10.1016/j.actbio.2017.06.036.

[57] K. Aoki, H. Haniu, Y.A. Kim, N. Saito, The use of electrospun organic and carbon nanofibers in bone regeneration, Nanomaterials 10 (2020) 562, doi:10.3390/nano10030562.

[58] H. Samadian, H. Mobasheri, M. Azami, R. Faridi-Majidi, Osteoconductive and electroactive carbon nanofibers/hydroxyapatite nanocomposite tailored for bone tissue engineering: In vitro and in vivo studies, Sci. Rep. 10 (2020) 14853, doi:10.1038/s41598-020-71455-3.

[59] S. Chen, Z. He, G. Xu, X. Xiao, Fabrication of nanofibrous tubular scaffolds for bone tissue engineering, Mater. Lett. 182 (2016) 289–293, doi:10.1016/j.matlet.2016.07.015.

[60] F.C. Vazquez-Vazquez, O.A. Chanes-Cuevas, D. Masuoka, J.A. Alatorre, D. Chavarria-Bolaños, J.R. Vega-Baudrit, J. Serrano-Bello, M.A. Alvarez-Perez, Biocompatibility of developing 3D-printed tubular scaffold coated with nanofibers for bone applications, J. Nanomater. 2019 (2019), doi:10.1155/2019/6105818.

[61] S. Chen, S.K. Boda, S.K. Batra, X. Li, J. Xie, Emerging roles of electrospun nanofibers in cancer research, Adv. Healthc. Mater. 7 (2018) e1701024, doi:10.1002/ADHM.201701024.

[62] J. Horne, L. McLoughlin, B. Bridgers, E.K. Wujcik, Recent developments in nanofiber-based sensors for disease detection, immunosensing, and monitoring, Sensors Actu. Rep 2 (2020) 100005, doi:10.1016/j.snr.2020.100005.

[63] A. Senthamizhan, B. Balusamy, T. Uyar, Glucose sensors based on electrospun nanofibers: A review fiber-based platforms for bioanalytics, Anal. Bioanal. Chem. 408 (2016) 1285–1306, doi:10.1007/s00216-015-9152-x.

[64] A.M. Al-Dhahebi, S.C.B. Gopinath, M.S.M. Saheed, Graphene impregnated electrospun nanofiber sensing materials: a comprehensive overview on bridging laboratory set-up to industry, Nano Converg 7 (2020) 27, doi:10.1186/s40580-020-00237-4.

[65] S. Bao, M. Du, M. Zhang, H. Zhu, P. Wang, T. Yang, M. Zou, Facile fabrication of polyaniline nanotubes/gold hybrid nanostructures as substrate materials for biosensors, Chem. Eng. J. 258 (2014) 281–289, doi:10.1016/j.cej.2014.07.078.

[66] W. Jia, L. Su, Y. Lei, Pt nanoflower/polyaniline composite nanofibers based urea biosensor, Biosens. Bioelectron. 30 (2011) 158–164, doi:10.1016/j.bios.2011.09.006.

[67] E. Ding, J. Hai, T. Li, J. Wu, F. Chen, Y. Wen, B. Wang, X. Lu, Efficient hydrogen-generation CuO/Co3O4 heterojunction nanofibers for sensitive detection of cancer cells by portable pressure meter, Anal. Chem. 89 (2017) 8140–8147, doi:10.1021/ACS.ANALCHEM.7B01951/SUPPL_FILE/AC7B01951_SI_001.PDF.

[68] Q. Guo, L. Liu, M. Zhang, H. Hou, Y. Song, H. Wang, B. Zhong, L. Wang, Hierarchically mesostructured porous TiO_2 hollow nanofibers for high performance glucose biosensing, Biosens. Bioelectron. 92 (2017) 654–660, doi:10.1016/J.BIOS.2016.10.036.

[69] S. Kajdič, O. Planinšek, M. Gašperlin, P. Kocbek, Electrospun nanofibers for customized drug-delivery systems, J. Drug Deliv. Sci. Technol. 51 (2019) 672–681, doi:10.1016/j.jddst.2019.03.038.

[70] A. Luraghi, F. Peri, L. Moroni, Electrospinning for drug delivery applications: A review, J. Control. Release. 334 (2021) 463–484, doi:10.1016/J.JCONREL.2021.03.033.

[71] X. Guo, J. Zhu, H. Zhang, Z. You, Y. Morsi, X. Mo, T. Zhu, Facile preparation of a controlled-release tubular scaffold for blood vessel implantation, J. Colloid Interface Sci. 539 (2019) 351–360, doi:10.1016/j.jcis.2018.12.086.

[72] M. Rychter, A. Baranowska-Korczyc, B. Milanowski, M. Jarek, B.M. Maciejewska, E.L. Coy, J. Lulek, Cilostazol-loaded poly(ε-Caprolactone) electrospun drug delivery system for cardiovascular applications, Pharm. Res. 35 (2018) 1–20, doi:10.1007/S11095-017-2314-0/TABLES/2.

[73] N.Ç. Bezir, B. Bozkurt, A. Evcin, B. Özcan, E. Klr, G. Akarca, O. Ceylan, Enhanced antibacterial activity of silver-doped chitosan nanofibers, in: AIP Conf. Proc, American Institute of Physics Inc., 2019, p. 030003, doi:10.1063/1.5135401.

[74] A. Keirouz, N. Radacsi, Q. Ren, A. Dommann, G. Beldi, K. Maniura-Weber, R.M. Rossi, G. Fortunato, Nylon-6/chitosan core/shell antimicrobial nanofibers for the prevention of mesh-associated surgical site infection, J. Nanobiotechnology. 18 (2020) 51, doi:10.1186/s12951-020-00602-9.

[75] M.T.P. Albuquerque, J. Nagata, M.C. Bottino, Antimicrobial efficacy of triple antibiotic-eluting polymer nanofibers against multispecies biofilm, J. Endod. 43 (2017) S51–S56, doi:10.1016/j.joen.2017.06.009.

[76] Y. Li, S. Gao, B. Zhang, H. Mao, X. Tang, Electrospun Ag-Doped SnO2 hollow nanofibers with high antibacterial activity, Electron. Mater. Lett. 16 (2020) 195–206, doi:10.1007/S13391-020-00203-6/FIGURES/9.

[77] F. Mohammadi, A. Valipouri, D. Semnani, F. Alsahebfosoul, Nanofibrous tubular membrane for blood hemodialysis, Appl. Biochem. Biotechnol. 186 (2018) 443–458, doi:10.1007/s12010-018-2744-0.

[78] V. Guarino, G. Ausanio, V. Iannotti, L. Ambrosio, L. Lanotte, Electrospun nanofiber tubes with elastomagnetic properties for biomedical use, Express Polym. Lett. 12 (2018) 318–329, doi:10.3144/expresspolymlett.2018.28.

[79] W. Wang, S.P. Gumfekar, Q. Jiao, B. Zhao, Ferrite-grafted polyaniline nanofibers as electromagnetic shielding materials, J. Mater. Chem. C. 1 (2013) 2851–2859, doi:10.1039/c3tc00757j.

[80] J. Joseph, A.G. Krishnan, A.M. Cherian, B. Rajagopalan, R. Jose, P. Varma, V. Maniyal, S. Balakrishnan, S.V. Nair, D. Menon, Transforming nanofibers into woven nanotextiles for vascular application, ACS Appl. Mater. Interfaces. 10 (2018) 19449–19458, doi:10.1021/acsami.8b05096.

[81] D. An, Y. Ji, A. Chiu, Y.C. Lu, W. Song, L. Zhai, L. Qi, D. Luo, M. Ma, Developing robust, hydrogel-based, nanofiber-enabled encapsulation devices (NEEDs) for cell therapies, Biomaterials 37 (2015) 40–48, doi:10.1016/J.BIOMATERIALS.2014.10.03.

Index

Page numbers followed by "*f*" and "*t*" indicate, figures and tables respectively.

A

Acid-base titration, 198
Activated sludge modified graphene-oxide, 98
Aligned nanofibers, 328
Alkali metal ion batteries (AMIB), 41
Alloy-type anode materials, 187
  charge-discharge process, 187
Amidoxime, grafting polymerization, 92
  chemical grafting, 92
  plasma technique, 92
Antimicrobial applications, 334

B

Biochar, 101
Bio-sensing applications, 332
Biowastes, adsorbents functionalization, 100
  biochar, 101
  extracellular polymer matrix, 101
  magnetic biochar composites, 103
  phosphate bacteria biochar, 102
Bis choloromethyl biphenyl, 6

C

Carbon
  dots, 52
  materials, 185
  nanofibers, 221
  nanotube oil-absorbing material, 167
Carbon nanotubes, 92, 221
  anode materials, 191
  based anode materials, 189
Carbon tubular nanofibers, 115, 124
  catalyst, 117, 124
  crosslinking method, 121
  electro-spinning, 118
  functionalized nanocomposites, 124
  mechanism and control, 119
  synthesis, 117
  uranium capture, 125
Chemical vapor dispersion (CVD)

technique, 117
Collagen, 324
Conversion-type anode materials, 189
Covalent organic frameworks, 1
Covalent triazin frameworks, 1

D

Dicholroxylene, 6
Dimethylformamide, 118
Divinylbenzene, 3
  highly crosslinked polymer networks, 5
Drug-delivery systems, 333

E

Electrochemical coupling, 140,
  see also Tubular carbon nanofibers
  (TCN)
Electrospinning
  process, 328
  technique, 323
  technology, 326
Electrospun materials, 327
Ethylene dichloride, 9
Extracellular polymer
  matrix, 101
  substance, 98

F

Fabricated polymer nanofibers, 192
Formaldehyde dimethyl acetal (FDA), 4, 21
Friedel craft acylation, 134
Functional carrier, 50
  fluorescent materials, 52
  magnetic nanomaterials, 85

G

Gel type precursors (GTP), 16
Graphene, 186, 219
  oxide, 219
  oxide-nano sheets, 97
Graphite, 183, 191

H

Hollow microporous organic capsules
    (HMOC), 21
Hollow porous pollen microspheres (HPP),
    187
Human umbilical vein endothelial cells, 328
Hummers' method, 219
Hydrothermal method, 240
Hypercrosslinked polymers, 1, 121
    applications, 20, 39
    capturing carbon dioxide gas, 23
    crosslinking bridging structure, 13
    crosslinking distribution, 12
    design and synthesis, 34
    dilution effect, 9
    drug delivery, 21
    hollow structured, 18
    hydrophobic skeleton, 22
    initial copolymer network, 10
    invention, 1
    macroporous and gel type polystyrene, 3
    mechanism, 35
    membrane gas separation, 21
    mesoporous, 18
    molecular weight effect, 15
    morphology, 16
    photocatalysis, 20
    pollutants removal, 22
    polycondensation, 6
    polymer resins, 16
    polystyrene network, 4f
    preparation, 1
    properties, 9
    reaction medium, 14
    research progress, 30, 32
    sensing, 20
    synthesis, 3
    third generation, 4
    tubular, 17

I

Imprinting factor, 49
Intercalation anode materials, 185

L

Liquid-phase oxidation, 135,
        see also Tubular carbon nanofibers
        (TCN)

Lithium-ion batteries, 183, 211
    anode materials, 184
    cycle stability, 183
    higher reversible capacity, 183
    mechanism, 185f, 203, 236
    storage, 203

M

Magnetic biochar composites, 103
Magnetic nanochain-based multishell
        absorbers, 302
    inorganic nanowire-based, 303
    magnetic Ni nanochain core, 303
Magnetic nanomaterials
    one-dimensional, 51
    two-dimensional, 52
    zero-dimensional, 50
Magnetic tubular fiber, 304
Mesoporous hypercrosslinked
        polymers, 18
Mesoporous silica nanoparticles, 334
Metal-oxides, 99
Microwave absorbers and electromagnetic
        stealth, 257
MnO_2-based anode materials, 212
MnO_2-based composite electrodes, 217
MnO_2-based nanostructured electrodes, 212
Molecular imprinting technique, 42
Molecularly imprinted polymers, 56
Molecules on imprinted polymers, 48
Monochlorodimethyl ether, 4
Multishell composite materials, 300
    carbon cloth core, 301
    carbon paper core, 301
    fiber-based multishell absorbers, 301
    polymer fiber core, 302
Multiwalled carbon nanotubes, 92, 190

N

Nanofibers, 332, 323
Nanofibrous membrane, 336
Nanostructures, 326
Nanotechnology, 323
Natural polymers, 324
Nerves cells, 329
Nickel-polyurethane based tubular
        nanofibers, 337

O

Oil-absorbing materials, 154
  adsorption force, 154
  adsorption principle, 157
  biomass-derived one-dimensional, 168
  chemical adsorption, 158
  classification, 159
  dipole interaction, 156
  dispersion force, 156
  electrostatic force, 156
  fibrous synthetic high, 160
  hydrogen bond adsorption, 158
  magnetic one-dimensional, 165
  natural material modified, 161
  one-dimensional, 161
  one-dimensional polymer nanofibers, 161
  one-dimensional polystyrene, 162
  physical adsorption, 157
  resin-based high, 160
  rubber-based material, 159
One-dimensional carbon nanomaterials, 250
  carbon nanotubes, 251
One-dimensional magnetic nanomaterials, 51
One-step mixing method, 222
Oxygen-plasma treatment, 137,
  see also Tubular carbon nanofibers

P

Phosphate bacteria biochar, 102
Photochemical grafting, 138,
  see also Tubular carbon nanofibers
Plasma-induced graft polymerization (PIP) technique, 57
Polydimethylsiloxane, 35
Polyethylene oxide, 324
Polymeric electrospun nanofibers, 323
Porous aromatic frameworks, 1
Porous biomass tubular carbon fibers, 175
Porous carbon nanofibers, 184

Q

Quantum dots (QD), 52
Quartz crystal microbalance (QCM) biosensor, 57

R

Raman spectroscopy, 196*f*, 229
"Rocking-chair" battery, 184

S

Selectivity factor, 49
Self-driven protein imprinting, 71
Self-driven surface imprinted
  magnetic tubular carbon nanofibers, 75
  tubular carbon nanofibers, 71
Smart responsive imprinted polymers (S-MIP), 65
Solid electrolyte interface (SEI) films, 187, 198
Solution processable hypercrosslinked polymers (SHCP), 21
Styrene
  bulk, 3
  DVB, 3
Styrosorbs, 3
Surface imprinted manganese dioxide-loaded tubular carbon fibers, 80
Surface-initiated reversible addition-fragmentation chain transfer aqueous precipitation polymerization, 67
Surface plasmon resonance, 56
Surface protein imprinting technique, 49
  composition, 65
  functional carrier, 50
  surface design and construction, 58
Synthetic nerve guidance conduit, 329

T

Tetrahydrofuran, 118
Thermo-sensitive MIPs, 65
Titanate nanotubes, 94
Titanium oxides, 186
Tubular
  carbon nanofibers, 192, 222
  conduits, 325
  nanofibers, 326, 327
  scaffolds, 327
  woven nano textiles, 328
Tubular carbon nanofibers, 41, 133, 276
  carboxylate, 141

covalent modification, 134
electrochemical coupling, 140
hydrophobically modified, 173
liquid-phase oxidation, 135
oxygen-plasma treatment, 137
photochemical grafting, 138
preparation and functionalization, 135
Tubular hypercrosslinked polymer, 169
Tubular hypercrosslinked porous polymer,
    17
Tubular magnetic carbon nanofibers, 279
Tubular polymer nanotubes, 121
Two-dimensional magnetic nanomaterials,
    52

U

Ultralight helical porous carbon fibers, 286
Up-conversion nanoparticles, 52
Uranium adsorption, 144
Urethral epithelial cells, 331

V

Vapor-grown carbon nanofibers (VGCNF),
    134
    structure, 134
Vertically aligned carbon nanofibers
    (VACNF), 138
Vinylbenzyl chloride, 30

W

Waste paper carbon (WPC), 96

X

XPS measurements, 229
XRD patterns, 228f
Xylene dichloride (XDC), 4

Z

Zero-dimensional magnetic nanomaterials,
    50

Lightning Source UK Ltd.
Milton Keynes UK
UKHW051928020223
416402UK00016B/158

9 780323 990394